ARTISTS ANODIZING ALUMINUM

The Sulfuric Acid Process

A Workshop Excursion with

David LaPlantz

A suite of brooches by David LaPlantz, fabricated, anodized and painted aluminum, shown 15% larger than actual size.

ARTISTS ANODIZING ALUMINUM:

The Sulfuric Acid Process

David LaPlantz

Press de LaPlantz, Inc.
Bayside, CA

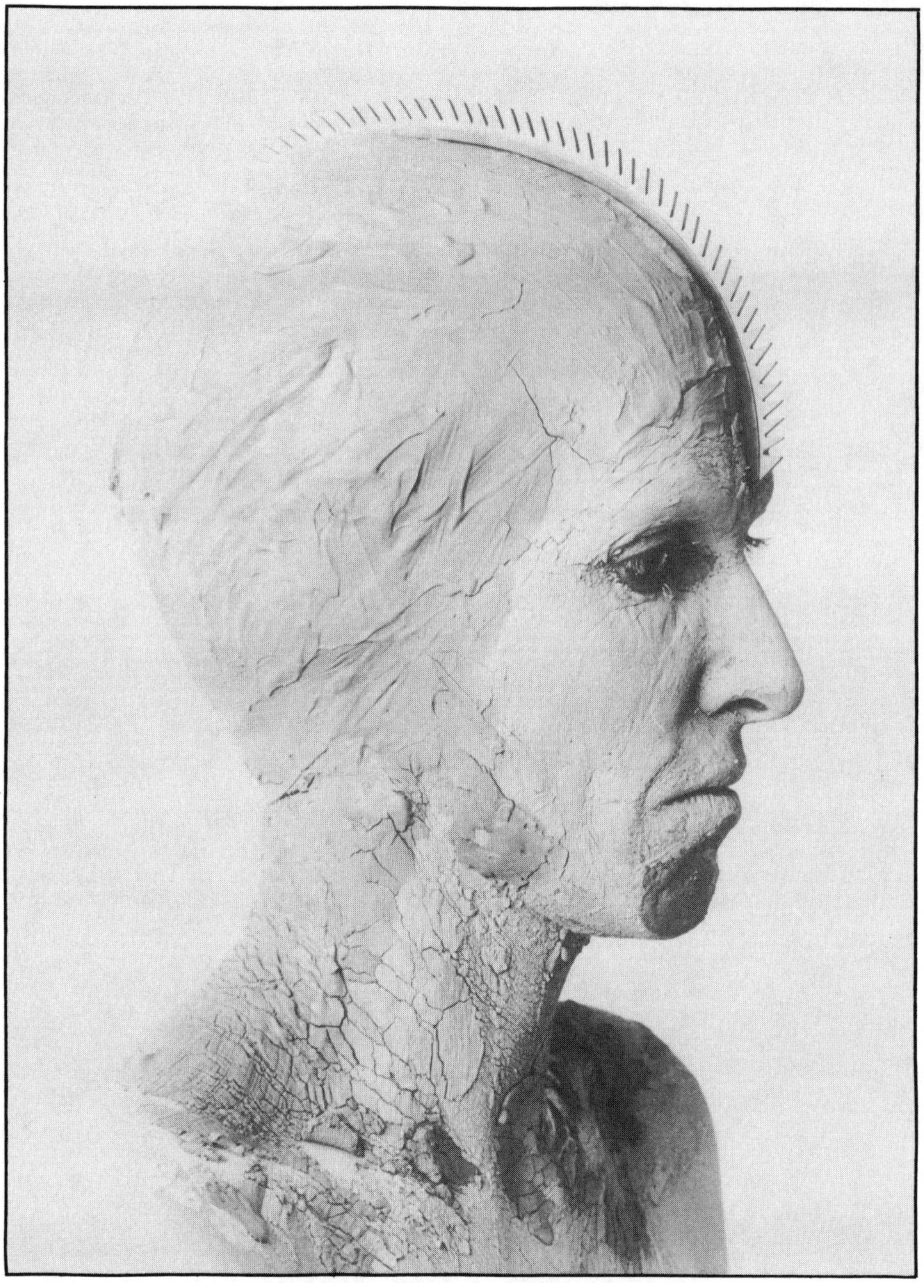

Head piece by Barbara Heinrich, anodized aluminum and brass, 1982, 1/2" x 9" x 1/8".

Inquiries should be addressed to: Press de LaPlantz, Inc, PO Box 220, Bayside, CA 95524.
Printed in the United States of America.

Library of Congress Catalog Card Number: 87-061000
ISBN: 0-942002-03-2

Table of Contents

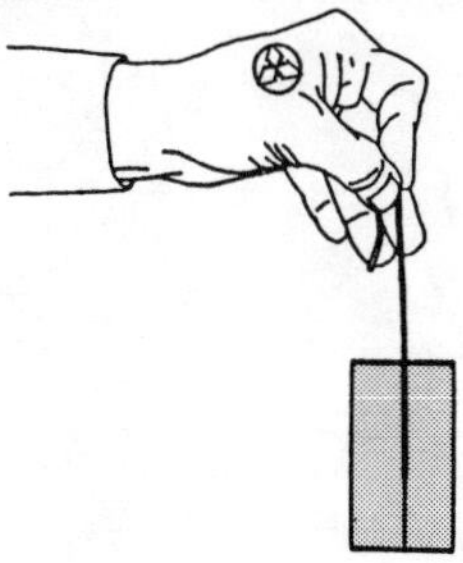
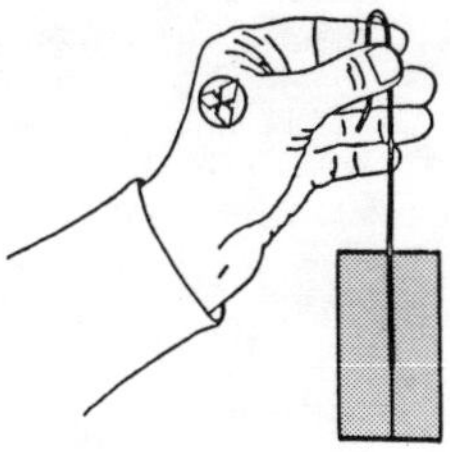

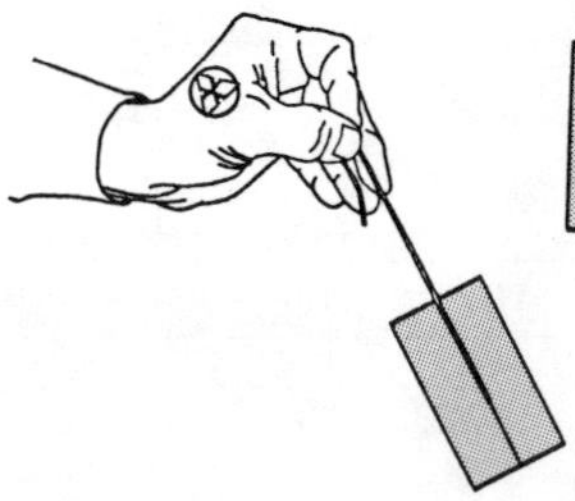
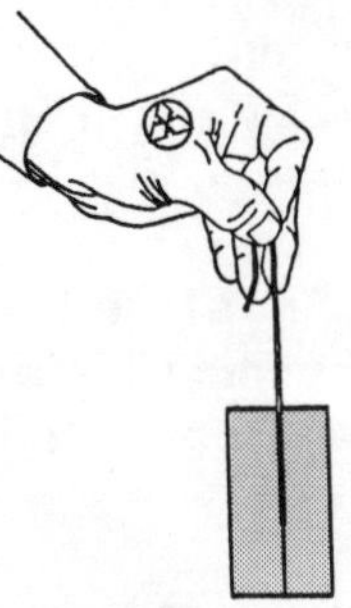
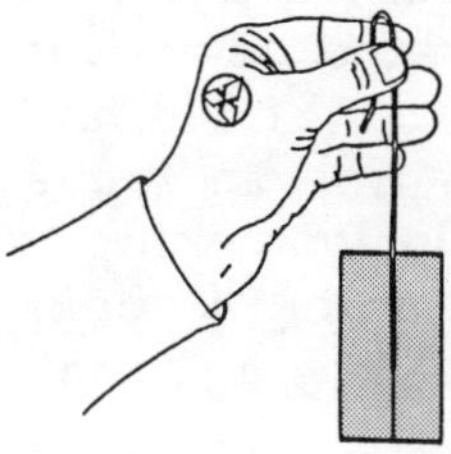
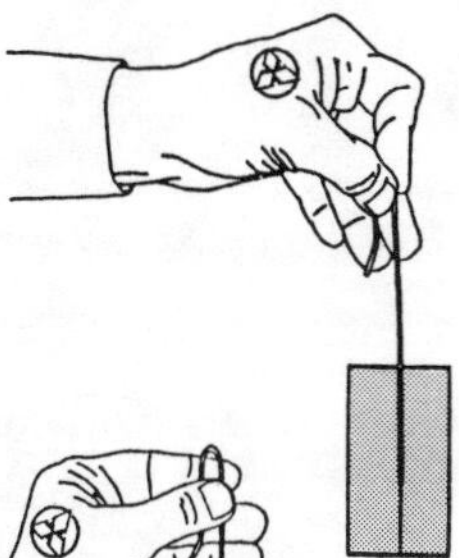
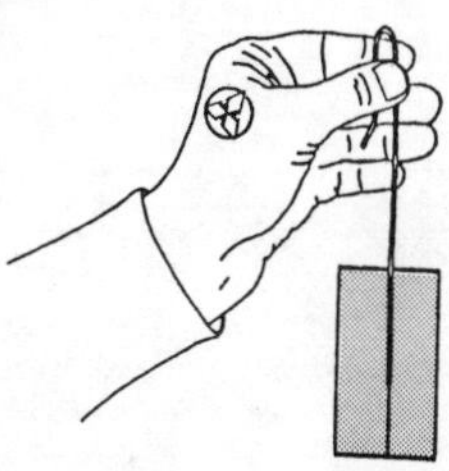

This book is dedicated with love to the memory of:
Ralph (Pop) Merle
Joseph John LaPlantz
Harold Dock
and to my loving, kind, and thoughtful Shereen.

Special thanks go to the following:

The writing of this book could not have been accomplished without the hand holding and incredible support from my wonderful wife, Shereen, and Bud Bolmer, my technical editor.

The production of this book could not have been realized without the generous support and hard labor of the Press de LaPlantz art staff: Patricia Andriese, Jane Sullivan, Tanya Boone and Kirt Frischknecht. They are the "Best!"

The writing of this book has been a long and difficult ordeal. I realize now, that I am not the type of person to sit at a computer fooling around with words. During most of this project, I have had to worry about words, captions, illustrations, "which line to leave in, which line to leave out." I don't mind the details in making a piece of jewelry. I do mind the picky details of producing a book. Next time, I'll leave it to someone else. I should be at my bench fooling around with some metal.

A special thanks must go to two, very special people in my life. The first thanks is to my loving wife, Shereen, who endured far more than anyone should have to endure, just to produce a book. She has held my hand, propped me up and put me back together each time I freaked out due to frustration, too little sleep, lack of understanding how to make this computer do what I wanted, etc. Thank you Shereen. Without your understanding and clear vision, when my vision was blocked, this book would have never been finished.

Help and support are two words vital to human existence. We all need it. We all need to practice it. During the time I was writing this book, I needed lots of both. Shereen was always there to provide the help, support and encouragement. Without Shereen, this book might still only be a dream, "like clouds in some coffee." For those of you who have endured long, drawn out projects like a book, you know exactly what I mean. I have wished, often, that I had never started this project. It has simply, gone on and on. It seemed at times that it was in charge of me, not the reverse. It grew larger and larger, almost out of control. The more I researched, the more I felt I needed to say. I am glad it is over. Shereen, to you, I say with all of my heart -- Thank You. This one's for you.

The second thank you is for P.W. "Bud" Bolmer, formerly the Head of Finishing Section, Kaiser Aluminum and Chemical Corporation, Center for Technology in Pleasonton, California. Bud has helped in so many ways so that the extremely technical explanations would be clear, correct and understandable. We spent many hours burning up the telephone lines, discussing the how and why of anodizing aluminum. Without Bud's assistance, much of this book would be non-existent.

The writing and producing of a book is a big job. At first, I figured that I could do it all by myself. What a naive thought! This book would not be in front of you, if it were not for the well appreciated long hours and hard labor from Patricia Andreiese, Jane Sullivan, Tanya Boone, Kirt Frischknecht, and Bil Coleman of the Press de LaPlantz team. Muchas Gracias to you all!

Totally unrelated, but on my mind, none the less, is to say that after finishing this book, I really must respect and stand in awe, and thank Oppi Untracht for everything he did for all of us in writing Jewelry Concepts and Technology. I thought 200 pages was work. I can't imagine the pressure of writing an 850 page book. I know how much I edited out. I can imagine that Oppi had enough extra material to write another book. Thanks Oppi!

Acknowledgements

I wish to thank the following people for helping me to write this book:

William Allen, Humboldt State University
Bud Bolmer, Retired Section Head of Finishing, Kaiser Aluminum
Robert Burrus, Met-L-Art, Inc.
Terry Burson, Milner's Anodizing
Clyde Davis, Humboldt State University
Ed De Large
Christina DePaul, University of Akron.
Mike Edwards, Humboldt State University
Bob Ensminger, Humboldt State University
Chuck Evans, Iowa State University
Arline Fisch, San Diego State University
Mark Fountain, University of Delaware
T. Franco, EAST/WEST DyeCom
William Frazer, Humboldt State University
Anne Graham, University of Delaware
Al Harcourt, University of California, Berkeley
Frank Koch, DeFelsko Corporation
Shereen LaPlantz
Tim McCreight, Portland School of Art
David C. Montgomery, Sandoz Colors and Chemicals
Silvio A. Rodriguez, C. H. Patrick and Co., Inc.
William B. Rusch, Jr., Anomatic Corporation
James Seiferheld, U.S. Specialty Color Corporation
Kirk Senseney, Milner's Anodizing
Mark W. Spencer, ISOMAT
Mrs. Spinosia, Aluminum Color Corporation
Norm Sprankle, Humboldt State University
David Tisdale
John Verhoeven, Iowa State University

Editorial Assistance
Shereen LaPlantz

Technical Editorial Assistance
Bud Bolmer

Design Concept/Layout
David LaPlantz/David D.Deco
Patricia Andriese/Jane Sullivan
Tanya Boone/Kirt Frischknecht

Illustrations
David LaPlantz/Jane Sullivan/Tanya Boone/Kirt Frischknecht

Photography
All photos by David LaPlantz unless otherwise identified.

Darkroom Work
Patricia Andriese

The Beginning. . .

My interest in color on metal started back in 1964. As an undergraduate student at Bowling Green State University, working with jeweler Hal Hasselschwert, I concentrated on making jewelry using transparent enamels on fine silver.

During the late 1970's, I was painting three-dimensional pieces that functioned as small change (coin) banks. After doing quite a few of the coin banks and other similar containers, I was looking around for the next "wave" I would ride. I was getting bored.

In the Fall of 1980, I brought Ed DeLarge to Humboldt State University to teach a titanium workshop. I was fascinated by the color potential of the titanium. I could see, however, that titanium might become another in the long list of the **darling** "techniques" or bandwagons that a large number of jewelers usually jump on. I was correct, and I am glad I did not leap on that bandwagon!

The same day that Ed left town, I was driving by the local trophy store. I knew that trophies usually had brightly colored metals on them. On a hunch, I walked into that trophy store and stepped into a whole **new** world. What a great day! I was introduced to colors on aluminum that were reminiscent of the transparent enamels I had worked with in the mid-1960's. That was all I needed. I was hooked! The aluminum that I found that day formed the foundation of the work I am still doing today. What a day that was......whew!!

After about four years of working with the painted trophy aluminum, I was introduced to the process of anodizing aluminum. I was at the University of Delaware teaching a workshop for Anne Graham. She had recently set up an anodizing area for a workshop David Tisdale presented to her students. Anne was kind enough to show me how to anodize. Thank you Anne.

As they say, the rest is history. At least, my history. Since that day, I have been researching, experimenting and developing my own personal approach to anodizing aluminum.

At first, I was only interested in "what" I could do with anodizing. I did not care about knowing "how and why" the process occurred.

In anodizing aluminum, I was drawn into what the process could provide for me. This **new** technique allowed me to communicate in a **different** way. Most techniques serve only as means to an end. Anodizing was just another means. It wasn't anything **that** special, or so it seemed at the time. The more I worked with the anodizing, the more I was hooked. What a wonderful ride. And, it's not over yet!

Recently my interest piqued in knowing the "how and why" of the anodizing process. That is why I am writing this book. After doing all of the research, it was time to show others the joy and the art possibilities in anodizing aluminum!

So, here I am writing a workshop experience in setting-up an anodizing studio and showing you how to anodize. I hope you enjoy this technique and that your ride is breathtaking. Let's go......

FRONT PAGES

What is Anodizing?
Glossary
Ex Libris

What is Anodizing?

Simply stated, the process of anodizing aluminum is an electrochemical treatment that converts the aluminum surface to an aluminum oxide.

Anodizing provides a protective and abrasion resistant film, or porous oxide layer, that may be left clear or dyed for decorative effects.

This is accomplished by making the aluminum positive (anode) or anodic in a sulfuric acid/water electrolyte. Hence, the term anodizing. Get it? The negative (cathode) may be aluminum, carbon, lead, or stainless steel. Lead is used most frequently. Direct current (D.C.) as supplied by a 12 volt battery or a D.C. power supply, that converts alternate current (A.C.), provides the current to "push" the chemical reaction along.

Anodizing is both a conversion process and an oxidation process.

1. In the conversion process, the aluminum surface is changed so that it has very different properties from the natural state of the metal.
2. The oxidation process creates a tough oxide layer similar to the rusting of iron or steel. "Anodizing might even accurately be described as an artificially-induced, self-terminating corrosion process," states Aluminum Technology, Book Five: Finishing Aluminum, published by the Aluminum Development Council, Sydney, Australia.

In industrial anodizing, three basic types of anodizing exist:
Sulfuric Acid
Oxalic Acid, and
Phosphoric Acid electrolyte solutions. Chromic acid and several others acids, fused salts, and some alkaline baths are also used.

The sulfuric acid method is the most commercially available and the easiest for artists to set up. This is why I am writing about the sulfuric acid anodizing process. If you are interested in reading more about the other anodizing methods, please consult the bibliography.

Part of learning is accepting that each individual puts a personal "spin" on what has just been learned. Some of the steps and formulas within this book can be varied, perhaps there really is "no one right way" to anodize. Of course, there is a big difference between how a commercial anodizer works versus the artist.

This book is my way of anodizing. It works for me, my results please me. The studio set-up and equipment are meant for the artist, on a small, affordable scale. The set-ups and methods are for the greatest efficiency.

I don't claim to be a chemist or an electrician, just an artist anodizing aluminum.

WHAT TIME IS IT ANYWAY? ANODIZING HISTORY LESSON TIME!

Between the end of World War I in 1917 and the stock market crash of 1929, the chromic acid, oxalic acid, and sulfuric acid processes for anodizing aluminum were patented.

The first patent was issued in England in November, 1924 to Bengough and Stuart. The second went to Flick and was assigned to ALCOA in the United States in February, 1925.

Following in the footsteps of the anodizing research were two patents for dyeing the anodic coatings.

What discoveries led to these patents? One story tells of coffee being spilled on anodized British Army cafeteria trays and the resultant staining. A second tells of a cat who knocked over a bottle of red ink on some freshly anodized sheets of aluminum. I can just see the "lights" that must have appeared in some people's heads when they saw the results of the coffee stains and spilled red ink! On occasion, fortune smiles on us, doesn't she?

When you think about it, the history of anodizing and dyeing aluminum is quite short. Just think what we will know about the subject in 50 to 100 years?

The following chart compares/contrasts aluminum anodizing and electroplating processes.

ANODIZING	ELECTROPLATING
REASONABLE PRECAUTIONS DUE TO CHEMICAL AND ELECTRICAL HAZARDS	**REASONABLE PRECAUTIONS DUE TO CHEMICAL AND ELECTRICAL HAZARDS**
ADEQUATE VENTILATION	**ADEQUATE VENTILATION**
D.C. power supply	D.C. power supply
higher voltage higher amperage	lower voltage lower amperage
bus bar used to support aluminum	bus bar used to support work piece +
aluminum is the anode	work piece is the cathode
aluminum **must** be firmly connected to racking structure	work piece requires a good connection, although resistance is not as critical
rack structure made of aluminum or titanium, never copper	wire to carry current may be copper
cathode is lead, aluminum, stainless steel, or carbon	cathode is the work piece, it can be most any conductive material *
sulfuric, chromic, oxalic, or phosphoric acid electrolytes	sulfuric acid, gold, silver, nickel, etc. electrolytes
metal must be grease free	metal must be grease free
coating forms from the metal and grows inward	deposited coating grows outward
aluminum, titanium, niobium, tantalum, and magnesium are metals that can be anodized	most metals, if pretreated properly, can be plated. non-metallic objects may be plated if coated with metallic, conductive layer of paint

+ work piece = object to be plated

* The anode is usually the metal being deposited (gold, silver, copper). It dissolves (it's anodized actually) by virtue of the plating current, and thus the concentration of metal ions in the bath remains constant. (The metal that's removed as plate at the cathode, is replaced by the metal dissolved at the anode.)

PROTECTIVE OXIDE

When aluminum is left out in the weather, a natural protective oxide is automatically developed. Anodizing, as mentioned earlier, is an artificially induced method for creating a thin, scratch resistant, transparent oxide layer on the surface of the aluminum. This anodic oxide is many times thicker and more protective than the natural oxide. Pure aluminum and its alloys can be anodized and dyed to produce wonderfully decorative effects.

Perhaps a good way to discuss the manner in which the oxide layer is formed is to compare and contrast the anodizing of aluminum to the electroplating process.

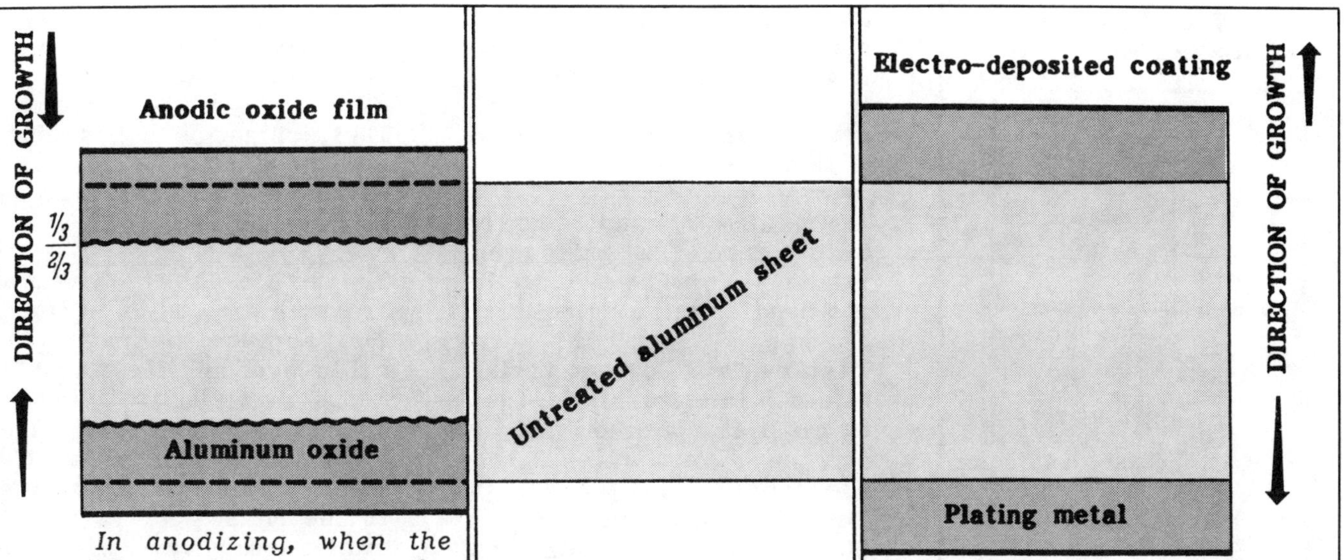

In anodizing, when the aluminum is made the anode (+), in an appropriate electrolyte, the conversion of the aluminum surface to an aluminum oxide occurs. The oxide layer grows in an inward direction. The oxide has a greater volume than the metal, so the surface of the aluminum grows outward by about one-third of the built-up film's thickness.

(After Dr. M. Schenk.)

In electroplating, a second metal is applied to the surface of the first metal, when the metal is made the cathode (-) in an appropriate electrolyte. The coating grows continuously in an outward direction.

ELECTROLYSIS

Electrolysis is the process that allows aluminum to be anodized. It is the production of chemical changes by the action of an electric current.

The electrolysis of water, in order to form hydrogen and oxygen gases (the components of water), is a familiar example of an electrolytic process. Other examples are batteries, both dry cells and storage, the corrosion (rusting) of steel and other metals, and most importantly, the anodizing of aluminum.

There are four requirements for an electrolytic process to take place. They are:

1. An electrolyte: A material that conducts electric current by means of ion movement. Just running electrons through a wire is not enough. For the electrolysis of water, a solution of sulfuric acid in water is a suitable electrolyte.
2. An anode: The positive electrode where the current enters the electrolyte.
3. A cathode: The negative electrode where the current leaves the electrolyte.

   The anode and cathode both must conduct electricity by movement of electrons. There must also be a chemical reaction that can take place at each.

   For electrolysis of water, the reaction at the anode is evolution of oxygen, while hydrogen is evolved at the cathode.
4. A driving force or potential (voltage) to make the two electrode reactions occur. For electrolysis of water, this requires an external power supply connected between the anode and cathode. Sometimes the electrode reactions take place spontaneously when the external circuit between the electrodes is closed. These electrolytic cells are batteries, and can supply electrical current.

SCHEMATIC DIAGRAM OF AN ANODIZING CELL

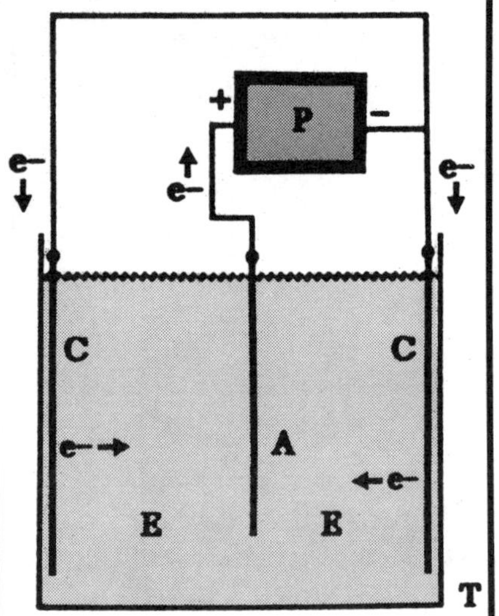

T = tank

C = cathodes

A = anode (aluminum)

E = electrolyte solution

e-→ = electron flow

P = power supply

A tank to hold the electrolyte and to mount the electrodes is convenient, since most electrolytes are liquid solutions of salts, acids and bases. This is, however, not a strict requirement for all cells.

When a D.C. current is passed through the electrolytic cell, its value is measured by an instrument known as an "ammeter." The ammeter is placed between the power supply and the anode. The current is read in units called "amperes."

When a steady current (I for amperes) flows through the electrolyte circuit (t for seconds) it is said that a charge "q = It" in coulombs passes through the circuit or electrolyte cell. The coulomb is a unit of charge equivalent to an ampere second or the ampere is equivalent to a coulomb/second.

Around 1893, Michael Faraday discovered the following principles in what eventually has been called electrolysis.

The amount of material released in electrolysis, that is, the mass of hydrogen or oxygen given off in the electrolysis of water, is directly proportional to the electric charge which passes through the cell; i.e. the product of current and the time.

When aluminum is anodized, the sulfuric acid solution can still be used as an electrolyte. The cathode can be an inert metal such as lead. The cathode reaction is still the evolution of hydrogen gas, just as in the electrolysis of water. However, the anode is now aluminum and it is not inert. When the external power supply is connected, the reaction at the anode now forms an oxide film on the aluminum, rather than evolving oxygen gas.

In anodizing, the aluminum is being converted to aluminum oxide. The conversion takes place due to the flow of electrons. For aluminum to become aluminum oxide, you have to give up electrons. Thus ionization takes place and the aluminum ions react with the water to form an aluminum oxide.

The oxide layer develops from the outside inward. The surface between the oxide layer and the aluminum keeps dissolving, building up, and redissolving. The aluminum oxide film has two layers. The first layer, next to the metal, is a very thin, continuous film called the barrier layer. The second layer, or outer layer, is porous and accounts for most of the total oxide thickness. The film grows at the metal-oxide interface. The first oxide formed stays on the surface as new oxide forms beneath it. Thus the oxide grows inward as more metal is converted. As the oxide grows, the thin barrier layer is continuously formed and then converted to a porous oxide. That is, the first barrier layer thickness remains constant, while the second porous layer thickness increases due to the conversion of the barrier layer.

A more technical explanation: The first (barrier layer) oxide that forms on the aluminum surface does not

THE TWO PRINCIPLE TYPES OF ANODIC OXIDE FILMS

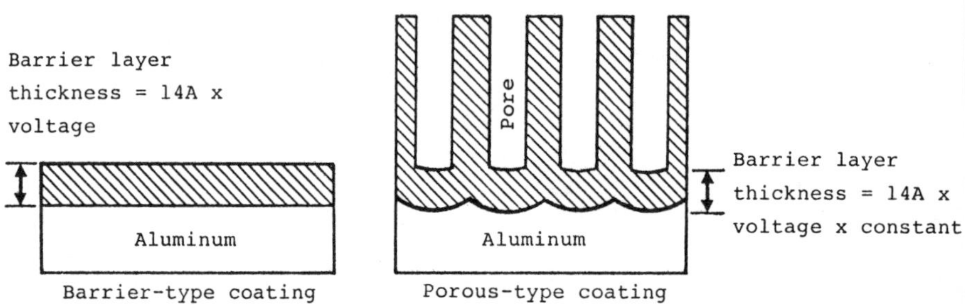

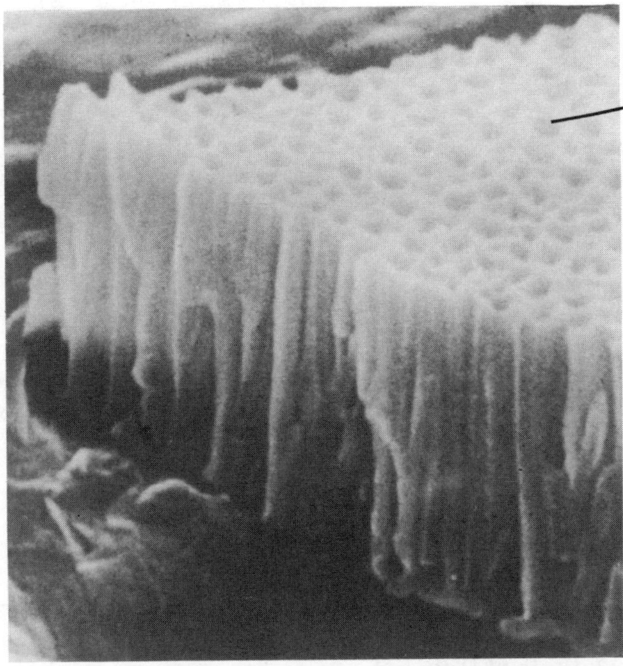

This photo is a scanning electron micrograph of a fracture section of an anodic oxide film formed in phosphoric acid at about 100 volts. It shows the columnar structure of the oxide and pore mouths at the oxide surface. The very thin barrier layer can be seen at the metal/oxide interface. The instrumental magnification was 20,000X and has been photographically enlarged to roughly 40,000X.

conduct electrons or ions, and resists the flow of electricity. If it were not for a very unusual effect of the sulfuric acid electrolyte combined with the influence of applied voltage, the oxide would stop the current flow. The oxide would then stop growing when it reached a limiting thickness. This oxide would be thin and not very useful for protection or decoration. In fact, if the electrolyte is boric acid or some other acids, rather than sulfuric, phosphoric, chromic or oxalic, the oxide does stop growing after a short time. For your information, these thin oxides are used in electrolytic capacitors. In the sulfuric electrolyte the barrier layer oxide dissolves -- not all of it, and not all over. The electrolyte drills tiny holes (pores) into the oxide.

If you imagine what a group of soda straws looks like, standing next to each other, you have the perfect mental picture of what the pores in the porous oxide layer look like. Compare your mental picture to the illustrations above and on page 18.

At the base of these pores the oxide is thin enough to pass some current, so the oxide continues to grow underneath at the metal-oxide interface. If the oxide grows too rapidly, the current will drop off, and the oxide growth slows down until the electrolyte can catch up with its pore formation. If the pores form rapidly, the oxide will grow faster. Eventually the pore formation and the growth of the oxide will reach a balance or "steady state." The net effect of all of this is: Relatively thick oxides can be formed. The oxide between the pores does not dissolve (at least not very rapidly). The result is a hard protective oxide layer with tiny little pores that extend down almost to the metal surface. These pores can be used to absorb dyes for color, while the bulk oxide protects the metal from corrosion and scratches.

The rate of oxide layer formation will equal the rate of barrier layer dissolution. An equilibrium develops between the current flow and the amount of dissolution at the barrier layer, the electrolyte interface. This determines the amount of current passed under any set of conditions.

If the voltage is increased, the dissolution will increase. This is because the dissolving action is not entirely controlled by the chemical reaction, but is assisted by the electric field.

An increase in: voltage, higher bath temperature, or a higher concentration of sulfuric acid affect the nature of the oxide layer developed on the aluminum. Anything that creates an increase in the dissolution rate of the barrier layer at the base of the pores will increase the current, and therefore, the rate of the oxide thickness' growth.

continued on page 19.

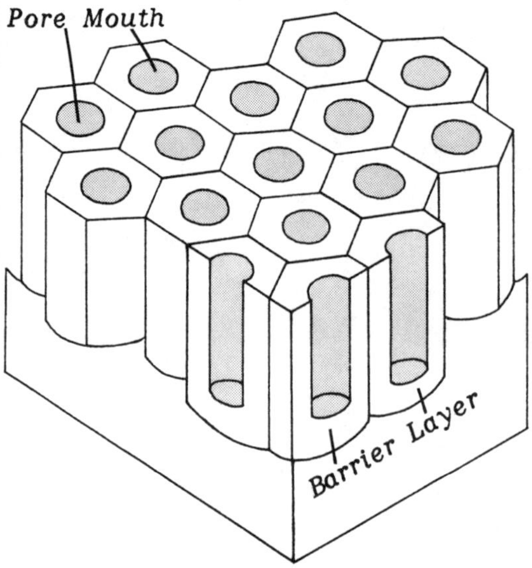

Structure of a typical cell of an anodic oxide coating showing basic units which determine its size. The cell size is twice the cell wall thickness, plus the pore diameter.

Actually, the walls of each pore are very thick when compared to the pore diameter. The outside is hexagonal, rather than round. The pores are packed together without leaving any void space, other than the pores, of course.

After anodizing, the aluminum may be left clear and sealed. The other option is to dye the freshly anodized film with an organic dye.

As the aluminum is dyed, the dye molecules penetrate the mouth of each pore and produce color on (in) the oxide layer. The final step would be the sealing of the oxide layer.

Sealing, in essence, is the closing or plugging of the pores. This involves the hydration of the anodic surface to form an alpha alumina monohydrate ($Al_2O_3H_2O$).

"When aluminum is made the anode in a suitable electrolyte, with a metal or carbon cathode, and an electric current is passed through the cell, the electrolyte releases oxygen ions which react with the aluminum ions to form aluminum oxide on the surface of the metal. The phrase **surface of the metal** is meant in a literal sense, since the reaction takes place with the parent metal --in other words, the coating grows inwards from the original surface of the metal. An important function of the electrolyte is to act as a solvent for the oxide, since it is only by dissolving portions of the growing oxide layer that the essential reaction can take place with the parent metal.

Anodic coatings on aluminum are therefore usually classified by the nature of the electrolyte, almost invariably an acid solution, since this indicates the degree of solvent action on the oxide layer, and the characteristics of the coating depending on this. With electrolytes with little or no capacity to dissolve the oxide, the anodic coatings are more or less rapidly self-terminating, hence thin and essentially non-porous. The type of coating is termed a **barrier** coating and has valuable properties for certain electrical applications.

Where appreciable solvent action is present, coatings are relatively porous except for the thin, dense portion at the metal interface which indicates where the solvent action and ion reaction terminated. These porous coatings, which account for the bulk of anodized aluminum, are very widely employed for decorative and protective purposes."-Aluminum Technology, Book Five: Finishing Aluminum published by the Aluminum Development Council, Sydney, Australia.

The drawing on page 16 shows the general nature of the two types of anodic coatings.

continued from page 18.

ELECTROLYTIC CELL

An electrolytic cell is required for electrolysis to take place. The electrolytic cell is composed of the following components:

ELECTROLYTE TANK:
: Container housing electrolyte.

ELECTROLYTE:
: The sulfuric acid/water solution.

ELECTRIC CURRENT:
: D.C. power supply

ELECTRODES:
: The aluminum (anode) and the lead (cathode)

ANODE:
: The positive charged electrode (aluminum) by which the current enters the electrolytic cell.

CATHODE:
: The negative charged electrode (lead) by which the current leaves the electrolytic cell.

As the current flows from the anode to the cathode, through the sulfuric acid electrolyte, bubbles of oxygen are given off at the positive charged aluminum (anode), while bubbles of hydrogen are given off at the negative charged lead sheet (cathode). The electrons flow (or ride on the ions as they move through the electrolyte) from the cathode to the anode.

THE ANODE REACTION

The overall anode reaction is:

1.

$$2AL + 3H_2O \rightarrow \underline{AL_2O_3} + 6H^+ + 6e^-$$

Meanwhile, the six electrons move through the electrical circuit to the cathode where

2.

$$6H_2O + 6e^- \rightarrow 3H_2\uparrow + 6OH^-$$

The cathode OH- and the anode H+ recombine in solution to reform water.

The complete cell reaction is the sum of 1 and 2.

$$2AL + 3H_2O \rightarrow \underline{AL_2O_3} + 3H_2\uparrow$$

The function of the acid is to dissolve or react with the solid Al_2O_3 to form pores so that the reaction goes on rather than allowing the Al_2O_3 to form an impervious film and stop the reaction. In this way, thick films of several microns (up to 3 mils) can be formed.

With a sulfuric acid electrolyte, the reaction to form the pores leaves roughly 15% pore volume; also, about 13% sulfate is incorporated (trapped, adsorbed or compounded) with the oxide. G. C. Wood in Oxides and Oxide Films, Volume # 2, J. W. Diggle ed., Marcel Dekker, Inc. New York, New York. (1973) suggests the following dimensions.

- Barrier layer thickness: 150 angstroms (9-11 angstroms/volt).
- Pore diameter: 150 angstroms (9-11 angstroms/volt).
- Pore population: 8 X 10^{10} pores/cm^2
- Pore volume: 14%.
- Barrier layer density: 3.2 g/cm^3
- Porous layer density: 2.7 g/cm^3

The above dimensions are typical for sulfuric acid films on aluminum alloy 1199 anodized in 15% H_2SO_4 at 77 degrees F., 12 amps per square foot, at 14 volts.

SCHEMATIC DIAGRAM OF AN ANODIZING CELL

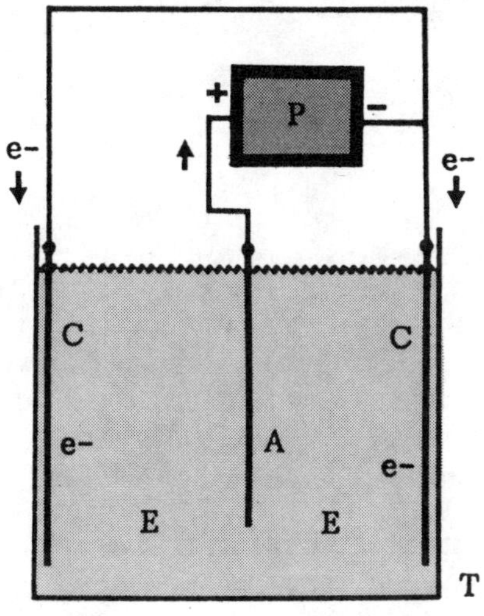

T = tank
C = cathode
A = anode (aluminum piece)
E = electrolyte solution
e- → = electron flow
P = power supply

ANODE REACTION:

$$2AL + 3H_2O \rightarrow \underline{AL_2O_3} + 6H^+ + 6e^-$$

CATHODE REACTION:

$$6H_2O + 6e^- \rightarrow 3H_2\uparrow + 6OH^-$$

COMPLETE CELL REACTION:

$$2AL + 3H_2O \rightarrow \underline{AL_2O_3} + 3H_2\uparrow$$

ANODIZING CONDITIONS:

Electrolyte: 15-18% Sulfuric Acid (15% - 22 oz/gal)
Voltage: 12-20 volts D.C.
Current Density: 10-15 Amps/Square Foot
Temperature: 60-80 degrees F. Optimum 68-72 degrees F.
Time: 20-40 minutes.

ANODIZED ALUMINUM IMPLIES.....

Some practical properties and implications of anodized aluminum include the following:

The pure oxide is water clear "glass." An application of this would be the use of clear anodized auto trim to protect the reflective sheet.

The oxide layer follows the contours of the mechanical and chemical finish on the aluminum.

The oxide is **brittle.** It will crack under mechanical or thermal treatments. If you take an anodized strip and bend it, you can actually hear it crack. This is a good experiment to try. Examine the bend area and look for visual cracks, elongated stress lines.

Once the aluminum has been anodized you cannot form the piece without possible distortion as mentioned above -- cracking. In the aluminum industry this distortion or cracking is a No-No. In art, why not!

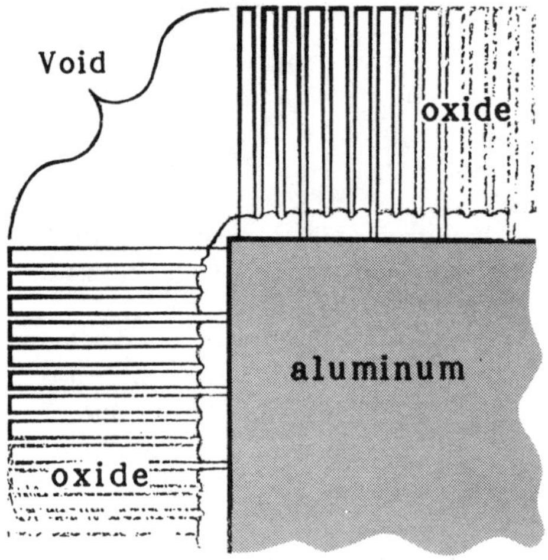

A void requires a generous radius for oxide coverage.

The oxide will grow perpendicularly to the metal surface, regardless of the orientation in the anodizing bath. Sharp edges will not cover with thick coatings. The oxide coating also reduces fatigue strength.

The functional applications of an anodized piece of aluminum include a hard and abrasion resistant surface.

It is important to understand that the dimensions of a piece increase when anodizing occurs (0.3 - 0.4 mil increase per mil of oxide).

Details of the structure of an anodized piece:

To resolve or **see** the structure of the oxide layer, an electron microscope is required. This item is well beyond the reach of our wallets so don't worry about running out to purchase one!

The pore diameter and barrier layer thickness vary with anodizing voltage, type of electrolyte and temperature. The dimensions increase with increased voltage.

The density and pore volume also vary with anodizing conditions and alloy. The pore volume decreases as voltage increases (larger pores, but fewer of them).

Sulfuric acid oxide coatings have long, narrow pores. For a visual analogy, think of the pores as the soda straws on page 17. To obtain a reasonable proportionality you would need mile-long straws with a wall thickness of 10 X pore radius.

Sealing

Once the aluminum has been anodized, whether clear or dyed, the oxide film must be sealed. As the oxide pores are sealed, the dye has an improved durability and is prevented from leaching out. The metal is also made resistant to staining from grease, oils, water, etc. because corrodents cannot get into the oxide pores.

Technically speaking, sealing is the hydration of the anodic oxide. The aluminum oxide is converted from Al_2O_3 to an alpha aluminina monohydrate $Al_2O_3 \cdot H_2O$.

This conversion causes the pores to plug, reduce in diameter, be closed off, or sealed. Bud Bolmer, my technical advisor, likens the sealing process to the plugging of a musket barrel with mud. Imagine the physical act of picking up some mud and plugging the end of a rifle barrel! A simple, but effective way to visualize and understand sealing.

Sealing does, however, soften the oxide coating. This softening effect is due to the hydration reaction. The amount of softening depends on the sealing time. Normal sealing should not present any problem for most applications. In industry, for certain critical abrasion resistance uses, coatings are left unsealed (if appearance in service is not important). A practical test for coating hardness is to rub the metal surface lightly with a slurry of a mild abrasive, such as pumice or Bon-Ami. The oxide coating should NOT be scratched. I recommend doing this on sample pieces, not the ones you really care about. If you run a few tests as you first start out anodizing, you'll learn to predict later results with some degree of accuracy. All you have to do is follow the procedures that give you the best results. You know, just like Formula Art.

We know that sealing plugs the pores. What is the process that's required to seal the oxide film? The simplest way to seal is: treat the oxide layer with hot (boiling) water. The oxide, as anodized, is anhydrous, or without water. The water treatment hydrates the oxide or causes a water molecule to enter the oxide structure. This reaction is represented by:

$$Al_2O_3 + H_2O \rightarrow Al_2O_3 \cdot H_2O$$

The hydrated oxide is called boehmite, or in this case, because the particles are too small to be resolved by X-ray, it's called pseudo-boehmite. The important thing for us to know is: The hydrated oxide is more voluminous (bigger) than the oxide was as anodized. When the hydrated oxide forms in the pores, it plugs them off. The mouths of the pores plug first, because that is where the water and heat are located. The longer the boiling, sealing water treatment lasts, the thicker the plug in the pore becomes.

Unfortunately, the whole surface of the oxide film is hydrated in water. This leaves a soft, powdery surface film called "smut." With dyed films, some dye bleeds out in a plain water seal. Bud feels the plain water seal leaves an unattractive smut on the metal's surface, although it is well sealed and resistant to further attack.

I tend to disagree. I have sealed the predominance of my pieces in boiling water and do not find the surface smut a problem. The degree of smut on my work is minimal. My colors are rich and vibrant; check the back cover and the interior color section. Remember, Bud is approaching this from an industrial anodizer's point of view. An artist does not always have the same worries as industry.

Bud suggests removing the smut with a wet cloth or a little mild abrasive, such as fine pumice. I dry my pieces and then apply a small amount of mineral oil with a clean cloth. This enhances the appearance of the anodized metal. The excess oil is then wiped off. It works for me. On pieces of jewelry that have been worn awhile, I periodically re-oil them to bring up the color vibrance. (I do wipe off the excess oil each time I oil a piece.)

In Bud's opinion, there is a better way to seal. Here are his thoughts. Anionic surfactants can be added to water to prevent the smut formation. Several of these are known and used in proprietary formulations. The problem here is that most of these inhibit pore plugging as well as smut formation. This latter problem is solved by the addition of nickel acetate (or some other nickel salt) to the sealing solution. Nickel and aluminum form a mixed oxide (co-precipitate) which deposits in the pores and provides an effective plug. Thus the surfactant-nickel acetate combination eliminates smut, yet prevents dye bleeding, while providing a chemically resistant oxide film. The sealing solution suggested above can be purchased pre-mixed, such as Sandoz Sealing Salt AS, from the two sources listed in this book. You may also mix your own sealing solutions. See below and on next page.

In addition to the water seal as discussed above, the Metal Finishing Guidebook/Directory suggests the following as sealing solutions:

WATER SEAL
Distilled or Deionized water (100 ppm SiO_2) Water purity counts.
Temperature: 200-212 degrees F.
Time: 30 minutes.

This is the seal I use. I'm not interested in subjecting my body to carcinogens, nickel acetate is a known carcinogen, so I stay away from it, advocating a long and healthy life. I recommend you do the same.

NICKEL/COBALT ACETATE SEAL
Nickel acetate: 5-5.8 g/L (0.67-0.78 oz/gal).
Cobalt acetate: 1 g/L (0.13 oz/ gal).
Boric acid: 8-8.4 g/L (1.1 oz/gal).
pH: 5-6
Temperature: 158-194 degrees F.
Time: 15-20 minutes.

DICHROMATE SEAL
$K_2Cr_2O_7$: 100 g/L (13.4 oz/gal).
$NaCo_3$: 18 g/L (2.4 oz/gal).
or NaOH: 13 g/L (1.7 oz/gal).
Temperature: 194-203 degrees F.
Time: 2-4 minutes.

Consider carefully if you want to use the nickel compound sealing solutions. Nickel is a heavy metal. The sealing solution with nickel can cause nickel itch, may cause cancer, and generates a hazardous waste and disposal problem. Many commercial anodizers use the nickel sealing solutions. For the results you'll require, I'll bet the water seal will work just as well as the nickel sealers. I do not want to confuse or scare you. I get good results with the water seal. It is the least expensive method of sealing. The hot water seal does not generate a hazardous waste.

Consider staying as far away as possible from cancer causing materials. The final choice, however, is yours. Choose wisely.

As discussed earlier, the simplest, most economic method of sealing is to use distilled or deionized (DI) water. Distilled or deionized water produces better sealing quality and corrosion resistance than tap water. Tap water is often unsuitable due to impurities and therefore should be avoided.

DISTILLED WATER VERSUS DEIONIZED WATER?

I am often asked which is better -- distilled or deionized water? So, let's ask the expert. Bud answered my question this way. Distilled and deionized water are the same for most practical purposes. The difference is in the way they are produced.

Distilled water is made by condensing the steam from boiling water. The dissolved salts (ionic materials) and other dissolved substances that have a different boiling temperature than water, either stay behind in the still, or escape as vapor.

Deionized water is made by passing the water through ion exchange columns. These columns absorb the positive ions from the water and exchange them for hydrogen ions. They also absorb the negative ions and exchange them for hydroxyl ions. The hydrogen and hydroxyl ions combine to form water. Some organic compounds which are not ionized in water sometimes remain in deionized water. They can often be removed by absorption on activated charcoal.

In distilled water, some organics are not separated cleanly, because they have boiling points very close to that of water. Also some ionic material is often physically carried over with the steam (like a mist). Deionizers are probably easier to operate from a maintenance and quality control standpoint than are stills. Thus, most commercial users of high purity water seem to use deionized water.

Whew, thanks Bud. What a mouthful of information! I always wondered what the difference was, if any, between distilled and deionized water and why most large anodizers utilized deionizers over stills.

DYES LEACHING

Some aluminum dyes bleed or leach out more in water than in nickel salt or other sealing additives. This is because the nickel salt sealers plug the pores quicker. I am not bothered by this leaching out problem. My pieces live up to my expectations. You may wish to experiment with both types of sealing and form your own conclusions within the framework of your particular requirements.

The water temperature is held at about 190-200 degrees F. in a stainless steel container with a lid to reduce heat loss. By maintaining a temperature just under boiling (212 degrees F.), the loss of water through evaporation is reduced.

A FLOW DIAGRAM

Some of the processes for the anodizing, dyeing and sealing of aluminum.

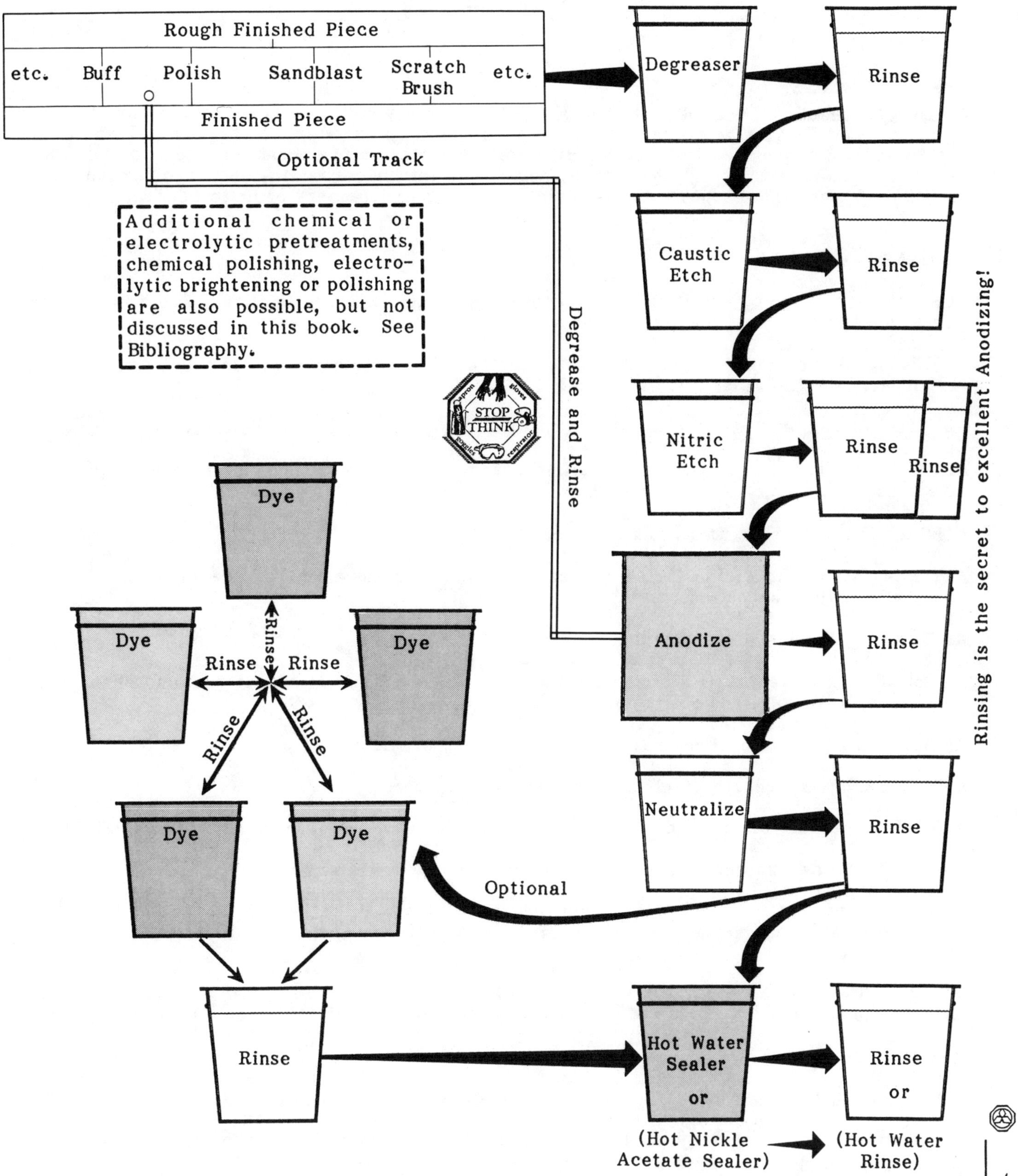

Glossary

Glossary/Vocabulary. I think most books put the glossary in the strangest place--in the back of the book! This glossary/vocabulary is located where you need it--in the front. I would like you to become acquainted with the **words** before you need to use them. My method of reading books is to start with the glossary **first,** because it introduces me to the "lingo" or language. It seems to make more sense to be acquainted with the vocabulary prior to becoming involved with the text.

GLOSSARY OF TERMS:

ACETIC ACID. A sour, colorless, liquid compound, having a sharp odor -- found in vinegar. Acetic acid is used to lower the pH of a solution.

ALIQUOT. To designate a part of a number that divides the number evenly and leaves no remainder. For example, 5 is an aliquot of 15.

ALKALI. One of a class of chemical compounds which combine with acids to form salts.

ALLOY. Any mixture of metals. Aluminum may be mixed with copper, silicon, magnesium, manganese, iron, zinc, or other metals.

ALPHA. The beginning, first, number one.

ALTERNATING CURRENT (A.C.). A current in which the movement of the electrons is back and forth in the conductor.

ALUMINA. An oxide of aluminum.

ALUMINUM. A silvery metallic element whose chemical symbol is Al. It is found abundantly, very light, malleable, and resists oxidation.

AMMETER. An instrument used to measure the rate of electrical current flow.

AMPERAGE. The strength of electrical current in amperes.

AMPERE. The unit of electric current, named for Andre Ampere, French physicist, 1775-1836.

ANGSTROM. One hundred-millionth of a centimeter, a measurement.

ANION. A negatively charged ion that migrates in the direction of the anode under the influence of electric current. The cation is the positively charged ion that migrates toward the cathode.

ANODE. The positive electrode of an electrolytic or voltaic cell. The current enters the electrolyte through the anode and positive ions accumulate while negative ions are discharged.

ANODIC. Of or from an anode.

ANODIZING (OF ALUMINUM). The anodizing of aluminum consists of converting the aluminum surface to an aluminum oxide. The aluminum work piece is made the anode, in a suitable electrolyte, hence anodic or anodized. The anode reaction forms aluminum ions which react with water to form the oxide. The purposes of anodizing aluminum include:

1. improve corrosion resistance; 2. increase hardness; 3. create a non-conductor of electricity; 4. most important to me, to create a surface that will accept dye for decorative reasons, **ART!** Some electrolytes used to anodize aluminum are: chromic acid, sulfuric acid, and oxalic acid. Phosphoric acid, several other acids, fused salts, and some alkaline baths are also used.

AVOIRDUPOIS (WEIGHT). The system of weights used in the United States for commodities of all kinds except drugs, precious metals and stones. To eliminate some possible confusion: The pound or unit weight is 16 ounces as opposed to the troy weight of 12 ounces.

BARRIER LAYER. That portion of the oxide layer closest to the parent layer of aluminum. Above the barrier layer are the pores that accept the dye after the aluminum has been anodized.

BASE. A substance capable of combining with an acid to form a salt.

BATTERY. A device that uses chemical energy to produce electricity. A 12 volt battery may be used as a power supply to anodize aluminum.

BUFFER SOLUTION. A liquid solution having a specific pH such as 4, 7, and 10 used to re-calibrate the pH meter. For optimum pH testing results, the meter must be standardized against the buffer solution periodically.

BUS BAR. A strong, copper bar that supports the work to be anodized. It carries the current from the power supply or rectifier to the electrolyte. I recommend bus bars of at least 1/4" X 1". Commercial anodizers often use 1" X 1" copper bars or larger depending on the size of the rack and weight of the work to be anodized.

CARBON. A non-metallic chemical element found in many inorganic compounds and all organic compounds. Graphite and diamonds are pure carbon. Graphite may be used as a cathode material in aluminum anodizing.

CATHODE. The negative terminal of an electrolytic or voltaic cell. Cathodes used for aluminum anodizing include lead, stainless steel, aluminum, and carbon. Aluminum cathodes should be removed when not anodizing.

CELL. In electricity, a receptacle containing electrodes and an electrolyte for decomposing compounds by electrolysis.

CONDUCTOR. A material through which an electrical current can readily travel.

CONVERSION. To change from one natural state to another, as in anodizing.

COULOMB. The quantity transferred by a current of one ampere in one second.

CURRENT. The transmission of electricity through an appropriate conductor, in this case, the electrolyte. The current is measured in amperes.

CURRENT DENSITY. The ratio of the current to the anode's surface in the electrolyte. The amount of current passing through the electrolyte is divided by the area of the anode (work piece). Most figures you will read about will be expressed in amperes per square foot, or asf.

DEGREASER. Anything that cleans the metal is a degreaser, ie. the complete removal of buffing compounds, grease, oils, etc. Degreasers are, strictly speaking, organic, but the most effective cleaners for aluminum are liquid detergents. The anodizing industry uses inhibited alkaline cleaners which operate around 130-160 degrees F. and are very good for removing grease. The inhibitors have been added to keep etching (metal attack) to a minimum.

DEGREASING. The part of the cleaning process that removes grease from all metals. This process can be accomplished with a liquid organic solvent (solvent degreasing) or a condensed solvent vapor (vapor degreasing).

DEIONIZED WATER. Water that has been processed to remove the ions.

DIELECTRIC. Non-conducting, an insulator--as in rubber and glass do not conduct electricity. Anodized aluminum is dielectric.

DIRECT CURRENT (D.C.) Electric current which moves continuously in one direction from positive to negative without any reversal. The electrons move from negative to positive.

DISTILLED WATER. Purified water by the distillation process.

DRAG-IN. Chemicals that are pulled into a solution due to improper rinsing from the prior bath or tank solution.

DRAG-OUT. Chemicals that are carried out and contaminate the next bath or tank solution.

DYES. A substance used to dye something, in this case aluminum, with a more or less permanent color. The dyes specifically made for aluminum are organic dyes.

ELECTRIC CURRENT. The flow of electrons along a conductor.

ELECTROLYSIS. The decomposition into ions of a chemical compound in solution by the action of an electrical current passing through the solution. Michael Faraday is given credit with developing this theory.

ELECTROLYTE. A solution or liquid through which current flows, by the movement of ions in an electrolytic cell, often an aqueous solution of acids, bases, or salts. Any substance which in solution is dissociated into ions and is made capable of conducting an electric current.

ELECTROLYTIC CELL. The structure in which electrolysis takes place. The components include: an anode, a cathode, electrolytic solution (electrolyte), a tank to hold the electrolyte, and electrical current to create the electron flow.

ELECTRON. The fundamental unit of electricity. It carries a negative charge.

EPA. Environmental Protection Agency.

ETCH. To dissolve or eat away the surface of a metal with an acid or base.

EXTRUDE. To force or to press out, as through a die.

EXTRUSION. The act of extruding or the state of being extruded.

FARADAY, MICHAEL. Faraday if often called the greatest experimental scientist of the 19th century. His experiments have provided us with the theory of electrolysis.

FASTNESS. The ability of dyes to resist fading. Fastness is measured on a scale of 1 to 8. A rating of 8 indicates no color change and is the highest degree of fastness. Generally for jewelry, lightfastness is not a big problem since the work is not constantly in direct sunlight. Fastness is quite important in architectural anodizing.

FUME HOOD. A general purpose three-sided compartment with a sliding glass front within which toxic or dangerous chemicals should be used. The interior usually has a sink and running water. Exhaust flow is provided by an electric fan.

HAZARDOUS WASTE. In anodizing, the following hazardous wastes are generated: sodium hydroxide or caustic soda (lye), nitric acid, sulfuric acid, heavy metals included in the dyes and sealing solution, in addition to rinse water. Many other hazardous wastes occur in life, but the ones mentioned above are those the anodizer generates.

HYDRATE. A compound formed by the chemical combination of water and some other substance.

HYDROGEN. An inflammable, colorless, odorless, gaseous chemical element. It is lightest of all known substances, its symbol is H.

INSULATOR. A material through which electricity does NOT pass readily such as wood, plastic, and don't forget that the surface of the aluminum, once anodized properly, is insulating. The bulk of the aluminum, however, is still a good conductor. A strong clamp on a thin oxide film will form a contact by breaking through to bare metal.

INTERFACE. The surface that lies between two parts of matter or space and forms their common boundary. The space between the barrier layer and porous oxide that comprises the anodic coating.

ION. An atom or group of atoms that carry an electric charge.

LEAD. A malleable bluish-gray metal, ductile and heavy, used as the cathode in anodizing aluminum.

MATTER. Anything which occupies space and has weight.

MICROMETER or MICRON. One millionth of a meter or one thousandth of a millimeter.

MIL. 0.001 inch or 25.4 microns or .0254 mm.

MONOHYDRATE. A hydrate containing one gram molecular weight of water per gram molecular weight of the combining element.

MSDS. Material Safety Data Sheet. This is vital information **you need to know** regarding the potential chemical hazards possibly in the products you work in your art. For instance, how toxic is acetone? Should you allow acetone to come in contact with your skin? Contact the vendor or the manufacturer of the product in question about each product used in your studio.

MSHA. Mine Safety and Health Administration.

MULTIMETER/MULTITESTER. A battery operated, high-sensitivity tester designed to measure A.C. and D.C. voltages, D.C. current and Resistance.

NATURAL GAS. A mixture of gaseous hydrocarbons, mainly methane, which occurs naturally in the earth, used as a fuel.

NEUTRAL. Neither acid or alkaline. On a pH scale, the number 7.

NIOSH. National Institute for Occupational Safety and Health.

NITRIC ACID. (HNO_3) A colorless, powerful and corrosive acid. The "smut" remover in anodizing aluminum.

OXIDATION. Anodizing is an oxidation process whereby the aluminum combines with oxygen to form an oxide layer on the aluminum's surface. Another good example of oxidation is the formation of rust on iron. Burning is an extremely rapid form of oxidation.

OXIDE. A compound of oxygen and one other element, in this case the metal is the other element and the resulting film is formed due to the reaction of the metal to the oxygen.

OXYGEN. A gas that is colorless, has no odor or taste. By volume it forms about half of the atmosphere. The chemical symbol is O. If you hadn't noticed, it is essential to life, combustion, and of course, anodizing, among other things.

pH. The measure of the acidity of a solution. A pH of 7 is neutral. pH values below 7 indicate a solution is acid, while pH values above 7 indicate an alkaline solution. Strong acids solutions have a pH below 1. Strong alkaline solutions have a pH above 12. The technical explanation is as follows--pH is a measure of the hydrogen ion activity of a solution. It is the negative logarithm of the activity. Its scale is from O (strong acid) to 14 (strong base). Acetic acid will lower the pH, while sodium hydroxide will raise it. The two common methods for measuring pH are the colorimetric and electrometric.

pH-COLORIMETRIC. For this method, pH papers are used. The color changes provide an estimation to about plus or minus 0.2 pH units.

pH-ELECTROMETRIC. For this method a battery powered pH meter is used. Carefully follow the directions that are provided.

PICKLE. An acid solution used to remove surface oxides from metal by chemical or an electrochemical action.

P.O.L. No, not the P.L.O., but P.O.L. This stands for Prest-O-Lite, the manufacturer of the valves and other connections for propane tanks.

POLYETHYLENE. A type of thermoplastic resin made by polymerization of ethylene; used in making translucent, lightweight, and tough plastic containers inert and impervious to acid/ chemicals.

PROPANE. A gaseous hydrocarbon of the methane family, derived from crude petroleum, used as a fuel.

PROPRIETARY. The word used to describe various products or processes made and sold by an individual or firm who have the exclusive rights of manufacture or sale. You will see this word used frequently in the various publications regarding anodizing, plating, etc.

RACK. In anodizing, the metallic device that attaches the aluminum to the copper bus bar. The rack is also the handle for the work. It provides a safe and proper way to support the work throughout the multiple liquid solutions required in anodizing. For single, one-of-a-kind works a simple aluminum wire will suffice. Multiple or irregular shaped works usually require more elaborate racks, either home made or factory constructed. Titanium is used frequently in the industry because it requires no stripping after each anodizing. Aluminum may be used, but requires stripping and rack maintenance, not to mention the amount of aluminum lost through each individual stripping. Aluminum racks must be discarded far earlier than similar rack devices constructed of titanium. Be certain to use aluminum wire of a similar alloy to the work for better current flow. Aluminum/copper alloy wires will reduce the current flow.

RCRA. Resource Conservation and Recovery Act.

RECTIFICATION. The conversion of an alternating current of electricity (A.C.) to direct current (D.C.).

RECTIFIER. The device which converts A.C. to D.C., thus the current will flow in one direction only.

RESISTS (ETCHING/BLEACHING, ETC.). The materials applied to the surface of the metal to act as a stop-out or to protect the metal from the corrosive action of an acid, as in etching/ bleaching, etc. Some resists include: paint, asphaltum, hard/soft grounds, fingernail polish, masking tape, electroplater's tape, peelable lacquers, etc.

RINSING. Almost every step in anodizing requires the rinsing action prior to the next procedural operation. Adequate and proper rinsing will insure consistently positive results (less rejects), minimizes contamination through drag-in and drag-out, and improves the overall quality of the work. I can't over stress the importance of good rinsing!

SEAL. The final step after anodizing (clear or dyed) which renders the oxide layer or pore structure closed. Boiling distilled/deionized water or nickel-acetate and distilled/deionized water with surfactants comprise the sealing solutions.

SODIUM HYDROXIDE. (NaOH) Better known as caustic soda or lye. This is a highly corrosive chemical used to etch or clean aluminum. Use this one with respect and utmost care and concern. Do not breath the fumes; cough, cough. Sodium hydroxide is also used to raise the pH of a solution.

SOLVENT. A substance with the cleaning power to dissolve grease, etc, (or resists, in etching).

SMUT. This term refers, generally, to any unsightly film left on the aluminum surface after an anodizing process step. The light gray to dark black color formed on the aluminum during the caustic etching. This color is formed due to the alloy content of the aluminum. Smut is also visible as the film left after the water sealing step.

SPECULAR. Mirror-like finish on metal.

SPLINE. The main shaft/rod that the rack components attach to.

STAINLESS STEEL. Steel alloyed with chromium, etc., creating a metal unaffected by rust or corrosion.

SULFATE (LEAD). The reaction of lead (cathode) and the sulfuric acid electrolyte. This is usually seen as a white colored film and can block the flow of electricity.

SULFURIC ACID. (H_2SO_4) A colorless, and strongly corrosive acid.

SURFACTANTS. A short term for surface active agents; chemical compounds that can alter the boundary conditions between liquid and gas, two or more liquids, or liquids and solids.

TANKS. The structures used to house the various solutions in anodizing, such as the electrolyte, water, acids, smut removers, dyes, sealers, etc. The housing for the sulfuric acid may be lead-lined steel, stainless steel or plastic lined steel. Plastics such as PVC, neoprene or polyethylene work well as tank liners. The lead-lined tanks provide the cathode, but you must be careful not to touch the lead with the work because of the shorting that will take place. Be careful! I use plastic garbage pails, buckets, etc. They are cheap and work well.

TAP WATER. The water that comes from the tap complete with necessary chemicals to purify. These chemicals may cause problems when mixed with acids, dyes, etc. Substitute deionized or distilled water for tap water when purity is required.

TEMPERATURE. The hotness or coldness of a substance as determined by the thermal energy it possesses.

THERMOMETER. An instrument for measuring temperature.

TRANSFORMER. A device that changes the voltage of an alternating current up or down.

VALVE ACTION. The ability of a metal's oxide to allow the current to flow only in one direction. Aluminum, niobium, tantalum, and titanium are valve metals.

VENTILATION. To bring in and circulate fresh air/to drive out foul, unhealthy air. An important way to stay healthy, consider ventilation in everything you do.

VOLT. In electricity, the unit of potential difference.

VOLTAIC CELL. A device which transforms chemical energy into electrical energy. This device consists of two dissimilar conducting materials immersed in a fluid which is also an ionic conductor.

VOLTMETER. The instrument used to measure the difference of potential.

VOLUME. Space that is taken up by matter.

Ex Libris

TECHNICAL INFORMATION

During my workshops, I usually drag out a pile of books that might be unknown to the participants or be of some special interest. Sometimes the books relate to the **other** aspects of life, rather than the topic at hand. Here's the pile of books for this workshop!

One of the best books I've seen that explains anodizing and especially "how to rack" is The Practical Anodizing of Aluminium by Hubner and Schlitknecht. Available from
MACDONALD & EVANS
8 John Street
London, England

Other outstanding books to complete your anodizing research **should** include the following:

The Surface Treatment and Finishing of Aluminum and its Alloys by S. Wernick, R. Pinner and P.G. Sheasby is a classic reference work on anodizing aluminum and so much more.
Don't miss Aluminum-Volume I,II and III edited by Kent R. Van Horn.
Both are available from:
ASM INTERNATIONAL
Metals Park, Ohio 44073

The Technology of Anodizing Aluminum by Brace and Sheasby is an excellent source of technical information on anodizing.
Also, Aluminum Workshop Practice by Hegman and Brace.
Both available from:
TECHNICOPY, LTD.
66 High Street, Stonehouse
Gloustershire GL10 2NA
England

Metal Colouring by Fishlock, published by Robert Draper Ltd., Teddington, England.

The Anodic Oxidation of Aluminium and Its Alloys by A. Jenny, translated by Winifred Lewis.

Die Praxas Der Anodischen Oxidation Aluminiums by Hubner and Speiser. This is available **only** in German from:
ALUMINUM VERLAG
GMBH Dusseldorf, Germany

Other titles of interest may be found in the bibliography. Look them up!

An invaluable handbook and a must for any jewelry, finishing operation, anodizing, or plating studio is the Metal Finishing Guidebook and Directory.

As the cover of this Guidebook states, "The most comprehensive reference source in the industry-everything the finisher needs to know!" I agree wholeheartedly. This yearly handbook contains well written and clearly diagramed information. Subjects include everything you've always wanted to know about metal finishing: from polishing to anodizing and plating, through filtration systems, racks, power supplies, etc. You will also find a complete directory of who sells what and where to locate it.

The Metal Finishing Guidebook and Directory table of contents may whet your appetite:

MECHANICAL SURFACE PREPARATION
CHEMICAL SURFACE PREPARATION
ELECTROPLATING SOLUTIONS
PLATING PROCEDURES
SURFACE TREATMENTS
ORGANIC FINISHING
CONTROL-ANALYSIS-TESTING
FINISHING PLANT ENGINEERING
TABLES AND DATA SHEETS
DIRECTORY
SUBJECT INDEX

How do you like that for a listing of information? You ought to see the subtitles in each of these categories. The Metal Finishing Guidebook/Directory is everything you need to know all wrapped up in one easy—to-read package.

I am totally sold on this book. Order one and you should not be disappointed. Actually, when you order the Guidebook you will also receive the "Metal Finishing" Magazine published monthly, except semi-monthly in January.

The subscription rates are \$28.00 a year/United States, other countries, \$75.00 a year. The Guidebook is **not** sold separately.

Order your Metal Finishing Guidebook/Directory and "Metal Finishing" Magazine from:
METALS AND PLASTICS PUBLICATIONS, INC.
One University Plaza
Hackensack, NJ 07601
201-487-3700

ALUMINUM PUBLICATIONS

The Aluminum Association, Inc. is an incredible source for information on aluminum. They sell many books and pamphlets on **everything** you will want to know about aluminum! Send for their extensive publications list.
THE ALUMINUM ASSOCIATION, INC.
Publications Department
818 Connecticut Avenue, N.W.
Washington, DC 20006
202-862-5100

To obtain another point of view regarding aluminum, write to the Aluminium Development Council in Australia. Their publications list is as impressive as the Aluminum Association, Inc. One such publication, Aluminium Technology, Book 5: Finishing Aluminium, has been an invaluable source of information and my constant technical companion since I first began anodizing aluminum.

In the post Crocodile Dundee era, we should give the folks **down under** the opportunity to be heard. What do you think, mate?

Write to our friends down under:
ALUMINIUM DEVELOPMENT COUNCIL OF AUSTRALIA, LIMITED
99 Elizabeth Street
Sydney, Australia NSW 2000

THE SEARCH

I suggest that if your local library does not have a copy of these, or any other book or periodical you need, that you investigate the inter-library loan system. This is an incredible way to complete your technical or visual research, no matter what subject you are interested in!

HEALTH AND SAFETY

The Michael McCann book, Artist Beware, is a must for everyone to read, not just artists! Everything you need to know regarding the materials you may work with is discussed in detail. Keep a copy in your studio at all times. This book was published by Watson-Guptill, New York, New York in 1979.

In 1985, McCann wrote the Health Hazards Manual for Artists, published by Nick Lyons Books, New York, New York. Check it out.

As a child, I discovered that catalogs were an excellent source of information. They were often as good as the encyclopedia, and contained better pictures! Write to the Lab Safety Supply Company for their 300 page catalog. The information and/or the pictures are worth it.

On the cover of the Lab Safety Supply catalog is a picture of a beautiful lake with trees, bushes, etc. Indeed, a tranquil setting. The caption in the lower left reads, "An extraordinary scene like this takes nature thousands of years to create. A hazardous materials incident in a town nearby could destroy it in seconds. By keeping in close compliance with new industry safety standards, you'll be doing all you can to make sure that never happens. Take a look inside for a complete selection of compliance materials and equipment."

They take their job seriously. I hope we will all do the same with our materials and chemicals in the studio.

Besides all of the information on the safety equipment available, you will find 20 pages devoted to other books of interest regarding chemistry, safety, etc.
LAB SAFETY COMPANY
P.O. Box 1368
Janesville, Wisconsin 53547
800-356-0783

STRANGE SOUNDING BOOKS

How is your perspective on life, your art, whatever? Zen and the Art of Motorcycle Maintenance by Robert M. Pirsig, published by the Morrow Publishing Company in 1974, may help to give your perspective on life a new slant. No, you don't have to like motorcycles to read this book. The only requirement is a zest for life and an eagerness to understand how the fabric of life is woven that creates our individual life histories. **Whew, I can't imagine that I just wrote that!**

The Thomas Register or the ThomCat is the most comprehensive listing of products and services in one source. The current Thomas Register numbers 21 volumes.
* Volumes #1 - 12 cover Products and Services
* Volumes #13 and 14 cover Company Profiles
* Volume #14, one half of which covers Product and Trademark Indexes

* Volumes #15 - 21 cover the Catalog File
Check out this overwhelming sourcebook at your local library. If the local library does not have the Thomas Register, ask at a nearby university/college library.

A friend of mine who just started a jewelry business used the Thomas Register to find possible sources to order materials from. He got all the phone numbers, many of which were 800 prefixes.

Many of the companies he called sent free samples. He found new and wonderful products he never knew existed! He has received lots of sample products to experiment with, his ideas have grown in **all** directions! All of these bonuses because he learned how to work the Thomas Register with the telephone company's services. The same success may happen to you.

If you're like me, the need to know certain facts, generic names of chemicals, processes and procedures is paramount. Here are a couple of those strange sounding books with some of the facts, processes and procedures you probably have wanted to know more about, but did not know where to look. The Scientific American Cyclopedia of Receipts, Notes and Queries edited by Albert A. Hopkins (1906, 25th edition published by Munn and Company, New York) and Henley's Twentieth Century Formulas, Recipes and Processes edited by Gardner D. Hiscox (The Norman W. Henley Publishing Company, New York, 1930). Look for these two books in your local library or used book store. I don't think you will be disappointed. The information inside of each of these books will keep you thinking for a **very long time.** I've seen copies of Henley's book for sale in the Publishers Central Bureau catalogs. Write for the current book catalog:
PUBLISHERS CENTRAL BUREAU
1 Champion Avenue
Avenel, New Jersey 07001

ARTISTIC STIMULATIONS

When you think about color, textures, patterns, and the forms that might stimulate your imagination, consider looking at fashion design in clothing, shoes, handbags, gloves, hats, watches, etc. Don't do the obvious. I agree that current fashion is what you have to consider if you are making jewelry as a fashion accessory. But, why not go back into the history of fashion? Everything comes around again! I subscribe to "Vogue" magazine and "W" from Womens' Wear Daily.
VOGUE
P.O. Box 55979
Boulder, Colorado 80321
W
Subscription Department
P.O. Box 56182
Boulder, Colorado 80321

When I first started doing the aluminum jewelry in 1981, I never thought I'd be subscribing to the above mentioned periodicals. At first, I subscribed to every fashion rag I could lay my hands on. After a year's worth of a **very full mailbox,** I settled on the two publications that I felt fed my imagination in the best possible way.

I look at fashion as a way to see what some people will be wearing next season. Let's face it, most of us will never weigh 98 pounds and wear the clothing seen in "Vogue" magazine. The styles, colors and patterns are great to look at to help stimulate ideas. I glance at the forms, check out the overlays, examine buttons and button holes, profile shapes of shoes, textures on hand bags, etc. Since jewelry is an accessory to clothing, why not have a preview of coming attractions?

COLOR TRENDS
Are you interested in knowing what color trends will look like in the future? If so, write to Color Trends for their yearly subscription rates.
COLOR TRENDS
8037 9th NW
Seattle, WA 98117

Aluminum is the American spelling.
Aluminium is the British Commonwealth spelling.

Brooch by David LaPlantz, fabricated, painted aluminum, engraved, actual size, 1981.

Pendant by David LaPlantz, fabricated and engraved, painted aluminum and Delrin, 3" diameter, 1983.

SAFETY

Safety Considerations
Waste Disposal

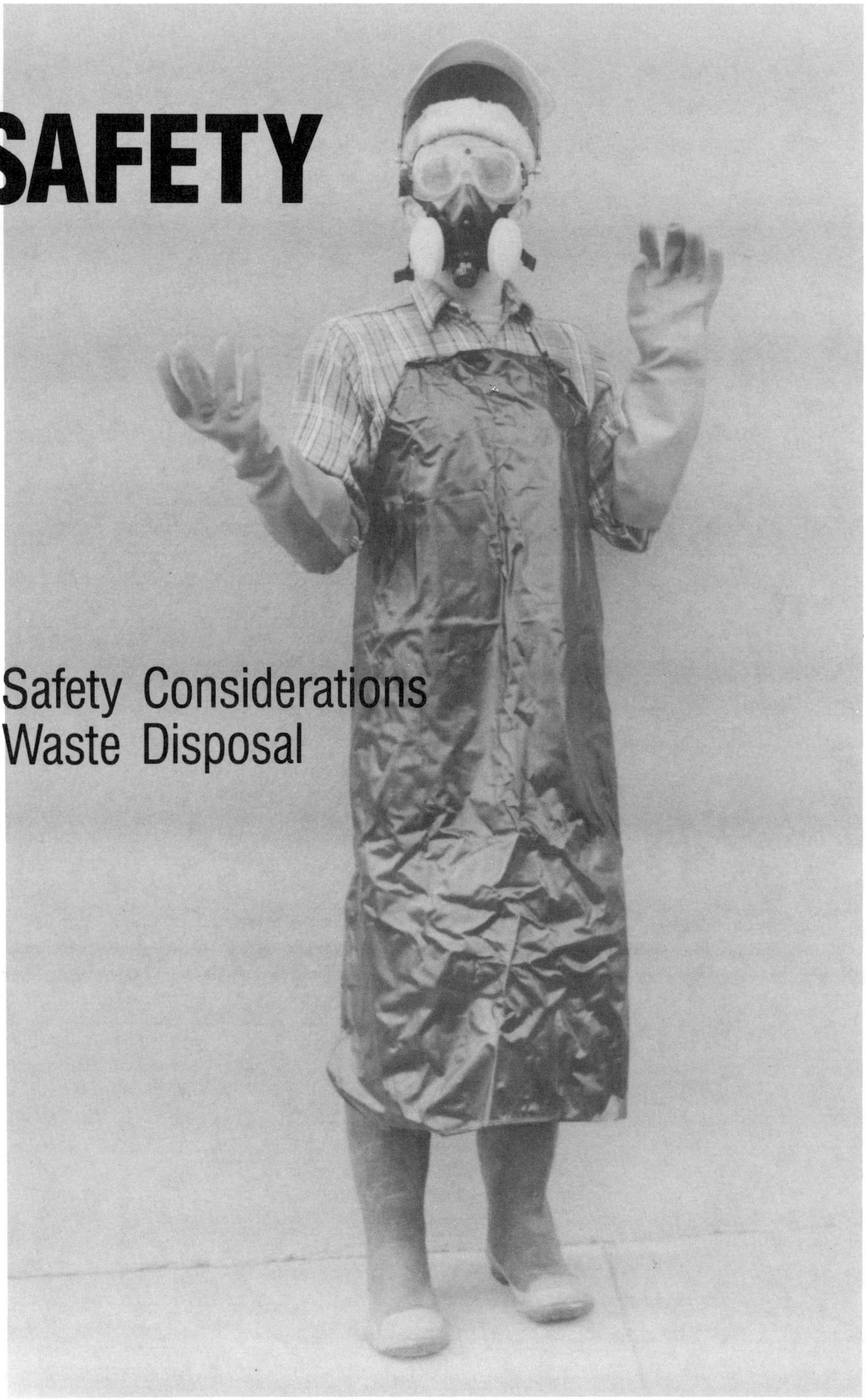

Safety Considerations

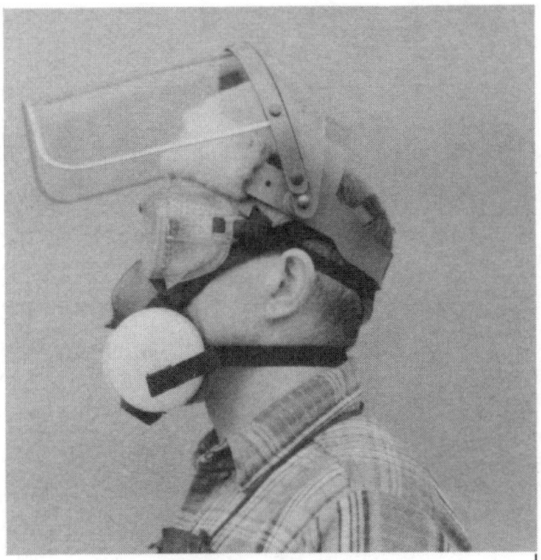

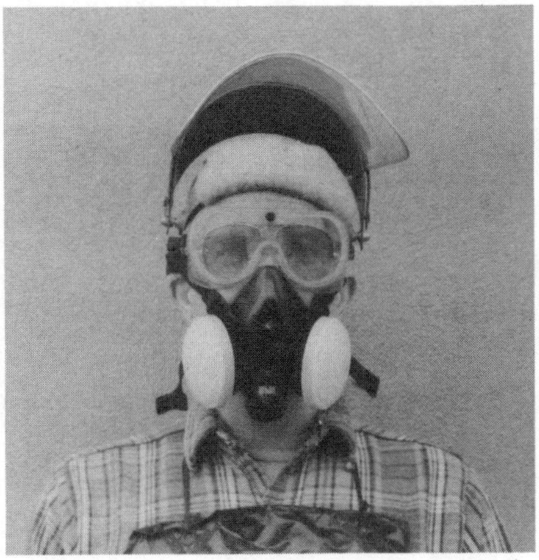

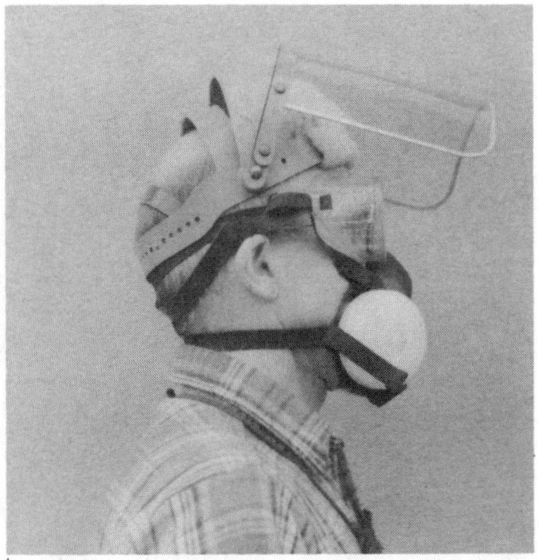

SAFETY CONSIDERATIONS

Let's be completely honest: to be totally safe from the effects of the chemicals used in aluminum anodizing, you should avoid or eliminate any contact with them.

With the aforementioned in mind, assuming you know how to safely handle, you will work with care and wear all of the proper safety equipment, then the chemicals can be used in a safe and sane manner.

Everything depends on how careful, cautious and concerned you are about continuing to live a healthy life.

The photos on this page, although taken with the serious side of my personality in mind, were also taken to produce some laughs. If you are laughing, please take a moment to think seriously about the safety gear required to protect you from the inherent dangers from the chemicals needed to anodize aluminum.

Don't be cavalier or stupid. No mater what chemical or materials you work with--know all you can learn about their dangers. Learn how to eliminate any danger to you, your family, your home, our land, water, the air and the planet in general.

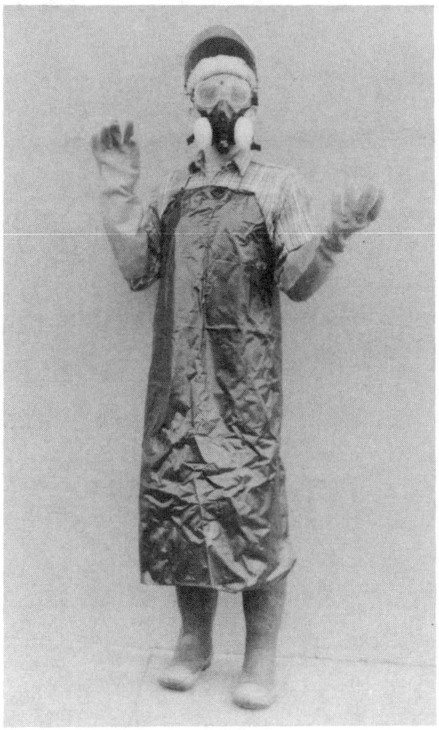

You have only one body. Once you damage your parts, that's it.

Therefore, think before you act. Think about the consequences/ramifications of each movement or action. The price may not be worth the action.

This is not a book written only on safety, but a workshop on a particular metalworking process. I have included as much as possible on each subject within the limits of this book. If you are unfamiliar with the proper use of any chemical, piece of equipment, etc., it is your responsibility to learn the information required from other, more proper, sources. Check the bibliography listings.

David LaPlantz demonstrating/wearing the safety equipment required for anodizing aluminum. This protective clothing includes: goggles, rubber gloves, plastic/rubber boots, and respirator--see pages 40-41.

MATERIAL SAFETY DATA SHEETS

In order to work more safely in your studio, obtain and read the Material Safety Data Sheets on all the products you use.

First off, you may be asking what is a Material Safety Data Sheet? A Material Safety Data Sheet (MSDS) is a report on the chemical toxicity of a given product.

A MSDS includes the following:

I. Product Identification and Manufacturer
II. Hazardous Ingredients
III. Physical Data
IV. Fire and Explosion Data
V. Health Hazard Information
VI. Reactivity
VII. Spill, Leak and Disposal Procedures
VIII. Special Protection Information
IX. Special Precautions and Comments

Read your MSDS with care. To obtain the needed information requires some creative detective work.

Contact the company you purchased the product from for an MSDS regarding that particular item.

HOW TO READ AND UNDERSTAND A MATERIAL DATA SHEET

To better understand the MSDS, contact your local Occupational Safety and Health Administration Office (OSHA) and request the form, "How to Read and Understand a Material Safety Data Sheet." This four page paper will be a big help in understanding the nature of the products you use. After obtaining MSDS on certain products used in anodizing and related processes, I am much more careful with **all chemicals.**

There is no reason to be half safe. Learn and understand what to and what not to do. Your life will be healthier and probably longer.

TOXICITY HOTLINE

If you have questions regarding toxicity, disposal procedures and other problems relating to chemicals in products, call the Chemical Manufacturers Association Hotline toll-free 800-262-8200 (in Washington, DC: 887-1315). A friendly voice will answer your question(s) or refer you to someone who will assist you.

An example of an MSDS is on the following pages.

The enclosed fume hood, as pictured, provides excellent ventilation. Therefore it's a great place to do anodizing and the related processes. The interior of the fume hood is well-lit, has a fan, running water, fire and chemical resistant walls, countertop, and sink. The plastic door slides up and down for protection. The door has been pulled down to the "glass safety mark" or optimum position to allow for a fresh air draw, while leaving enough space for the worker's arms and hands to fit inside the hood. Thus, if a spill occurred, the plastic door would protect the worker from harm.

Glass Safety Mark

The cost of a fume hood is out of the reach for most of us, at least when we are starting out. If, however, you are working in a school or laboratory situation where a fume hood exists, be sure to use it to the maximum.

MATERIAL SAFETY DATA SHEET

ALUMINUM BORDEAUX
1-3756-0-133

REVISED DATE: 04/01/86
PAGE: 1

COMPANY NAME REMOVED

HMIS

HEALTH	1 *
FLAMMABILITY	1
REACTIVITY	1
PERSONAL PROTECTION	E

A. PRODUCT IDENTIFICATION

TRADE NAME: ALUMINUM BORDEAUX R CONC — CAS NO: PROPRIETARY
SYNONYMS: C.I. ACID RED 194
CHEMICAL FAMILY: METAL COMPLEX OF AZO DYE
FORMULA: PROPRIETARY

B. HAZARDOUS COMPONENTS

COMPONENT	CAS NO	%	TLV
CHROMIUM AS AN INTEGRAL PART OF DYE MOLECULE	7440-47-3	>0.1	0.5 MG/M3 OSHA

C. PHYSICAL PROPERTIES

BOILING POINT ('F): NOT AVAILABLE
VAPOR PRESSURE (MM HG): NOT AVAILABLE
VAPOR DENSITY (AIR = 1): NOT AVAILABLE
SOLUBILITY IN WATER: SOLUBLE
APPEARANCE AND ODOR: RED POWDER, MILD ODOR
SPECIFIC GRAVITY: NOT AVAILABLE
PERCENT VOLATILE (BY WEIGHT): NOT AVAILABLE
EVAPORATION RATE: NOT AVAILABLE
PH (% IN WATER): NOT AVAILABLE

D. FIRE & EXPLOSION HAZARD DATA

FLASH POINT ('F): NOT AVAILABLE   FLAMMABLE LIMITS: LEL UNKNOWN UEL UNKNOWN
EXTINGUISHING MEDIA: WATER X ALCOHOL FOAM X CO2 X DRY CHEMICAL X
SPECIAL FIRE FIGHTING PROCEDURES:
PREVENT RUNOFF TO SEWERS AND BODIES OF WATER FROM FIRE FIGHTING INVOLVING THIS PRODUCT AS THIS PRODUCT CONTAINS CLEAN WATER ACT PRIORITY POLLUTANTS AND WILL ALSO BE A RCRA HAZARDOUS WASTE MATERIAL.
UNUSUAL FIRE AND EXPLOSION HAZARDS:
ORGANIC DUSTS HAVE POTENTIAL TO BE EXPLOSIVE WITH STATIC SPARK OR FLAME INITIATION. MAINTAIN GOOD HOUSEKEEPING FOR CONTROL OF DUST.

E. REACTIVITY DATA

STABILITY: STABLE   CONDITIONS TO AVOID: NONE KNOWN

HAZARDOUS POLYMERIZATION : WILL NOT OCCUR
CONDITIONS TO AVOID: NONE KNOWN

INCOMPATABILTY:
STRONG OXIDIZING AGENTS
HAZARDOUS DECOMPOSITION PRODUCTS :
THERMAL DECOMPOSITION MAY PRODUCE OXIDES OF CARBON, NITROGEN, AND SULFUR. ASH WILL CONTAIN CHROMIUM.

F. PERSONAL PROTECTIVE EQUIPMENT INFORMATION

RESPIRATORY PROTECTION:
USE NIOSH APPROVED RESPIRATOR OR DUST MASK.

PROTECTIVE GLOVES:
BUTYL RUBBER, PVC OR NEOPRENE

EYE PROTECTION:
SAFETY GLASSES OR CHEMICAL SPLASH GOGGLES

VENTILATION:
LOCAL VENTILATION RECOMMENDED - MECHANICAL VENTILATION MAY BE USED.

OTHER PROTECTIVE EQUIPMENT:
NONE KNOWN

* G. HEALTH HAZARD DATA *

THRESHOLD LIMIT VALUE: NOT ESTABLISHED LD50 (ORAL-RAT) >5000 MG/KG BODY WEIGHT
DERMAL IRRITATION-RABBIT: MILDLY IRRITATING EYE IRRITATION-RABBIT: MILDLY IRRITATING

EFFECTS OF OVEREXPOSURE:
CHROMIUM: CHROMIUM AND CERTAIN COMPOUNDS OF CHROMIUM HAVE BEEN REPORTED TO CAUSE DAMAGE TO THE LUNGS RESULTING IN CUMULATIVE DAMAGE TO THE LUNGS AND FROM RESULTS OF TESTING WITH LABORATORY ANIMALS ARE SUSPECTED OF CAUSING CANCER. DERMATITIS HAS BEEN REPORTED FROM REPEATED, PROLONGED CONTACT WITH CHROMIUM COMPOUNDS. SENSITIZATION MAY OCCUR ON REPEATED CONTACT. THERE HAVE BEEN NO REPORTS OF ANY OF THESE EFFECTS FROM THE USE OF THIS PRODUCT.

KNOWN EFFECTS ON OTHER ILLNESSES:
NO KNOWN EFFECTS ON OTHER ILLNESSES.

LISTED CARCINOGEN: NONE OSHA NTP X IARC X OTHER

* H. EMERGENCY AND FIRST AID PROCEDURES *

SKIN:
WASH THOROUGHLY WITH SOAP AND WATER FOR 15 MINUTES. IF SKIN IRRITATION OCCURS, SEEK MEDICAL ATTENTION.

EYES:
FLUSH IMMEDIATELY UNDER RUNNING WATER FOR FIFTEEN MINUTES. IF REDNESS OR IRRITATION OCCURS, SEEK MEDICAL ATTENTION.

INHALATION:
REMOVE SUBJECT TO FRESH AIR. SEEK MEDICAL AID IF LUNG IRRITATION PERSISTS OR IF BREATHING BECOMES DIFFICULT.

INGESTION:
IF INGESTED, SEEK MEDICAL ASSISTANCE.

NOTE TO PHYSICIAN:
NONE KNOWN

* I. SPILL AND DISPOSAL INFORMATION *

STEPS TO BE TAKEN IN CASE OF SPILL:
CONTAINS HEAVY METALS. SPILLS SHOULD BE SWEPT OR SHOVELED UP AND COLLECTED FOR DISPOSAL. CLEAN UP BY REMOVAL OF CONTAMINATED SOIL OR BY FLUSHING WITH A LIMITED QUANTITY OF WATER IF APPROPRIATE. PLACE MATERIAL OR COMTAMINATED SOILS IN AN SUITABLE DISPOSAL CONTAINER.

WASTE DISPOSAL INFORMATION:
THIS PRODUCT BECOMES A RCRA HAZARDOUS WASTE WHEN IT BECOMES A WASTE MATERIAL DUE TO CHROMIUM CONTENT. RECOMMENDED DISPOSAL IS BY INCINERATION WITH PROPER AIR SCRUBBING DEVICES AT APPROVED FACILITIES. MAY ALSO BE LANDFILLED IN APPROVED FACILITIES.

RCRA HAZARDOUS WASTE: NO YES X RCRA # D007 CERCLA: NO YES X RQ NONE

FOLLOW ALL LOCAL, STATE, AND FEDERAL REGULATIONS.

* J. OTHER REGULATORY INFORMATION *

TSCA: WE CERTIFY THAT ALL COMPONENTS OF THIS PRODUCT ARE REGISTERED UNDER THE REGULATIONS OF THE TOXIC SUBTANCES CONTROL ACT.

CWA: THIS PRODUCT CONTAINS THE FOLLOWING PRIORITY POLLUTANTS AND THEIR CONCENTRATION:
CONTAINS PRIORITY POLLUTANT CHROMIUM AT GREATER THAN 0.1%.
FDA / USDA: THIS PRODUCT IS REGISTERED WITH FDA / USDA NO X YES FOR USE PER SECTIONS:
NONE

DOT: REGULATED YES NO X RQ NONE UN/NA NO. NONE

MATERIAL SAFETY DATA SHEET

ALUMINUM BORDEAUX
1-3756-0-133

REVISED DATE: 04/01/86
PAGE: 3

COMPANY NAME REMOVED

SHIPPING NAME: TEXTILE DYEING OR PRINTING, PIGMENTED, OR REDUCERS OR EXTENDERS, TEXTILE DYEING OR PRINTING PASTE. NMFC 154800.

HAZARD CLASSIFICATION: NOT D.O.T. REGULATED

* K. S P E C I A L P R E C A U T I O N S *

HANDLING AND STORAGE INFORMATION:
KEEP CONTAINERS CLOSED. AVOID GETTING ON SKIN OR IN EYES WHEN HANDLING PRODUCT.

OTHER PRECAUTIONS:
NONE KNOWN

IN ACCORDANCE WITH GOOD PRACTICES OF PERSONAL CLEANLINESS AND HYGIENE, HANDLE WITH DUE CARE AND AVOID ANY UNNECESSARY CONTACT WITH THIS PRODUCT.

THIS INFORMATION IS BEING SUPPLIED TO YOU UNDER OSHA "RIGHT TO KNOW" REGULATION 29 CFR 1910.1200 AND IS OFFERED IN GOOD FAITH AS TYPICAL VALUES AND NOT AS A PRODUCT SPECIFICATION. THE INFORMATION CONTAINED HEREIN IS BASED ON THE DATA AVAILABLE TO US AND IS BELIEVED TO BE TRUE AND ACCURATE. NO WARRANTY, EXPRESSED OR IMPLIED, REGARDING THE ACCURACY OF THIS DATA, THE HAZARDS CONNECTED WITH THE USE OF THE MATERIAL, OR THE RESULTS TO BE OBTAINED FROM THE USE THEREOF, IS MADE. SANDOZ CHEMICALS CORPORATION ASSUMES NO RESPONSIBILITY FOR DAMAGE OR INJURY FROM THE USE OF THE PRODUCT DESCRIBED HEREIN.

VENTILATION

Ventilation is of utmost importance in the anodizing process. The fumes/gasses produced by the caustic soda, nitric and sulfuric acids, etching and bleaching solutions, nickel acetate sealer and other chemicals require being cautious.

Caustic soda gives off hydrogen gas when reacting with the aluminum during the etching. During strong etching, the mist over the etch tank also carries some caustic. If breathed, the caustic causes a coughing or choking sensation.

Nitric acid gives off hydrogen gas and oxides of nitrogen.

Sulfuric acid gives off hydrogen gas during the anodizing process. Hydrogen is a very combustible gas. It can become an explosion hazard if not vented.

When mixing the dyes and nickel acetate with water, the dust is hazardous.

Do not be casual around the anodizing, dyeing and sealing processes. Your health and later life may depend on the care and concern you exhibit now.

After considerable research into the subject of ventilation, it became obvious that I could devote about one-third of this book to just that topic. Since ventilation is one aspect of anodizing, and there is so much to cover on the entire subject, I decided to do the next best thing.

The **best** book that I have found on the subject of ventilation is entitled, strangely enough, Ventilation by Nancy Clark, Thomas Cutter, and Jean-Ann McGrane.

ORDER IT FROM:
Center for the Safety in the Arts
5 Beekman Street
New York, New York 10038
212-227-6220

This 117 page, well written book is jam-packed with everything you need to know about ventilation for the artist, craftsperson, and others involved in the arts.

This symbol will appear throughout the book. It will warn you when to wear the proper safety equipment. Who knows, when wearing this equipment you might make Blackwell's "Top Ten" list.

Sulfuric, nitric and hydrochloric acids will cause severe burns to skin and eyes, with potential loss of vision. Nickel acetate may irritate your eyes.

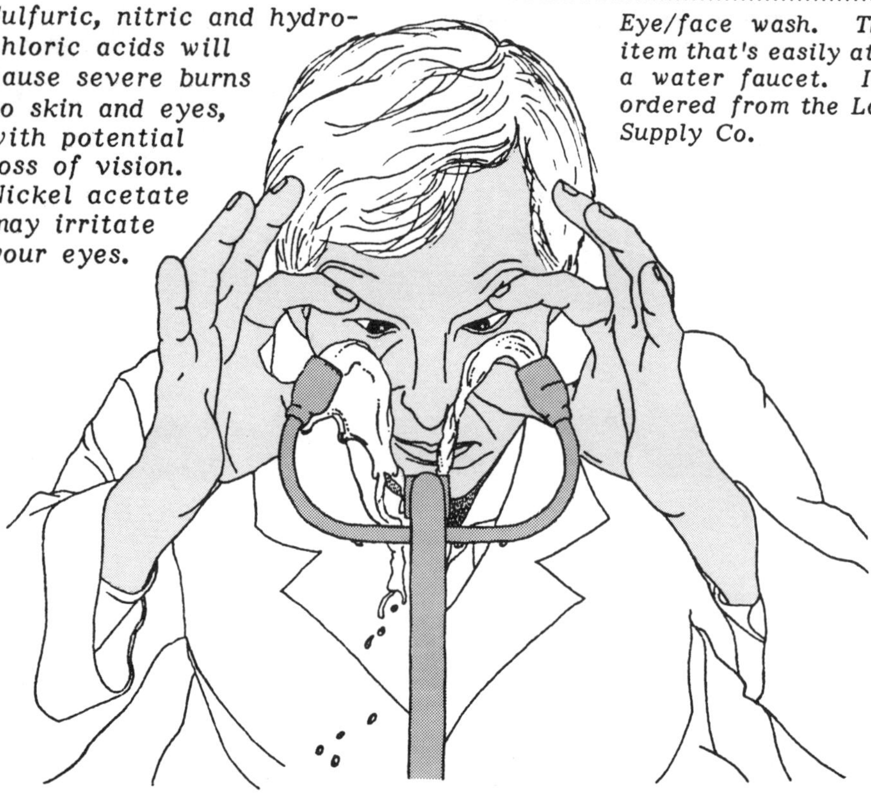

Eye/face wash. This is an item that's easily attached to a water faucet. It may be ordered from the Lab Safety Supply Co.

CHEMICALS SPLASHED IN YOUR EYES

What do you do if you have an accident and splash acids, caustic soda, nickel acetate, or the organic aluminum dyes in your eyes?

Flush your eyes with copious amounts of cool tap water, then apply cold compresses. Next, you need to be transferred to a physician, preferably at a hospital emergency room for further examination and treatment.

GOGGLES

All of the Material Safety Data Sheets (MSDS) that I have read regarding acids, dyes and sealing solutions containing nickel acetate, recommend wearing chemical splash goggles as the proper eye protection, rubber gloves, an apron, boots, and the appropriate approved respirator.

In the photos, I am wearing chemical splash goggles with a full-face plastic shield. It may be overkill. But I'd rather not have any chemical splash on any part of my face, especially in my eyes.

The chemical splash goggles keep the splash out of the eyes. Be aware: Splash goggles are different from the regular goggles you are used to wearing. Splash goggles have no vent holes on the side.

The full face shield protects the rest of the face. Full face respirators provide more complete protection. This will be discussed in more detail in the respirator section that follows.

RUBBER GLOVES

The rubber gloves I am wearing are long enough so I can reach into my tanks, fishing out the items that invariably drop off their racks. This always happens at the most inopportune times -- like when you can't locate your gloves in a hurry! Your skin **will** be affected by all of the chemicals in anodizing. Synthetic latex and vinyl (polyvinyl chloride) offer resistance to most solvents, acid and alcohols. Be certain to check with the dealer or manufacturer as to **exactly** what chemicals the gloves **will resist.**

Liquid sodium hydroxide and glacial acetic acid, used to raise or lower the pH value, cause burns to the skin. When the spill occurs you will not notice the burning sensation/the damage until later. Wear the ... gloves!

PLASTIC/RUBBER APRON

This apron is exactly like the kind we used to wear in chemistry class. It is long enough to protect from the spill that should never happen. There **is** a spill waiting for each of us. Count on it! Rubber covered cotton and the new non-woven material made from polyethylene fibers provide the needed protection.

RUBBER BOOTS

The rubber boots will protect you from spills, and possible electrocution. You never know when you might be standing in a spill, be unaware of it, and turn on a piece of power equipment, and...... My boots also have steel toes. They came with the boots. If a heavy item falls on my feet, I am safe. Smashed toes make walking painful!

RESPIRATOR

This is the most valuable tool to promote and provide healthy lungs for your future breathing. Invest in a happy, healthy future, buy the best respirator you can afford. Your lungs will thank you every day.

The following information on respirators and the proper cartridge usage has been taken from the Safety Equipment and Supplies Catalog.
Lab Safety Supply Company
P.O. Box 1368,
Janesville, Wisconsin 53547
800-356-0783

The two types of respirators available are half face and full face. The full face provides maximum protection. Be sure to purchase a respirator that is NIOSH/MSHA approved. NIOSH stands for National Institute for Occupational Safety and Health. MSHA stands for Mine Safety and Health Administration.

The respirator you purchase should feature a standard cartridge holder, which permits the facepiece to be adapted to a variety of uses by simply changing the cartridge-filter combination.

Two types of cartridges are available:
1. Chemical cartridges are filled with treated activated carbon which absorbs gases. The adsorptive capacity is limited, however. When the wearer detects any taste, odor or irritation, immediately leave the area and change the cartridges.
2. Mechanical filters provide protection by trapping the particles. When breathing resistance becomes excessive, the filter should be changed.

WILSON BRAND 1200 SERIES RESPIRATORS

A full face respirator is for hazardous areas where eye and face protection are needed along with respiratory protection.
FULL FACE RESPIRATOR:
Nickel acetate (Sealer):
HEPA Filter #S3202
Nitric acid: Air line/oxygen*
Phosphoric acid:
Air line/oxygen*

* Air can be supplied from a plant compressed air system providing grade D breathing air or by a clean air pump.

A half mask respirator covers the nose and mouth, providing respiratory protection only.
HALF MASK RESPIRATOR:
Caustic soda: Dusts/Mists #S3200
Sulfuric acid: Dusts/Mists #S3200
Dyes: HEPA Filter #S3202
Aluminum oxide dust and silica: HEPA filter #S3202. (Silica is a by product of using glass brush -- see Pretreatment chapter.)

WILSON BRAND 1200 SERIES CARTRIDGES/FILTERS

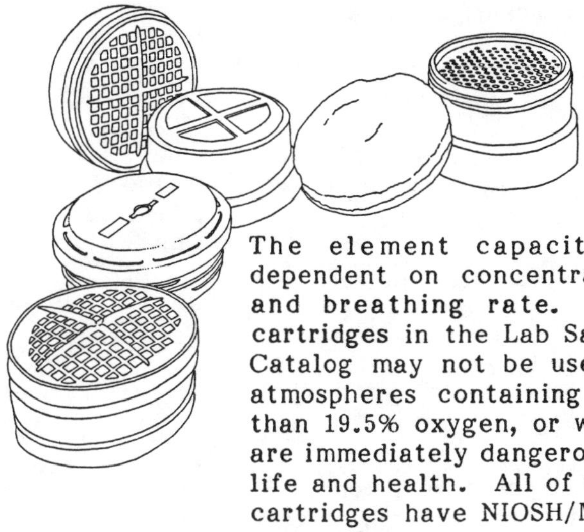

The element capacity is dependent on concentration and breathing rate. The cartridges in the Lab Safety Catalog may not be used in atmospheres containing less than 19.5% oxygen, or which are immediately dangerous to life and health. All of their cartridges have NIOSH/MSHA approval.

The manufacturer of these respirators suggests that when nickel acetate and phosphoric acid are heated, the air line respirator be used.

Do not be alarmed by the above comment regarding the air line suggestion. If you have an adequate ventilation system that draws away the fumes, this oxygen/air line system is not required. Besides, the cost of such a system would probably keep you from anodizing aluminum altogether.

BAKING SODA

I keep boxes of baking soda in my studio. There is always a two pound box next to the area where the anodizing takes place. The baking soda will neutralize the sulfuric acid in case of a spill.

SAFETY CONSIDERATIONS FOR PROPANE TANKS

Practice safety at all times. If you have questions about your appliance system or LP-gas system, contact your local LP-gas dealer at once. Do not wait, ask now! Do not store LP-gas tanks or cylinders indoors or in enclosed areas.

Always store cylinder(s) in the upright position.

Do not expose LP-gas containers to fire or heat above 120 degrees F.

Always store with the service valve closed and plugged.

Do not attempt to repair LP-gas containers, valves or regulators.

Always tighten POL connection-left hand thread. (POL stands for Prest-O-Lite, the manufacturer's name). Do not tamper with safety release valve device which is opposite the outlet connector.

Valve information may be obtained from the appropriate manufacturer.

Always secure tank to a wall with a chain so that it cannot fall over.

Check all tank and line connections periodically to be sure they are **all** tight. When testing for leaks, use soapy water-**not matches.**

When using tank, slowly open POL service valve all the way.

Close cylinder valve when not in use.

When you take your LP-gas tank to the LP-gas dealer for filling, transport it in the proper upright position, with the valves closed and a POL plug inserted. Secure the tank against falling or rolling.

If the Lp-gas tank is dropped or damaged in any way, take it to your local LP-gas dealer immediately for a check-up. A leaking tank is dangerous!

If the cylinder is equipped with a fixed liquid level gauge (10% valve) be certain it is tight.

SAFETY CONSIDERATIONS FOR ELECTRICAL APPLIANCES

Read all instructions before operating.

To protect against electrical shock, do not immerse cords, plugs, base of unit, or complete appliance in water or any other liquid. Do not stand in water when turning off, on or plugging/unplugging the electrical appliance. Shocking results may occur!

Unplug from outlet when not in use and before cleaning.

Do not operate any appliance with a damaged cord or plug. If you do not feel confident about repairing, return to authorized service facility for examination, repair, or adjustment.

Do not let the power cord hang over edge of table or counter, or to touch hot surfaces.

Be certain the control knob is turned off before inserting or removing plug from wall outlet.

Basic considerations regarding the use of electricity must always prevail -- think safety at all times. Think first, act second. Your life hangs in the balance.

IMPORTANT SAFEGUARDS FOR IMMERSION HEATERS

Never plug the unit into an electrical outlet until the heat probe has been placed in the liquid.

Always use the rubber safety handle when unit is in use. Avoid any contact with the metal probe when it is hot.

To avoid electrical hazards, never allow cord or plug to get wet. The metal heat probe is the only part of the unit that should ever come in contact with water or liquid.

Unplug the unit when not in use.

Allow the heat probe to cool thoroughly before cleaning or storing.

Discontinue use if the cord becomes frayed or damaged.

Center For Safety In The Arts

If you want to know who to contact regarding health hazards, workshops and other health related consultation services contact the people at:

CENTER FOR SAFETY IN THE ARTS
5 Beekman Street, Suite 1030
New York, New York 10038
212-227-6220

They have an incredible list of publications and data sheets that just might help to save your life, if not, prolong it. Check them out. Write for their information list.

The following has been taken verbatim from the information provided by the Center for Safety in the Arts.

The Center for Safety in the Arts (CSA)* is a national clearing house for research and education on hazards in the arts, including visual and performing arts, museums and educational facilities. CSA is partially supported with public funds from the National Endowment for the Arts, the New York State Council on the Arts, the New York State Health and Safety Training and Education Program and the New York City Department of Cultural Affairs.

CSA has received grants from the March of Dimes Birth Defects Foundation, Exxon Corporation, C.S. Fund, the Buckeye Trust, National Museum Act and from numerous individuals.

CSA PROVIDES THE FOLLOWING:

Art Hazards Information Center. The Information Center answers about 50 written and telephone inquiries daily on art hazards. These include questions on hazards of specific materials, ventilation, respirators, and other precautions, referrals to physicians trained in occupational health, etc. CSA distributes over 70 books, pamphlets, articles and data sheets to disseminate information on art hazards.

Art Hazards News. CSA publishes a four page newsletter which appears 10 times per year. It covers many topics including new hazards, precautions, government legislation and regulations, lawsuits, and a calender of events. Presently, 4500 copies of each issue are printed. Of these about 1500 free copies are sent to arts organizations and publications; over 1500 are sent to subscribers. The remainder are distributed at lectures, workshops and conferences, and through the Arts Hazards Information Center. Reprinting of articles from the newsletter is encouraged, and many arts publications do so on a regular basis. Written permission through CSA must be obtained prior to publication.

A one year subscription to Art Hazards is $15.00.

Lecture Program. CSA provides speakers for lectures and full-day workshops for art schools, public school districts, and professional organizations for art teachers, artists, conservators and theater craftspeople.

Art Hazards Course. CSA offers a five day course on art hazards the second full week of every June in New York City. This course is intended for art teachers and administrators, theater professionals and others interested in art hazards. CSA also offers these courses to sponsoring institutions.

Consultation Program. CSA will conduct health hazard surveys for art schools, museums, public schools, art centers, theaters, etc. CSA will also provide consultation services to institutions building a new arts facility or renovating an existing facility.

Do you see what I mean? Wait until you see the publications and data sheet listing! Don't hesitate! Write for yours today! Your life may hang in the balance. I've always wanted to use that phrase.

* formerly the Center for Occupational Hazards

Waste Disposal

In anodizing aluminum hazardous/toxic wastes are produced. This waste cannot and should not be haphazardly dumped or disposed of down the drain, or thrown out onto the land. The chemicals are dangerous to the land and the underground water.

So what do you do? First off, don't panic. Second, decide the optimum size of your tanks (pints, gallons, etc.) in relation to the size of the aluminum you wish to anodize. For most of us, whose tanks will not be larger than three to five gallons, we are considered a small generator of toxic waste.

The following introduction is from Understanding the Small Quantity Generator Hazardous Waste Rules: A Handbook for Small Business, produced by the United States Environmental Protection Agency. I have included this information so that you understand the law and proper disposal procedures.

This thirty-one page booklet will tell you everything you need to know regarding the proper dumping of your hazardous waste. I have highlighted certain portions from this informative publication.

In 1976, Congress passed the Resource Conservation and Recovery Act (RCRA) which directed the U.S. Environmental Protection Agency (EPA) to develop and implement a program to protect human health and the environment from improper hazardous waste management practices. The program is designed to control the management of hazardous waste from its generation to its ultimate disposal -- from "cradle-to-grave."

EPA first focused on large companies, which generate the greatest portion of hazardous waste. In recent years, however, public attention has been focused on the potential for environmental and health problems that may result from mismanaging even small quantities of hazardous waste.

In November 1984, the Hazardous and Solid Waste Amendments to RCRA were signed into law. With these amendments, Congress directed EPA to establish new requirements that would bring small quantity generators who generate between 100 and 1000 kilograms (kg) of hazardous waste in a calender month into the hazardous waste regulatory system. EPA issued final regulations for these 100 to 1000 kg/mo generators on March 24, 1986. Most of the requirements are effective September 22, 1986.

Write to and request your free copy of Understanding the Small Quantity Generator Hazardous Waste Rules: A handbook for Small Business, EPA/530-SW-86-019 from
United States Environmental Protection Agency
Office of Solid Waste and Emergency Response
Washington, D.C. 20460

There are three categories of hazardous waste generators. For most of us who are doing anodizing on a small scale, we fit into the category of a generator of no more than 100 kg/month of hazardous waste. The following will explain the three categories.

KEY: 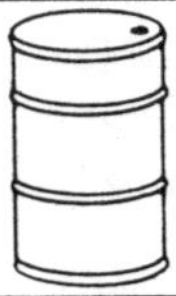 *= 1 barrel = about 200 kilograms of hazardous waste which is about 55 gallons*

CATEGORIES OF HAZARDOUS WASTE GENERATORS

I. Generators of No More Than 100 kg/month

If you generate no more than 100 kilograms (about 220 pounds or 25 gallons) of hazardous waste and no more than 1 kg (about 2 pounds) of acutely hazardous waste in any calender month, you are a conditionally-exempt small generator and the federal hazardous waste laws require you to:

1. Identify all hazardous waste you generate.
2. Send this waste to a hazardous waste facility, or a landfill or other facility approved by the state for industrial or municipal wastes.
3. Never accumulate more than 1000 kg of hazardous waste on your property. (If you do, you become subject to all the requirements applicable to 100-1000 kg/month generators explained in Hazardous Waste Rules Handbook.)

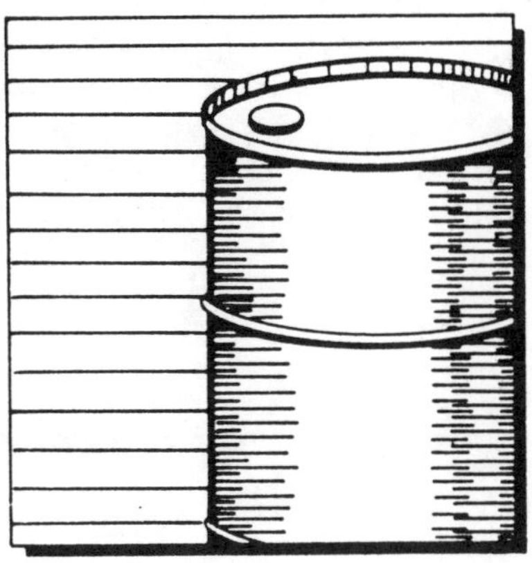

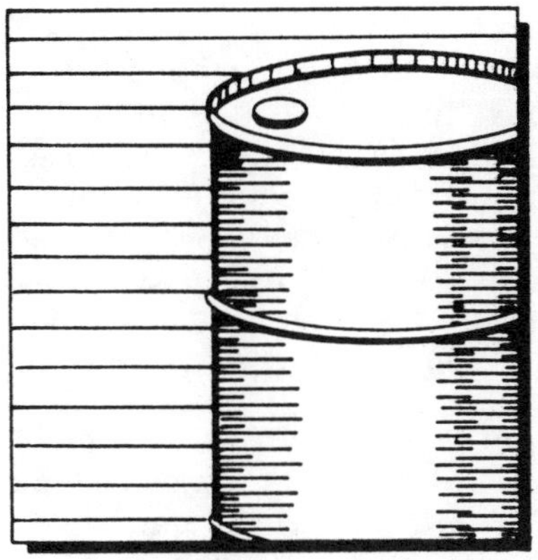

II. 100-1000 kg/mo Generators

If you generate more than 100 and less than 1000 kg (between 220 and 2,200 pounds or about 25 to under 300 gallons) of hazardous waste and no more than 1 kg of acutely hazardous waste in any month, you are a 100-1000 kg/mo generator and the federal hazardous waste laws require you to:

1. Comply with the 1986 rules for managing hazardous waste, including the accumulation, treatment, storage, and disposal requirements described in the Hazardous Waste Rules Handbook.

III. Generators of 1000 kg/mo or More

If you generate 1000 kg (about 2,200 pounds or 300 gallons) or more of hazardous waste, or more than 1 kg of acutely hazardous waste in any month, you are a generator of 1000 kg/mo or more and the federal hazardous waste laws require you to:

1. Comply with **all** applicable hazardous waste management rules.

Dispose of hazardous waste only in the appropriate manner and location.

Make sure that you check with your local, state, and federal agencies before doing anything with your hazardous waste. The transporting of hazardous waste could get you in some deep trouble. I know that in some larger metropolitan areas of California that hazardous waste collection sites are set up twice a year to collect the waste from small generators like you and me. All you have to do is to transport the waste to the collection site.

The California Waste Exchange

California Waste Exchange
Department of Health Services
Alternative Technology
714/744 P Street
P.O.Box 942732
Sacramento, CA 94234
916-324-1807

The California Waste Exchange (CWE) is a program set up by the State Government to promote the use, reuse, and exchange of industrial waste. CWE is administered by the Toxic Substances Control Division of the California Department of Health Services. The CWE was created as a service to the industry. The cost of raw material, the disposal of wastes (especially hazardous wastes), and the energy costs for processing have all escalated dramatically in the last few years. By promoting recycling the CWE hopes to assist in keeping costs down and thus to keep industry in California healthy and vigorous. The mutual advantages are obvious-reducing wastes going to Class I landfills (a short supply exists), saving natural resources and protecting the environment. State law requires that hazardous waste generators attempt to recycle their waste before disposing of them. Recycling is to everyone's benefit.

Even though most industries routinely try to reclaim as much material as possible from their own waste, the idea of recycling your waste off-site or using someone else's waste as feed stock may be new to you. Let us assure you that both recycling and waste exchanges are now proven concepts. Waste exchanges have been successful in Europe and Japan for approximately twenty years. In the United States, the number of exchanges has increased from four in 1976 to 29 in 1981. Since 1976 the CWE has successfully recycled numerous waste streams -- in one year over 50 waste streams were recycled that had previously been disposed.

What wastes can be recycled? Solvents (all types) are the materials which can most readily be reclaimed. In California, there are now a number of commercial solvent recyclers who can assist you in recycling your solvents. Solvents that cannot be cleaned up may be incinerated for their thermal content. Surplus material (out of date materials, laboratory reagents, etc.) are another important group of recyclables. Other examples of recycling exist. Pickling acid from steel industry can be converted into fertilizers or air pollution abatement solutions. Calcium hydroxide can be used as a soil conditioner, in the cement industry, or as a means of neutralizing acids. Metals can be recovered from plating solutions. Oils may be re-refined or used as a fuel. Empty containers may be reconditioned or recycled as steel scrap. Many other recycling or reuses are possible. We have named just a few.

The CWE programs offers the following services to industry:

1. Directory: CWE staff publishes a Directory of Industrial Recyclers. The Directory is to make the industry aware of presently available commercial recycling opportunities.
2. Newsletter/Catalog: CWE staff publishes a quarterly Newsletter/Catalog. The Newsletter is intended to keep people informed of developments in the filed of hazardous waste management (I.E., law, regulations, technology, etc.). The Catalog list wastes wanted and wastes available so that industry can buy, sell, or exchange their wastes.
3. Direct Assistance: CWE staff, when requested, will assist a company to recycle their waste. This can include plant inspection, process review, or meeting with plant representatives to inform them of regulatory or technological issues.

This costs nothing! There is no charge for any of these services. The State is funding this program as a public benefit to promote recycling. Recycling and CWE will Work! The CWE success will require some people to review their old "throw-away ethic" and change it to a new conservationist mode of thinking.

What does your State require??

EPA addresses & Phone Numbers

EPA AND STATE HAZARDOUS WASTE CONTACTS FOR ASSISTANCE
RCRA/Superfund Hotline
1-800-424-9346
(In Washington, D.C.:382-3000)

EPA Small Business Ombudsman
Hotline 1-800-368-5888
(In Washington, D.C.:557-1938)

National Response Center
1-800-424-8802
(In Washington, D.C.:426-2675)

U.S. EPA REGIONAL OFFICES

EPA Region I
Connecticut, Massachusetts, Maine, New Hampshire, Rhode Island, Vermont
State Waste Programs Branch
JFK Federal Building
Boston, Massachusetts 02203
617-223-3468

EPA Region II
New Jersey, New York, Puerto Rico, Virgin Islands
Air and Waste Management Division
26 Federal Plaza
New York, New York 10278
212-264-5175

EPA Region III
Delaware, Maryland, Pennsylvania, Virginia, West Virginia, District of Columbia
Waste Management Branch
841 Chestnut Street
Philadelphia, Pennsylvania 19107
215-597-9336

EPA Region IV
Alabama, Florida, Georgia, Kentucky, Mississippi, North Carolina, South Carolina, Tennessee
Hazardous Waste Management Division
345 Courtland Street, N.E.
Atlanta, Georgia 30365
404-347-3016

EPA Region V
Illinois, Indiana, Michigan, Minnesota, Ohio, Wisconsin
RCRA Activities
230 South Dearborn Street
Chicago, Illinois 60604
312-353-2000

EPA Region VI
Arkansas, Louisiana, New Mexico, Oklahoma, Texas
Air and Hazardous Materials Division
1201 Elm Street
Dallas, Texas 75270
214-767-2600

EPA Region VII
Iowa, Kansas, Missouri, Nebraska
RCRA Branch
726 Minnesota Avenue
Kansas City, Kansas 66101
913-236-2800

EPA Region VIII
Colorado, Montana, North Dakota, South Dakota, Utah, Wyoming
Waste Management Division (8HWM-ON)
One Denver Place
999 18th Street, Suite 1300
Denver, Colorado 80202
303-293-1502

EPA Region IX
Arizona, California, Hawaii, Nevada, American Samoa, Guam, Trust Territories of the Pacific
Toxics and Waste Management Division
215 Fremont Street
San Francisco, California 94105
415-974-7472

EPA Region X
Alaska, Idaho, Oregon, Washington
Waste Management Branch--MS-530
1200 Sixth Avenue
Seattle, Washington 98101
206-442-27778

THE ALUMINUM

Alloys
Painted Aluminum

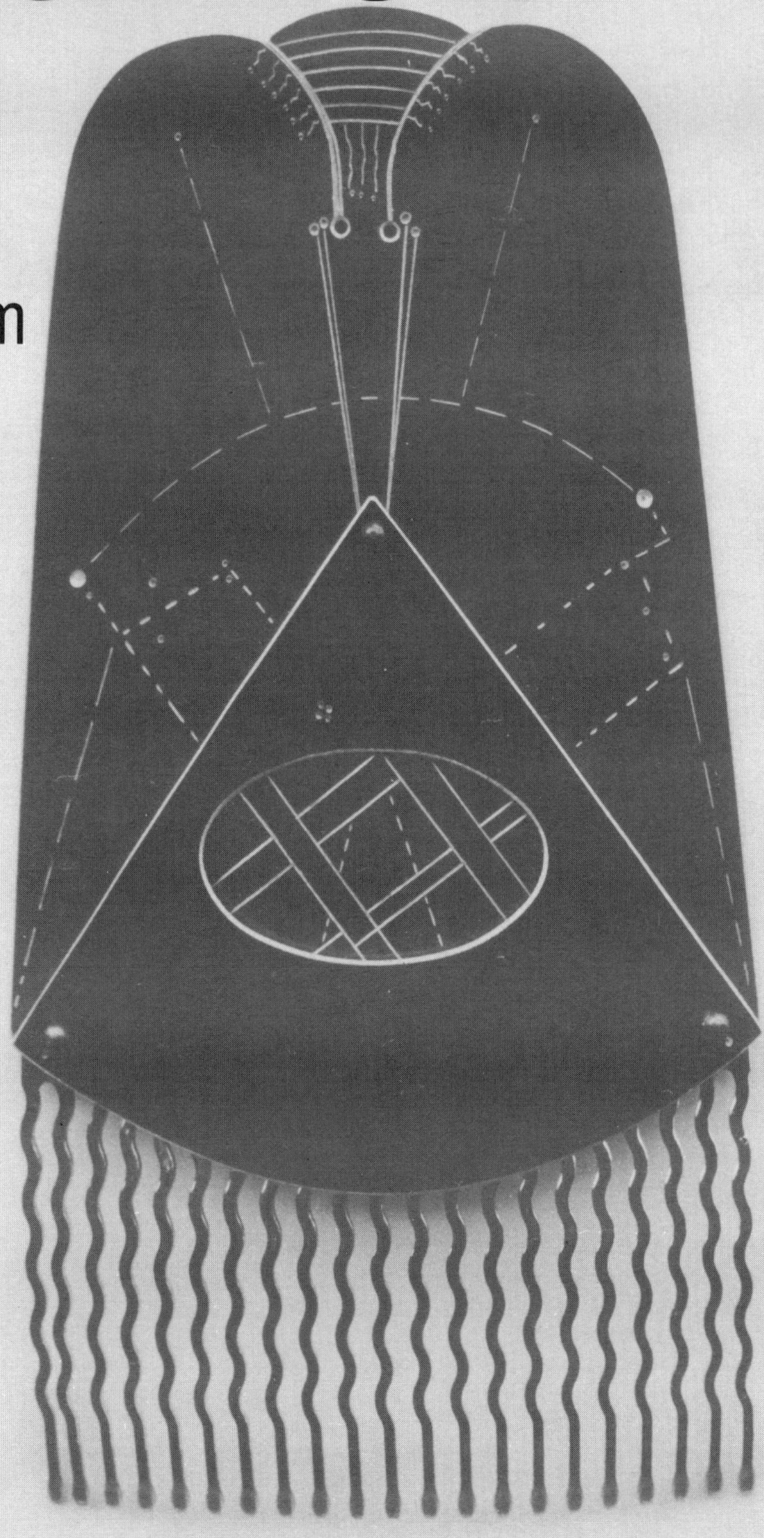

Hair comb by David LaPlantz, fabricated, painted aluminum, engraved, 1984.

Alloys

Frequently you will read/hear that this or that alloy is best for anodizing. I could not agree more, but, why not try anodizing and dyeing as many alloys as you can lay your hands on? Keep a sample ring of anodized and dyed aluminum along with these stats:

- how each piece was prepared prior to anodizing
- length of anodizing time
- what dye (cold or hot) was used
- immersion time
- alloy number
- plus any information that might be important.

This is how breakthroughs in any technique are discovered. Besides, do you really want to play follow-the-leader? Is Formula Art your goal? Do you have the courage to test your limitations? Who knows what **you** might discover?

This is my sample ring of available aluminum stock shapes.

Personally I enjoy finding aluminum shapes and alloys that excite me by what they might do for me. The magic is in finding something new, at least new to me. Each alloy, if anodized and dyed in the same dye for the same time period will probably yield slightly different results color-wise. Imagine if you were working on a piece of jewelry and your goal was to have a variation of one color; you could use two or three different aluminum alloys and realize the three values of the color. This might produce an effect much different than using the same alloy and immersing the pieces for various time periods in the same dyebath. If you happen to live in an area that is NOT blessed with lots of major industry, you may not have access to great scrap yards or the possibility to go to the aluminum dealer to buy their end-cuts and left overs. What do you do?

I live in a rural area of California. Locally, I have a steady supply of K and S Brand aluminum sheet and tubing. My local hobby store and at least three hardware stores stock the same fantastic display rack with all of the aluminum, brass and copper I will ever need. Well, almost.

At least half a dozen hardware/home improvement stores nearby stock those nifty aluminum rods, tubing, strips, etc. in six to eight foot lengths. Some of the aluminum is anodized, some un-anodized. I buy both types. I like the anodized because it has a much better surface finish and has the durable oxide layer already on the aluminum. Sometimes I use the anodized just as it is and saw the stock to fit the need of the piece. Other times, I drill, file, grind, sandblast, or whatever through the oxide layer, and then re-anodize the piece which produces color in the exposed areas while the remainder of the pre-anodized areas remain silver in color. Sound rich to you? You bet it is -- I love this simple, but elegant way to ornament my work. Sometimes the simpler treatments yield the best results. This **is** a matter of opinion. What do you think?

Aluminum boiler plate.

The drawings on page 51 illustrate **some** of the possible shapes of aluminum available to you. These are end cut or edge views, with the exception of the overhead view of the perforated screen. Check them out at your local hardware or hobby store. I have drawn them actual size so that you will be able to relate them to the scale of your work. Enjoy. Let your imaginations soar.

I recommend that you obtain a copy of C-6: Aluminum and Its Alloys from the Aluminum Association (address in Ex Libris chapter). This booklet is a complete reference for, as the title suggests, aluminum and its alloys.

The following has been excerpted from the above book.

The principal metals added to aluminum are copper, manganese, silicon, magnesium and zinc. Other elements are also added (in small amounts) for grain refinement and to develop special properties.

Aluminum and its alloys are divided into two categories. They are formed either as wrought or cast aluminum. Since my main research thus far has been with the wrought, I will concentrate my discussion in that particular category. Wrought forms would include sheet and plate, foil, extrusions, bar and rod, wire, forgings, impacts, drawn or extruded tubings, etc.

Each wrought aluminum alloy is designated by a number. This number describes the alloy as either wrought or cast and the alloying element(s). Wrought alloys are given four digit numbers. Cast alloys are identified by a three digit number followed by a decimal.

DESIGNATION SYSTEM--
WROUGHT ALUMINUM ALLOYS

ALLOY SERIES	MAJOR ALLOYING ELEMENT
1xxx*	99.00% minimum aluminum
2xxx	copper
3xxx	manganese
4xxx	silicon
5xxx	magnesium
6xxx	magnesium and silicon
7xxx	zinc
8xxx	other element
9xxx	unused series

*An exception is for the 1xxx series alloys, where the last two digits indicate the minimum aluminum percentage.

Here is how to read the wrought aluminum alloy numbers:

The first digit is the alloy series or principal alloying element(s). The second digit, if different than 0, denotes a modification in the basic alloy. The third and fourth digits form an arbitrary number which identifies the specific alloy in the series. For example:
Aluminum alloy 6063
6 = 6000 series means magnesium and silicon are the major alloying elements
0 = no modification in the basic alloy
63 = the individual alloy, but is of no major significance.

Alloy 1060 contains a minimum of 99.60% aluminum, for example.

In general, the alloys which contain manganese, magnesium, and silicon, plus pure aluminum are the best for making art. These alloys anodize well and produce the best colors. Their alloy designation numbers would be: for pure aluminum: 1000 and for alloyed aluminum the 5000 and 6000 series. The most common of the 6000 series would be 6061 and 6063. The 1000 series produces the most vibrant color, for example: All the dye manufacturers use the 1000 series for their color chip charts.

I have omitted any discussion on temper designation, mechanical properties, workability, machinability, corrosion behavior, etc., because the main topic of this book is anodizing aluminum, not the working of aluminum. Refer to the publication C-6: Aluminum and Its Alloys for further information.

These are mainly end cuts/edge views of available aluminum stock shapes. All the aluminum shown on this page is alloy 6063.

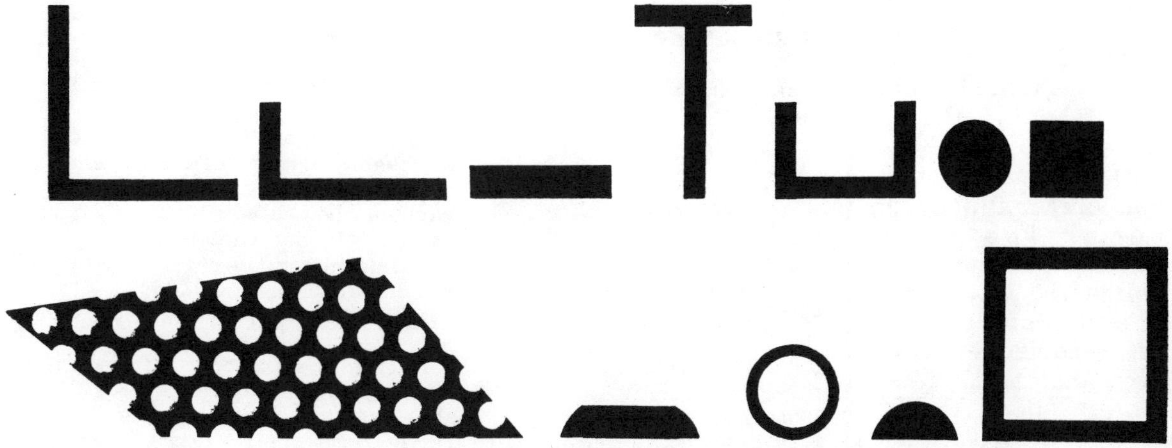

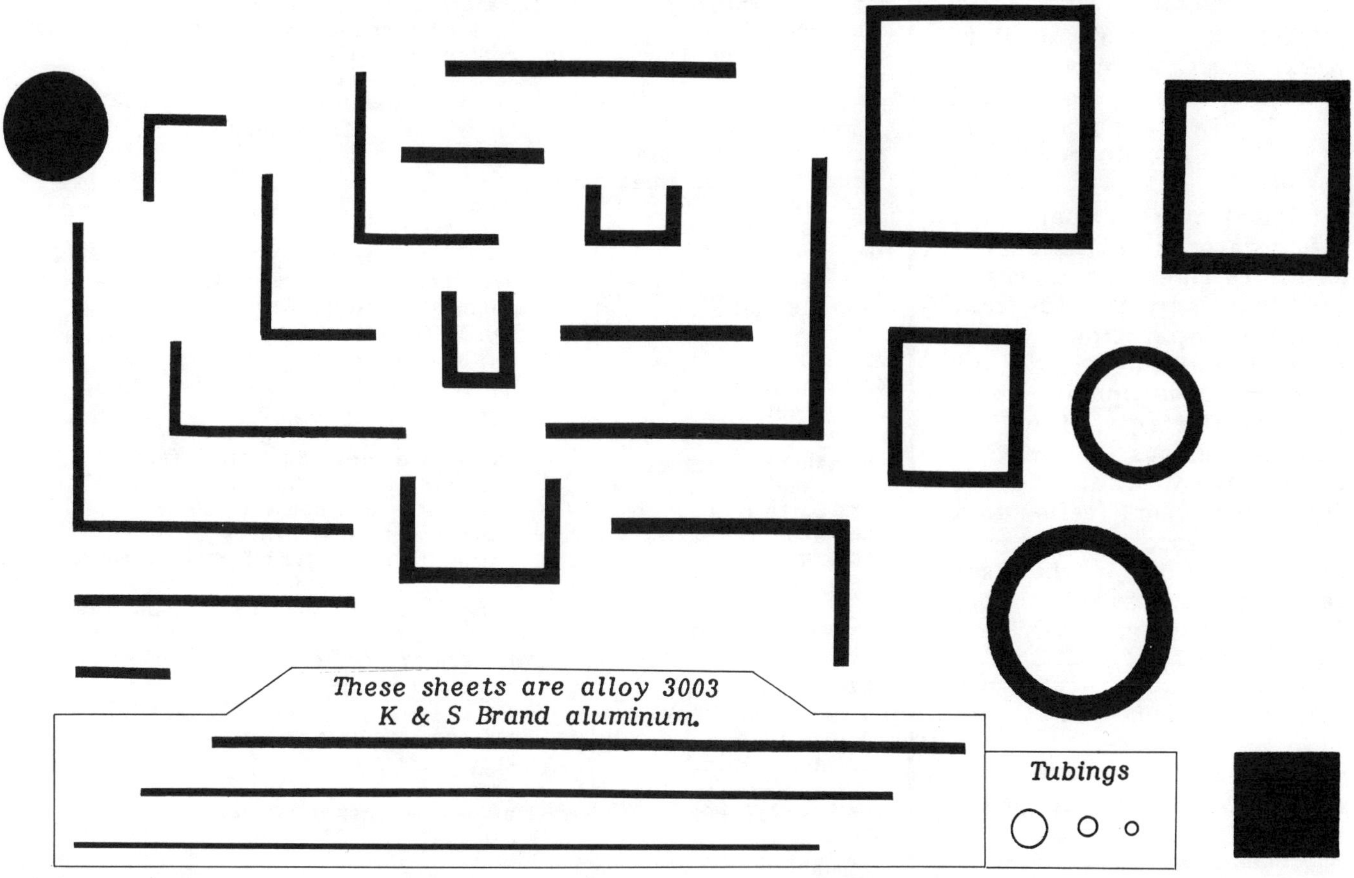

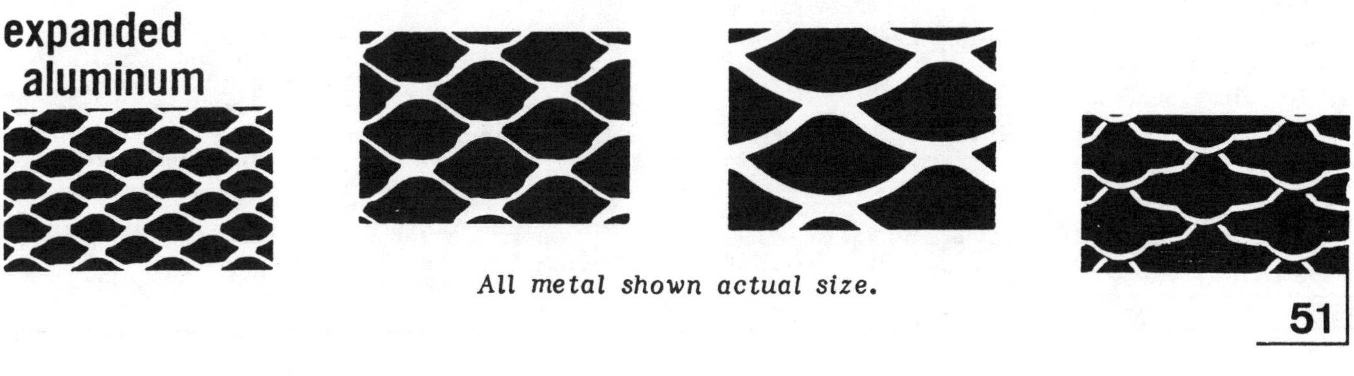

All metal shown actual size.

Aluminum is light weight (less filling?), cheap -- you bet, easy to work, comes in all varieties of interesting shapes and forms (sheet, rod, tubing, bar stock, plate, extrusions, etc.). It is readily available and capable of being anodized and dyed any color. The only limitation is you and me. What a delightful challenge and quite a burden, at the same time. Someone once said, "If you can think it--it is possible." Now, that's confidence. Care to test this theory?

Undoubtedly after you start working with aluminum you'll start to be more interested in aluminum scrap. For example: An old road sign catches your eye. Are you intrigued by some part of a refrigerator, like the shapes on the refrigerator door, the egg container, the ice tray and its separations? You then anodized them, dye them and come up with really wonderful effects. Then, you become curious as to what alloy they are. The following chart tells what alloys are used to make some products. These numbers are fairly basic and consistent for alloy designations.

- Radio Shack: #150-0035, aluminum wire: alloy 1350-T14.
- "K + S" brand aluminum sheet and tubing: alloy 3003-H14.
- S. Bates knitting needles: #0-3: alloy 2017-T4 #4-8: alloy 5056-H18 (Susan Bates uses Sandoz and Pylam dyes) #9-15: alloy 3003-H18.
- Boye Needle Company knitting needles: alloy 5052.
- Highway signs and "No Parking" signs: alloy 5052 and 6061.

Have you ever wondered what aluminum alloys normal day-to-day use items might be? Some of the answers are located below.

ALLOY NUMBER	ITEM
1xxx (99.99%)	jewelry (small quantity magnesium added)
5405	jewelry
1145 foil	disposable oven liners, burner drip pans
3004 sheet	removable burner reflectors, drip pans
6063, 6463 extrusions	appliance moldings and handles
3003, 6063 tube	vacuum cleaner wands
3003 sheet	vacuum cleaner canisters
1100, 1145	air conditioner fin stock
5050, 5052	air conditioner fan blades
5x57-O, H25 sheet 6063-T42 extrusions	refrigerator compartment doors, dip trays, door racks
1100, 3003, 7072	refrigerator fin stock, refrigerant lines, and defroster tubes
5005, 5357 sheet	sheet refrigerator shelves
1345 hard-drawn temper wire	refrigerator shelf wires
6063-T42	refrigerator shelf frames
1100, 3003 sheet	ice trays
3003, 3004, 5052	cooking utensils, food-handling equipment
3003, 5357 deep drawn	coffee pot shells
1100	household foil
5050	pen and pencil parts
3000-H18 tube	large size knitting needles
1100, 3003, 5052, 5056 wire temper full hard/half hard	smaller size knitting needles
5056-H14 wire (H111 or H112)	zippers
3004, 5050 sheet	eyelets and ferrules
3003, 5052, 6063	aluminum ferrules as furniture legs
5005-H19 wire	branches-artificial Christmas trees
5050-H19 foil	needles-artificial Christmas trees
3003	canister sets
5357-H36	clothes hangers
6063-T6, 6063 T832	mop, brush, car washing handles
3003 sheet	covers on hair dryer blower housings/heating mechanisms
5050 sheet	houseware fans and heater fans
1100 sheet	portable electric heater reflectors
3003 pattern sheet	mail boxes and house numbers
3000 and 5000 series sheet	storage cabinets/lockers
6061-T6, 6063-T6 extrusions 6061-T4	commercial furniture for restaurants, offices, and institutions
3003, 3105, 5052 3003, 3005 welded tube	camping, fishing and hunting supplies
6061-T6	firearms (receivers)
2024-T4 extrusions 6262-T9 bar stock	open gun sights
6061-T6 tube	telescopic sights
2014-T61 forging	revolver frame
2024	arrows
6061-T6, 7178-T6	badminton racket frames
7178-T6, 7075-T6 sheet laminated to wood	skis
7178-T6, 2014-T6, 6063-T832 tube	ski poles
3003, 3105	above ground swimming pool sidewalls
5050 sheet	large institutional in-ground pools
6061, 6063	pool ladders, diving boards, cleaning equipment, pool furniture
6063 tubing	camera tripods
1100 sheet, 3003 extrusion	containers for packaging, storing film
3003, 5052	radio chassis
3003, 6063 tube	outdoor television antennas
2014-T6	pipe wrenches (18-60 inch length)
6063-T832, 3003-H18 tube	lawn/garden equipment handles
3003 hard drawn tube	pole saws, tree trimmers (telescoping handles)

Painted Aluminum

TRUE CONFESSIONS-THE TABLOIDS NEVER LIE

"Extra! Extra! Read all about it! Artist buys and works with pre-anodized aluminum!"

Yes, it's true! I confess. I buy and work with pre-anodized/dyed aluminum sheet, rod and tubing. Surprised? Why not work with pre-anodized metal? It's nice to get an idea and be ready to jump on it immediately. Why wait until I anodize the shape? I like the option of sawing out my ideas from someone else's colored metals. By mixing and matching mine with theirs, I can extend my ideas further. There is no reason **not to use** pre-fabricated parts, pre-anodized or painted aluminum. Why not? If it works, use it! And does it **ever work!**

The anodized rod that I use comes in the form of commercially made knitting needles. Yes, I said knitting needles! Knitting needles come in many wonderful colors and sizes. The choice of rod size starts at about 1 millimeter and stops at 5 millimeters. Above the 5 millimeter size, the knitting needles are made from extruded tubing--what a bonus! The largest tubing is about 16 millimeters. Think about the possibilities?! Now, locate your nearest knitting store. The fun awaits.....

Pre-anodized and dyed aluminum sheet can be purchased from Duragraphic Systems. Check out their engraving stock aluminum.

DURAGRAPHIC SYSTEMS
P.O. Box 2831
Plainfield, New Jersey 07062
201-651-0667

New Hermes, Incorporated sells pre-anodized and dyed aluminum sheet also. Their offices are located throughout the United States.

NEW HERMES, INCORPORATED

East Coast
20 Cooper Square
New York, New York 10003
212-777-3080

Southwest
12243 E. Northwest Hwy.
Dallas, Texas 75228
214-328-9835

Midwest
3642 West 128th Place
Alsip, Illinois 60658
312-371-3744

West Coast
11711 Monarch Street
Garden Grove, California 92641
714-898-9265

Southeast
2620 Park Central Blvd.
Decatur, Georgia 30035
404-981-2562

Florida
5680 N.W. 161st Street
Hialeah, Florida 33014
305-625-1751

New Hermes also sells painted aluminum along with other interesting materials for the trophy industry and to make art with, write for a complete catalog.

The Accu Cutter Company sells painted aluminum along with other materials and supplies for the trophy/engraving business. Write for their catalog.

ACCU CUTTER COMPANY
3730 East Industrial Way
Building 28
Riviera Beach, Florida 33404
305-848-8888

I got my start, in colored jewelry, working with the painted, trophy metal. It is a great material to work with. Although the painted aluminum surface is not as durable as the anodized aluminum, it has its positive points. I **still** use it today. Consider buying the pre-anodized and/or painted aluminum and see how they work for you!

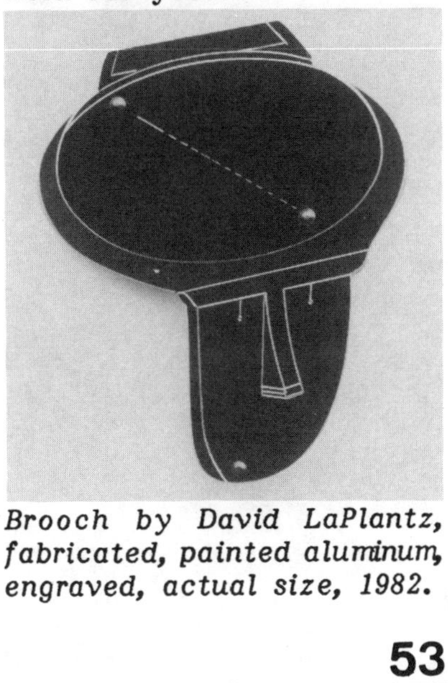

Brooch by David LaPlantz, fabricated, painted aluminum, engraved, actual size, 1982.

THE PREPARATION

- Studio Arrangement
- Possible Containers
- Anodizing Set-up
- Dye Bath Set-up
- Sealing Set-up

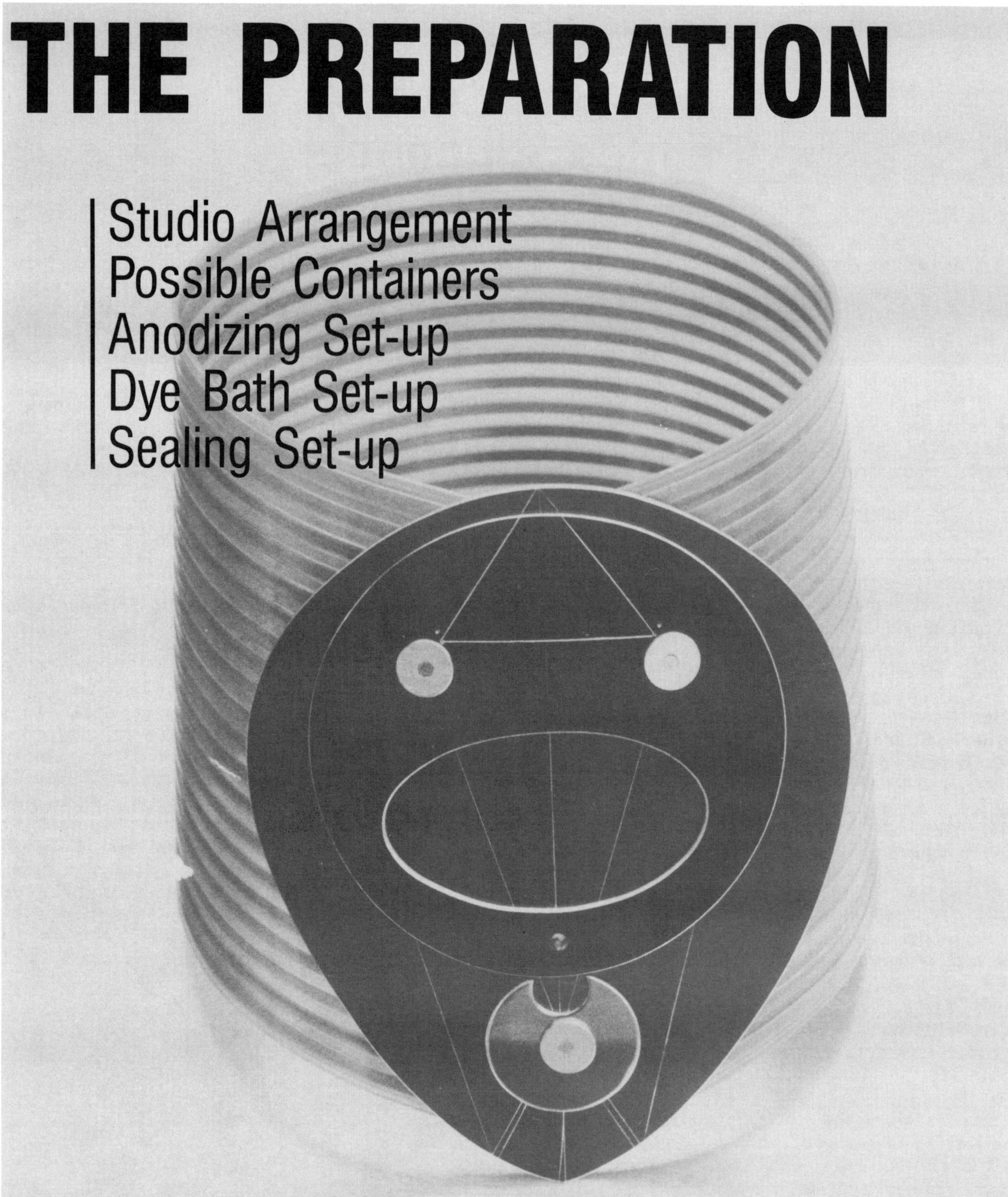

Bracelet by David LaPlantz, fabricated and engraved, anodized aluminum with electrical ribbon wire, 3 1/2" high, 1983-4.

Studio Arrangement

Let's talk about the layout of your anodizing studio space. You may wish to work in the same space as your current studio. Consider the following when deciding on your space requirements:

1. The general size of your work
2. If you anodize one piece at a time or many.

These considerations dictate the size tanks you will need. If you're like most people, the size of your tanks will range from 1 quart, through 1/2 gallon, 1 gallon, and up to 5 gallon containers with handles and tight fitting lids.

For example, let's set up an anodizing space for a 5 gallon anodizing tank, power supply, and the 1 gallon containers for the acids and rinse baths, an area for 6 to 8 -- 1 gallon dye baths, and a sealing area. For this you will require at least 2 -- 3 foot by 6 foot separate bench areas.

Be certain to take into consideration:

- ventilation (a must)
- lighting (natural and/or electric)
- water source (sink)
- electrical outlets (grounded)
- flooring/drainage, etc.

VENTILATION

If you locate your anodizing area near a window, you can exhaust the fumes out the window. For ventilation information, I certainly hope you've bought the Ventilation book recommended in the Safety chapter.

QUICK AIR THOUGHTS

A cross draft situation is nice if your studio is laid out to utilize air coming from behind you and out the window which is in front of you. A common mistake is locating the gas producing chemicals near a window and considering that enough ventilation. When the window is open and air is blowing in, the fumes are NOT exhausted, they are blown into the artist's face. Be certain that the fumes blow, or are drawn, away from you. A large fan behind the anodizing tank will draw the fumes away from you and out of that open window. YES, don't forget to open the window! It sounds funny, but so often that window doesn't get opened! Think before you begin anodizing.

I recommend using plastic sheeting to cordon, or seal off your anodizing studio for two reasons:

1. To prevent dust and contaminates from entering the anodizing area.
2. To stop chemical mists and vapors generated in anodizing from migrating throughout the rest of your studio/home.

BRIGHT LIGHT THOUGHTS

Lighting from a skylight would be the best type of light, and it is free. Important dye decisions and color comparisons are best made in direct sunlight, as opposed to artificial light, which alters color.

WATER

It would be nice if the anodizing area contained running water and a sink. If not, you're part of the bucket brigade! The newer deep plastic sinks are recommended. A nice feature of a sink, as you will see particularly in the Milner's chapter, is the ingenious use of a spray hose to totally rinse the aluminum. Spraying can forcibly rinse all the nooks and crannies in your piece. Remember: Rinsing is the secret to excellent anodizing.

ELECTRICAL OUTLETS

Be certain that your anodizing area has enough electrical outlets to run the power supply, lights, etc.

FLOORING/DRAINAGE

I recommend that you stand on a slat wood sub-floor in the areas directly in front of the anodizing, dyeing and sealing processes. This will protect you against electrocution, spilled chemicals, dyes, and sealing solution, etc. Don't be half safe. Consider your future and those who love and depend on you to be there.

The slat sub-floor is constructed from 2 X 4's and 1 X 6"s. The 2" X 4" boards serve as the foundation. They are laid flat. The 1" X 6" boards are used to provide the flooring. Suggested board lengths are 2" X 4" X 36 inches and up as required; 1" X 6" X 36 inches and up as required. The 2 X 4's are laid 2 inches in from the ends of the floor boards and spaced out at 12 inch intervals. The safe distance between the 1 X 6's floor boards is about 1 1/4 inches.

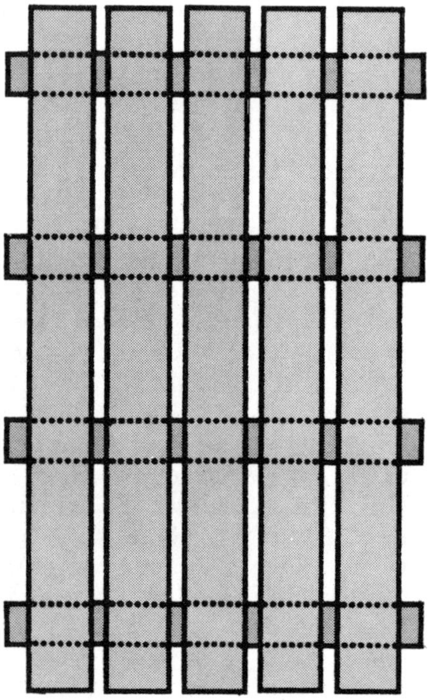

Larger openings between the boards will cause tripping hazards. You will see a similar slat sub-floor at Milner's Anodizing during a field trip later in this workshop. Watch for it.

STUDIO ARRANGEMENT

The anodizing, dyeing, and sealing of aluminum is a multi-stepped process and always makes me feel like a real (mad) scientist. The entire methodology may remind you of some of those process laden experiments performed in chemistry or physics classes.

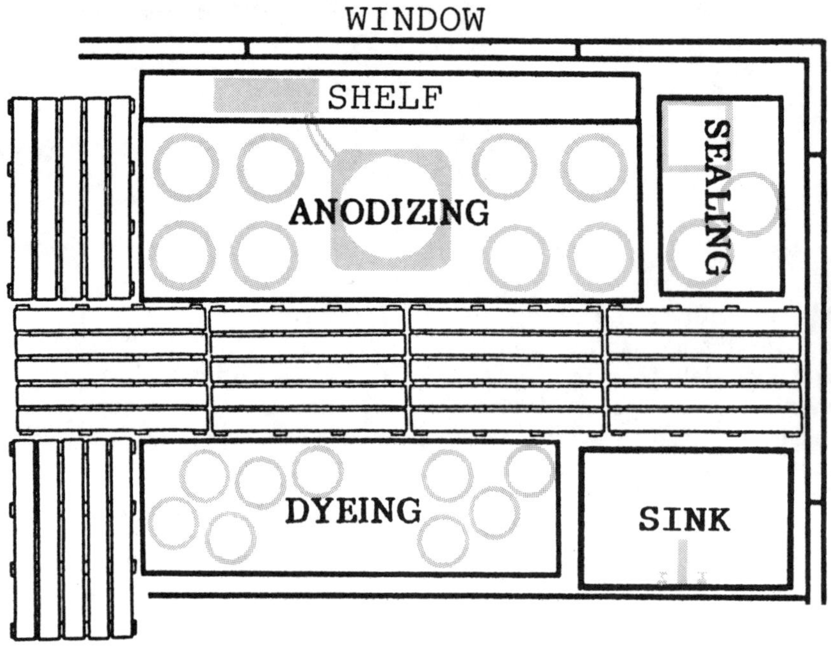

Do you want to walk miles each day to do this multi-stepped process? Of course not. So, set up your studio, be it for anodizing, jewelry, or whatever, with logic and steps in mind. Consider the well-designed kitchen, the sink, stove, refrigerator, and countertop work areas are placed close to each other in the obvious order of use. If not, you will literally walk miles each day to prepare your food, not to mention clean-up.

In the drawing above I have laid out a **possible** arrangement for the anodizing, dyeing and sealing processes. My anodizing area looks like the drawing. It works for me. Perhaps this same floor plan will work for you. Consider your needs, the direction you like to work; left to right, right to left, etc. **Find your best way to work.**

The layout of my anodizing studio allows me to move laterally no more than 6-8 feet in either direction. After anodizing, all I have to do is to turn around and I am ready to dye. Once the pieces are dyed and rinsed, in only a couple of steps I can seal my work.

I do more than one thing at a time in my studio. After all, time is precious. For some of us, time is also money. Therefore, if you can be dyeing, for instance, while you are putting new pieces in the anodizing bath, your time will be better spent. A good and efficient work cycle, as well as studio layout, will allow you to produce more and still have time to enjoy your life.

WORK BENCHES

Your work bench surface should be made from a material that will be unaffected by the acids and not be bothered by dampness. Formica kitchen countertops will work. A stainless steel work surface would be nice, but expensive.

Scrounge around in your local salvage yards, check out the used section of a kitchen supply store, restaurant auctions, etc. The kitchen supply stores are constantly buying and selling used equipment. Haunt such businesses and keep them aware of your needs. If you have a major college or university nearby, check out their auction or salvage sales. Don't discount your local high school. The high schools in my area have surplus items for sale each year. Be on guard, those special items you want crop up at the strangest times and places.

The bench area for the acids (nitric and sulfuric, and the caustic solution) and dyes requires some means to contain those liquids, should a leak or spill occur. Old stainless steel sinks with countertops usually have a raised edge to retain the spills. There is nothing worse than having to clean up spilled nitric, sulfuric, or caustic soda from the floor. In lieu of the stainless steel countertop, I'd recommend constructing your bench as follows:

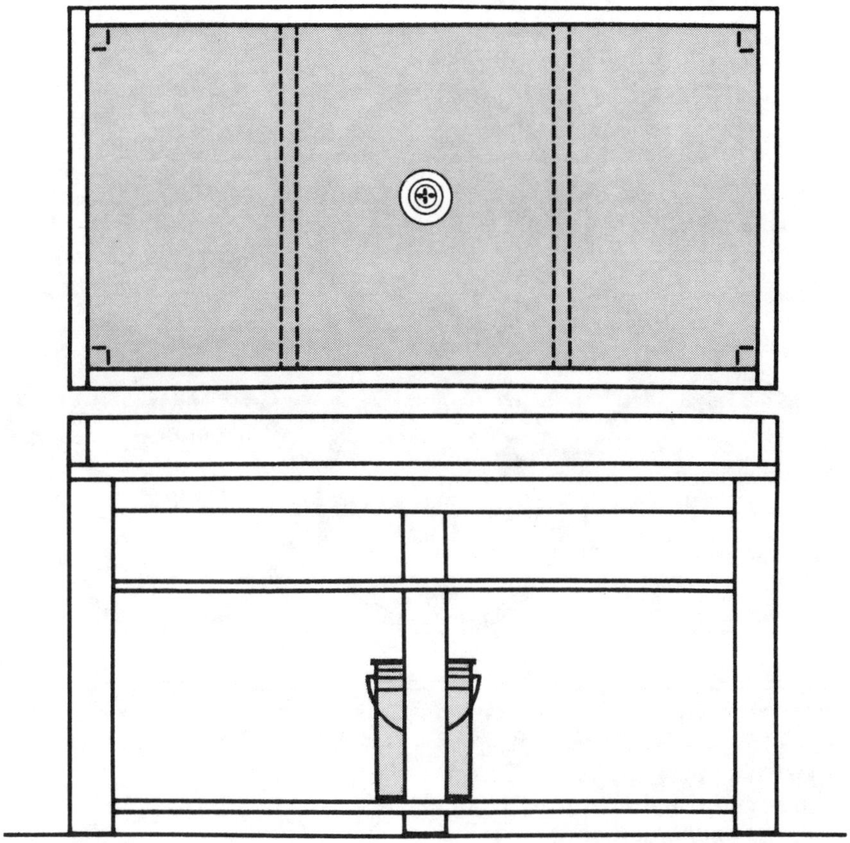

1. The 6 legs are made from 4 X 4's, cross braced, with a shelf or two between the top and the floor. Consider the size of liquid containers, and other items you might store on the shelf(ves). The bench may be made stationary or have wheels. This one shows easily attached casters from the hardware store. Wheels would allow you to move your anodizing outside into excellent ventilation and light.

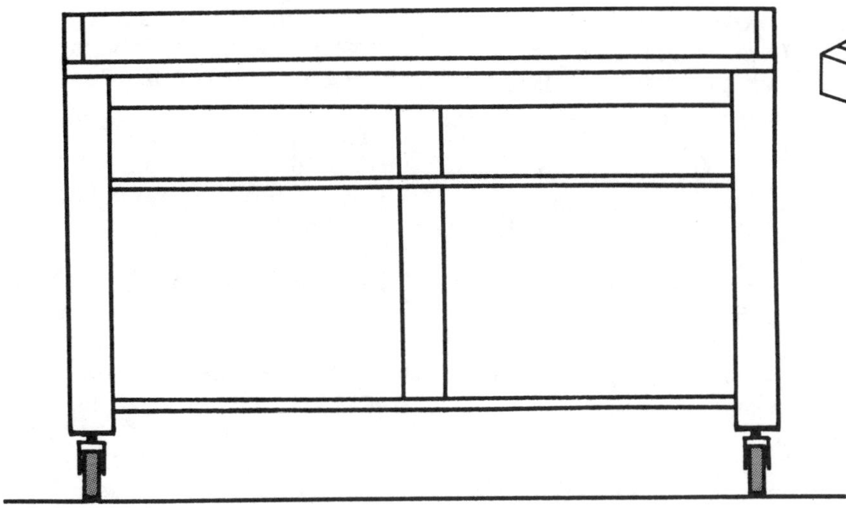

2. Be certain that the bench top is made from one solid piece of 3/4-1 inch plywood. I prefer my bench tops to be 3 feet deep by 6-8 feet long with 2 -- 2 X 4 cross braces spaced evenly under the plywood. You may or may not wish to utilize the entire 4 feet of the 4 X 8 X 3/4 inch plywood sheet. I prefer not to have to reach as far to the back of my bench. Consider what works best for you and construct your bench accordingly.

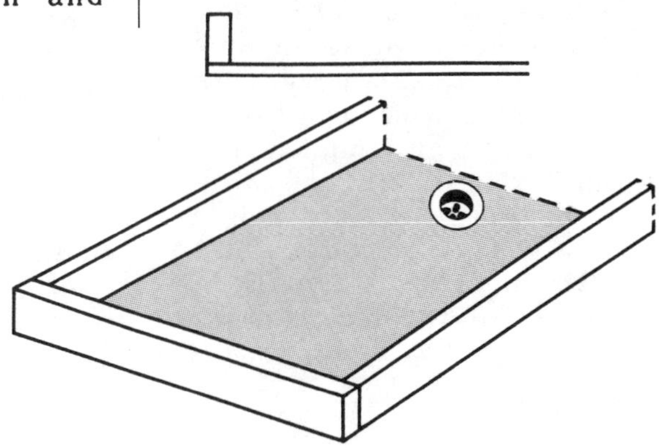

BASIC ORGANIZATION OF THE TANKS

Organize the tanks (plastic containers/buckets) in the formation they will be used in the anodizing process.

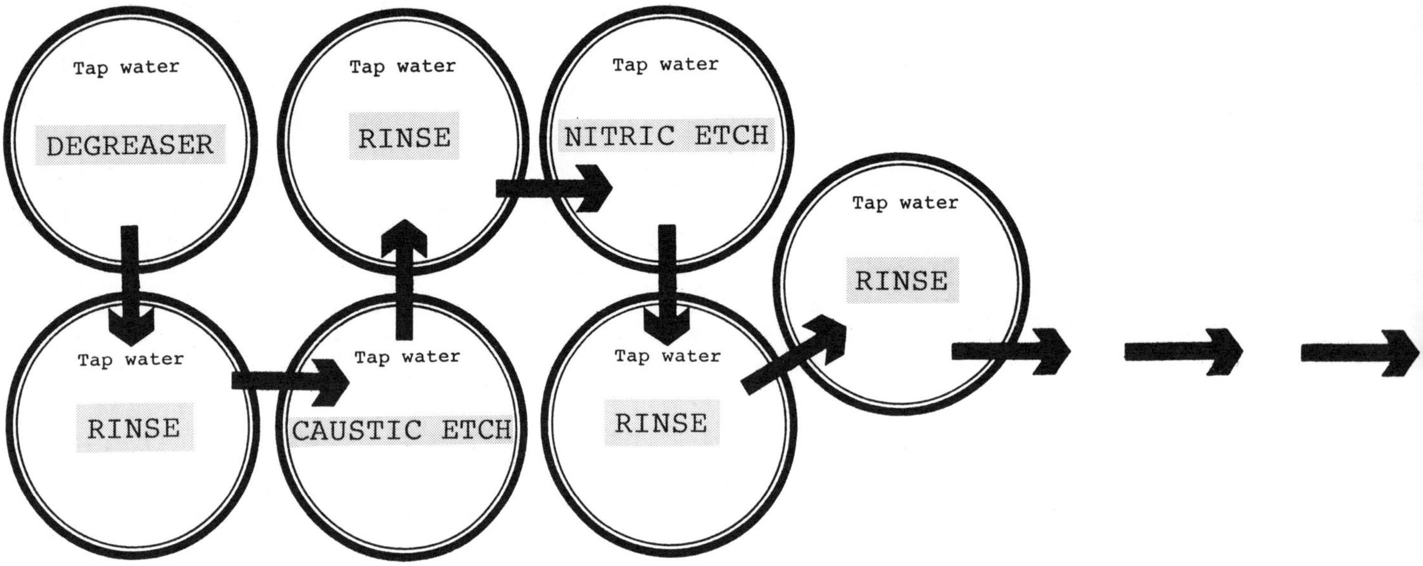

3. It is wise to construct a retaining wall or rim to prevent any spilled liquids from running onto the floor. I'd recommend using a 2 X 4 for this wall/rim. Prior to nailing the 2 X 4 wall on top of the bench, be sure to use silicon caulking beneath the 2 X 4, sealing it to the bench top. You may also wish to run a bead of caulking around on the inside of the wall/rim as an extra precaution seal. Using the tapered applicator provided with the caulking, apply a bead of the silicon to the inside edge of the molding rim. **Use your index finger** with a cloth covering to spread out the sealer. Allow the silicon caulking to cure, usually overnight or follow directions.

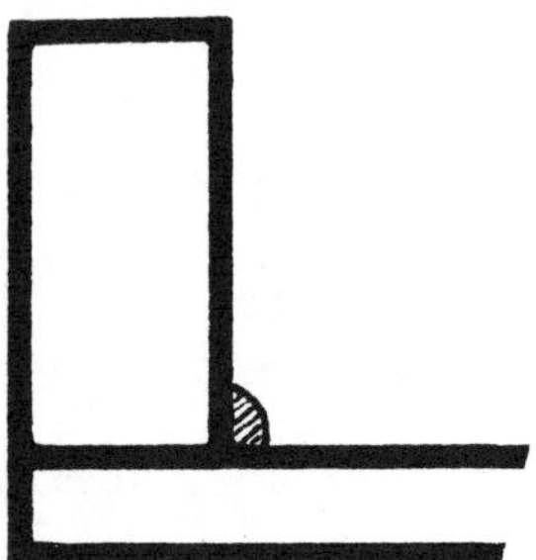

Now that you have a reservoir to contain a potential chemical spill, how do you clean up that spill? Why not install a sink drain in the middle of your bench? This drain can extend through the plywood top and have a plastic valve connected to the pipe. A bucket placed below the drain could then collect the spill or facilitate clean up.

You may now paint the bench top. Use a paint that is impervious to acids. Check with your local paint dealer. A couple of coats will insure a bench top that will not be affected by your chemicals and provide a reservoir to contain your liquids should a spill occur. Of course, now that you have prepared for the worst, none will ever occur! Aren't you glad that you know how to prevent the worst from ever happening? Yes, simply be prepared?

ACCIDENTS TO UNINVITED GUESTS

If you have small children and/or pets, be sure to have a way to close off your anodizing studio. The chemicals used are not kind to licking, small hands, eyes, etc. Your studio is just like a small chemistry laboratory -- respect and consider the dangers little ones/anyone might get into, if left unattended, for **any** period of time. Their life and future is in your hands!

You can **never** be too careful. Don't think that just because you are an adult nothing can happen to you. Accidents are waiting for each of us -- a split second of non-thinking or careless behavior is all it takes.

*I prefer to work left to right. The following illustration is not meant to be a model, but rather a guide. Figure out **your** best organization. Let's start out with an overhead view of the basic set-up:*

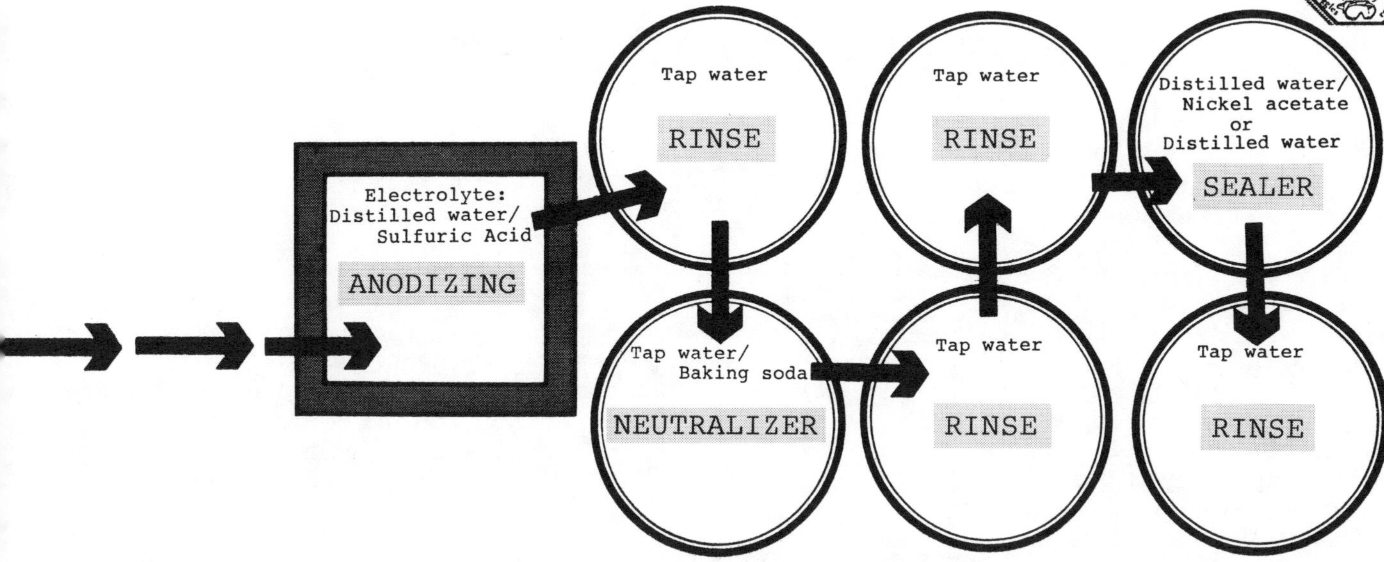

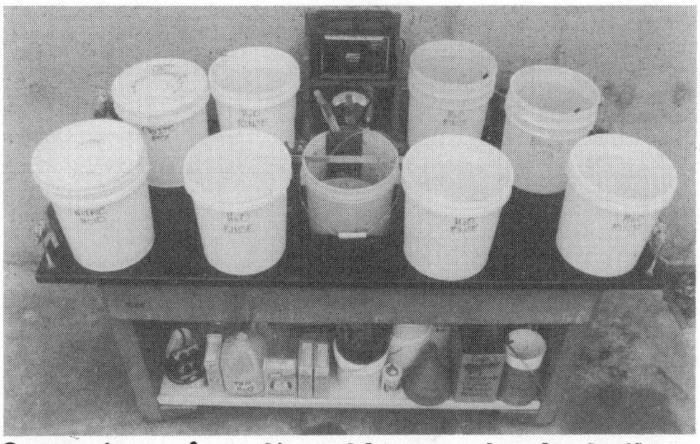

Over view of studio table, on wheels, built as suggested, with tanks in place. You'll notice that fewer tanks are on this table than are recommended, I'm learning to rinse more and more to eliminate drag-out.

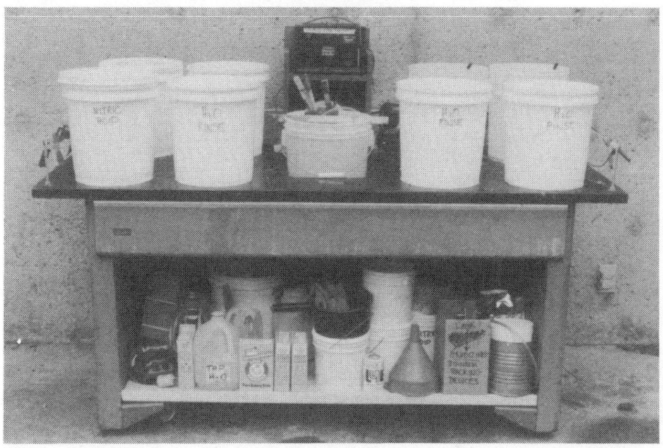

The table top is a chemistry table top with a pre-cut sink hole that the anodizing tank is resting inside. Notice the well used storage shelf below.

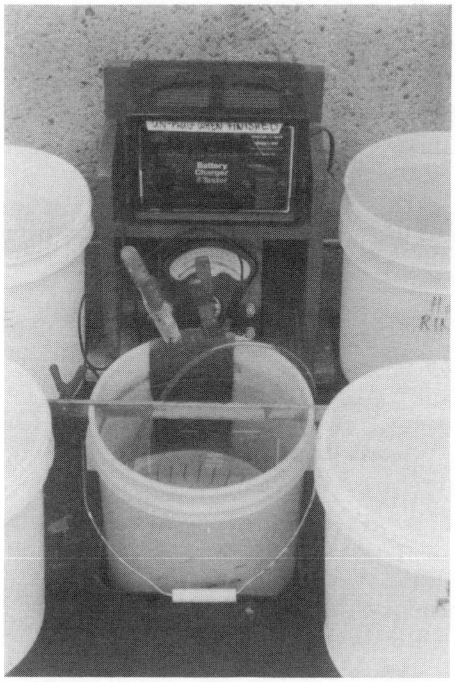

Notice that the power supply is on a shelf for accessibility to switches and knobs. Raising the power supply is also a safety feature. There is an ammeter wired in series between the power supply and the bus bar. Notice that the spring clamp is helping to hold the lead to the tank.

Possible Containers

Many plastic buckets or similarly shaped containers are needed to hold the various chemicals, rinses, dye baths, etc. required to anodize and dye aluminum. These containers should be made of polyethylene or polypropylene since they are inert to the chemicals used in anodizing. If you depend on buying everything new, get ready to spend lots of bucks. If you don't mind scrounging and/or working with some used items, your dollar will travel further.

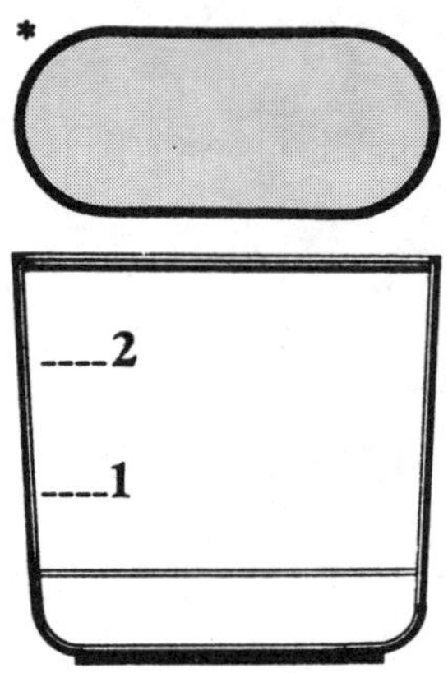

This 2 1/2 gallon container is actually a plastic, waste paper, vanity-type can for the bathroom. Be sure to choose your favorite designer color to spark up the studio!

Vital statistics:
(Inside dimensions)
5 pint: 5 5/8"(dia.) X 6 1/2"(deep).
1 gallon(short): 7"(dia.) X 7 1/4"(deep).
1 gallon(tall): 3 3/4"(dia.) X 9 1/2"(deep).
5 gallon: 11"(dia.) X 13 1/2"(deep).

*** overhead view**

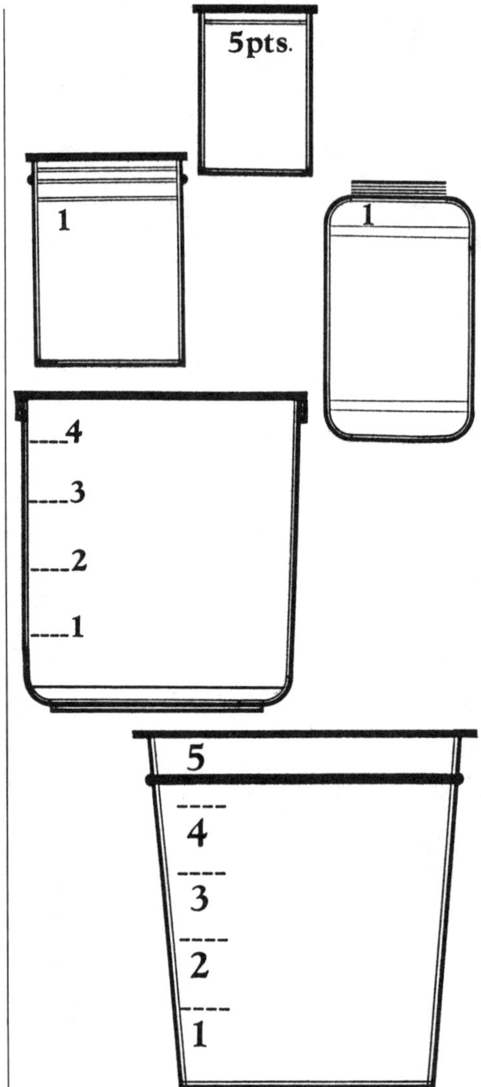

These plastic containers are mainly from restaurant/kitchen supply/food sources. The 5 pint, 1 gallon short, 1 gallon tall, 4 1/2 gallon, and 5 gallon containers all come with lids, and are round in shape. They used to hold mayonnaise, pickles, mustard, cottage cheese, etc. Handles are provided on the short 1 gallon, 4 1/2 gallon, and 5 gallon buckets. This is a must for carrying, pouring, and disposal.

I get many of my buckets from the university dining hall on campus and from local restaurants. Ask around, you can set up your anodizing studio quite cheaply! Also, don't forget to check with your local high school or junior college dining halls. I am always looking for cheap plastic buckets and odd-shaped containers used in a restaurant and/or in the average home setting. **Free** sounds better than **price**, to me.

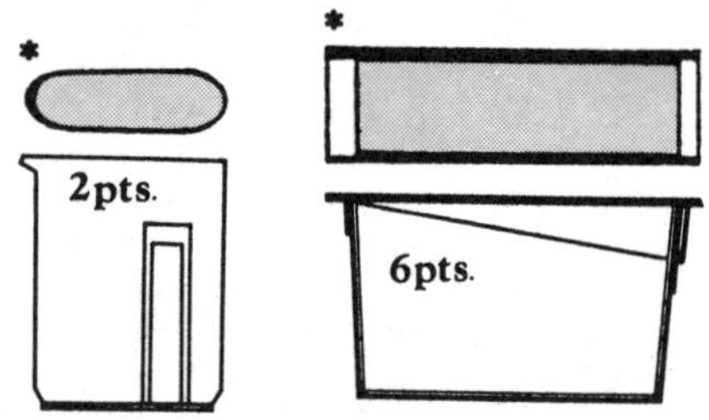

This 2 pint container is a refrigerator door pitcher. The 6 pint container is probably for ice. Both are slender, and/or long/tall, and may provide the exact volume to house dye for special shapes.
2 pint: 2 3/4"(across) X 6"(wide) X 7 1/4"(deep).
6 pint: 4"(across) X 10"(wide) X 6"(deep).

This smaller Rubbermaid brand plastic container is a see-through storage box with a snap-on lid. It holds 19 1/2 quarts and measures 10"(across) X 15"(wide) X 8 1/2"(deep).

*

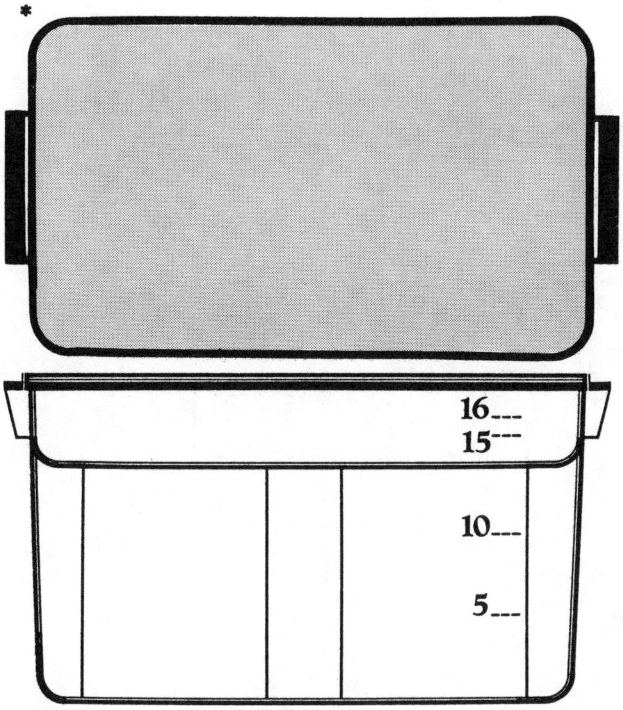

This 16 gallon Roughneck container, with lid, might serve as a larger volume electrolyte tank. The Roughneck series of containers come in a variety of sizes, all with lids. Check them out and see which sizes best fit your requirements.
16 gallon: 12"(across) X 18 1/2"(wide) X 16"(deep).

If you have limited amounts of dyes, say, in 1 gallon containers, and/or you have a piece larger than the existing dye container, what do you do? A voice in the back of the room always says, "Why not get more dye?" That costs money and we want to save money. In this instance I look into my stock of odd-shaped plastic containers. Consider garbage buckets, waste paper containers, anything that catches your eye and forces your imagination to figure out how you could use it. Walk up and down the aisles in the food and discount stores where the plastic ware is stocked. Consider the shapes. What size pieces can I work with considering the containers I have now? What size work do I plan in the future? Does it have a lid? Do I need a lid? Is it easy to clean and rinse? Does it have a handle, or a pouring spout? When filled with a liquid, will I be able to lift it without difficulty? Is the plastic wall thick enough to support the liquid once the container is filled? Consider all of the possibilities and ramifications before you lay down your hard earned bucks!

The voice in the back of the room still has not received a logical answer to the question. Let's try this one. When you only have so much dye and do not want to (or can't) mix up more, why not use a flexible plastic container? Then you can squeeze the container between your knees, or somehow decrease the width/diameter, while raising the dye level. Then your piece will be totally submerged in the dye. I have used this method many times with great results. Who says you need more dye in larger capacity dye bath containers? With a little creative thinking you can work out almost any problem.

What a great idea! The design of this bucket provides you with 2 different rinsing containers, for example. If you work with long, thin forms or have limited work space, this is the container for you. The only drawback is that it is difficult to empty one side without spilling the other. I know, I know -- use a pump or siphon (hint, hint). Two gallon (double well): each side -- 5"(across) X 12"(wide) X 10"(deep).

*

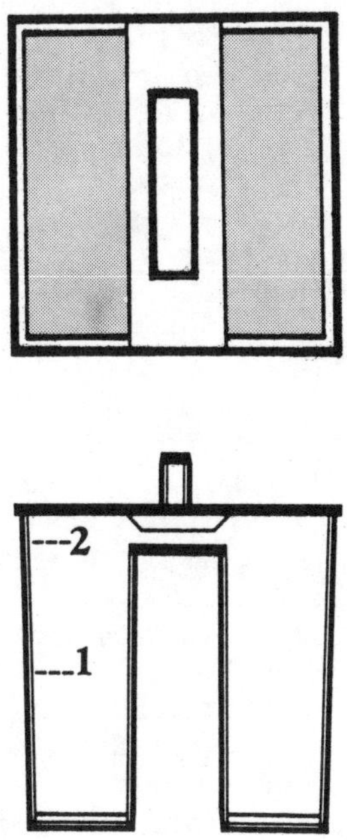

*** overhead view**

Anodizing Set-up

The equipment and supplies you will need and where to buy them.

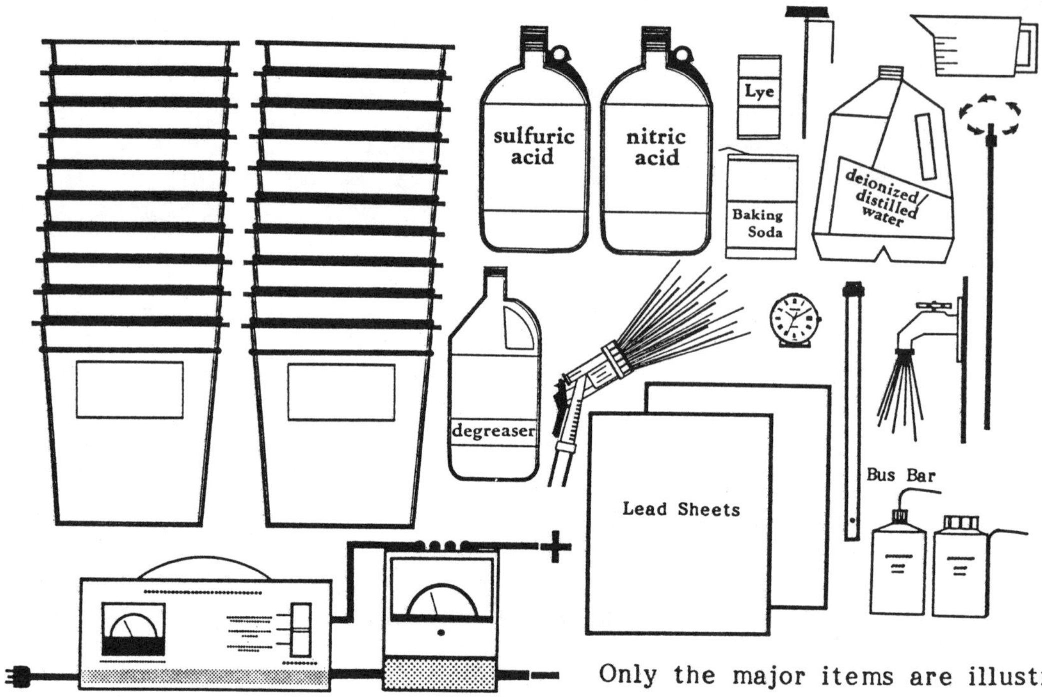

Only the major items are illustrated.

Grocery Store:
- 18-20 Plastic Containers (1 quart, 1, 2 or 5 gallon capacity)
- 8 Plastic Lids for Containers
- Distilled/Deionized Water
- Pyrex or Stainless Steel Measuring Cup
- Caustic Soda (Lye)
- Baking Soda
- Soap/Degreaser

Hardware Store:
- Timer (Clock)
- 12-26 Gauge Copper Wire
- Power Supply, Rectifying (changing) A.C. to D.C. -- 12-20 volts, 5-10+ amps current capacity. (See also Jewelry Supply Houses, Discount and Automotive Stores) A battery charger works for me.
- Clamps and Cable

Sheet Metal Contractor, Scrap Metal Yard:
- Lead Sheets

(1 qt container-5" wide X 8" long X 1/8" thick)
(1 gal container-6" wide X 10" long X 1/8")
(2 gal container-6-8" wide X 14" long X 1/8")
(5 gal container-12" wide X 16" long X 1/8")

- Copper Bus Bar/Brass Machine Screw, Washers and Nut (1/4" X 1 1/8")
- Aluminum, alloys:
  - 1100, 3003, 5052, 6061, 6063, etc.

Plastics Supply House:
- Plastic Rod for Stirring; add black electrician's tape to designate dry end.

Aquarium Store:
- Floating Aquarium Thermometer

Chemical Supply House:
- Chemistry Wash Bottles
- Nitric Acid, Reagent Grade
- Sulfuric Acid, Reagent Grade

Knitting/Weaving Store:
- Knitting Needles-#1 or #2, longest possible, any color (Racks)

Janitorial Supply House:
- Soap/Degreaser

Water Company
- Tap Water/Spray Rinse

Optional Items:
- See Testing Chapter:
  - Multimeter/Multitester
  - Oxide Thickness Tester
  - pH Meter
  - Ammeter
- See end of this chapter:
  - Air Agitation/Stirring System
  - Cooling System

Obviously many of the above items may be purchased at other locations. This list provides a start.

PREPARING ANODIZING TANK: BUS BAR

Prepare two slots in the top of the tank to support the copper bus bar. The bus bar should be located in the middle of the tank. Mark the middle of the bucket.

Two rectangular slots will be cut, one on each side of the bucket.

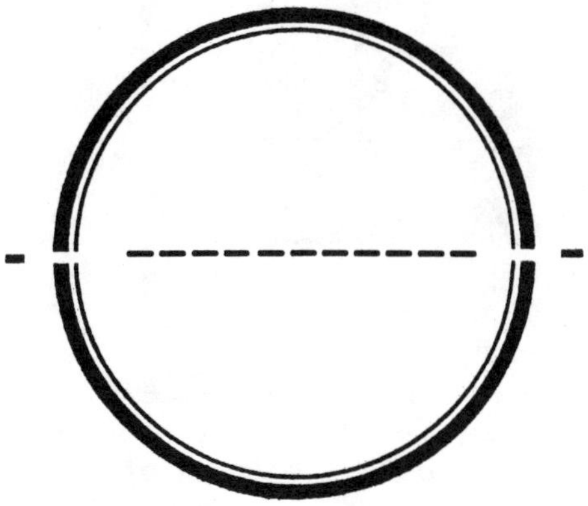

Measure the height and thickness of the bus bar. My bus bar measures 1/4 inch thick X 1 inch high X 16 inches long.

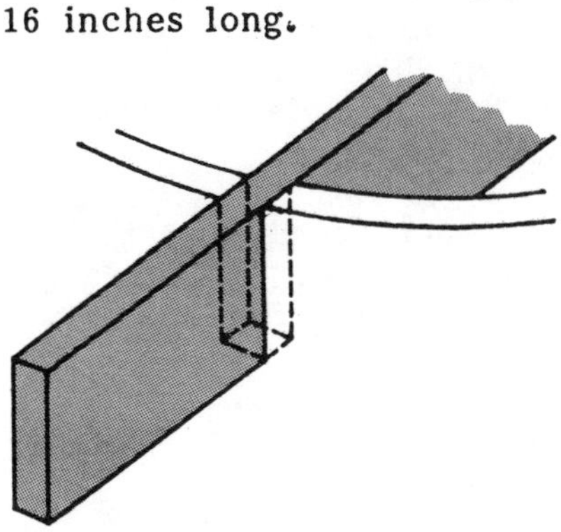

Each slot will be equal to the height and thickness of the bus bar. A sharp pocket knife or an X-acto blade will do. Watch your fingers, cutting plastic without slipping is an art in itself. Or, should I say, slipping for some of us, is, an art in itself?!

TIP:

If your anodizing tank is not easily cut with a knife, try drilling two holes at the corners of the rectangle, then saw from the top edge of your container down to the holes with a coping saw. The bottom cut goes from hole to hole to finish off the rectangle.

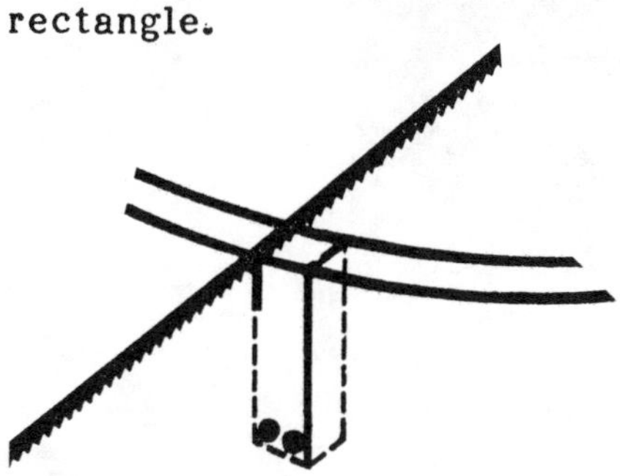

Remember, the plastic fragments from the drilling or sawing are not healthy. Beware of the plastic dust or small particles, they are dangerous! Wear goggles and appropriate respirator. Be certain to vacuum you, your clothes and your tools when finished. Work outside on some of these items when possible.

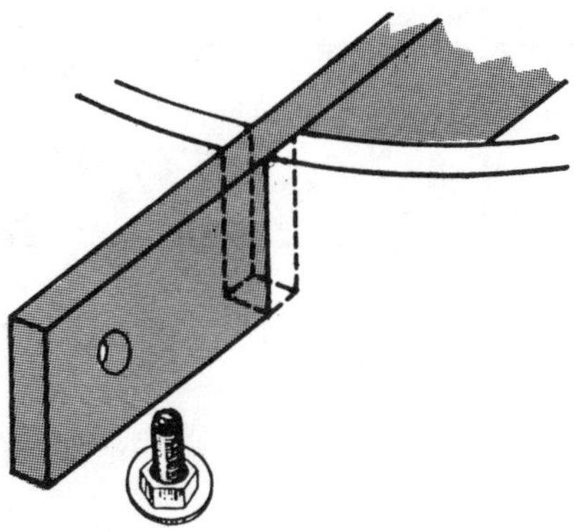

The connection for the positive lead from the D.C. power supply is a 1/4 inch X 1 1/8 inch brass machine screw, washer and nut. This "connection screw" will extend about 1/2 inch beyond the bus bar.

Drill the 1/4 inch hole for the connection screw about 1 inch from the end of the bus bar. Insert the screw, slip on the washer and tighten down the nut, finger tight.

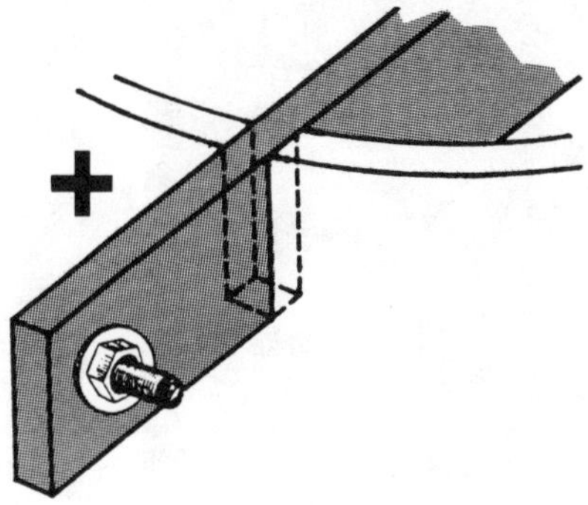

Overhead View

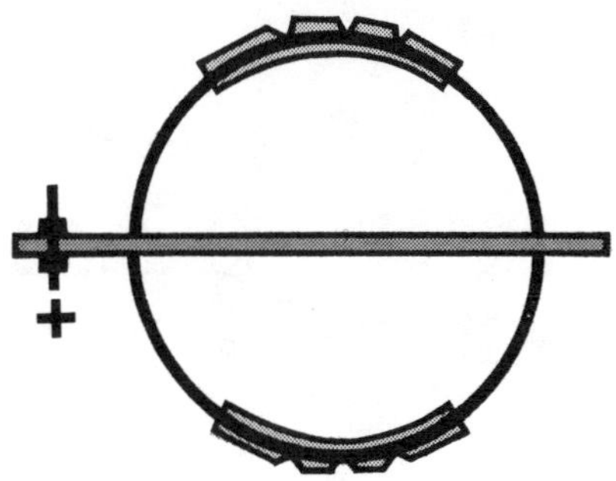

STOP THINK — apron, gloves, goggles, respirator

PREPARING ANODIZING TANK:
LEAD CATHODE

Cut two lead sheets (cathodes) with metal snips to size. Some people recommend cathodes equal in size to the piece you're anodizing, while some anodizing tanks are completely lead lined. Since I work with several different sizes/pieces at once, I personally use two cathodes that are each 12 inches wide X 16 inches long X 1/8 inch thick for my five gallon tank.

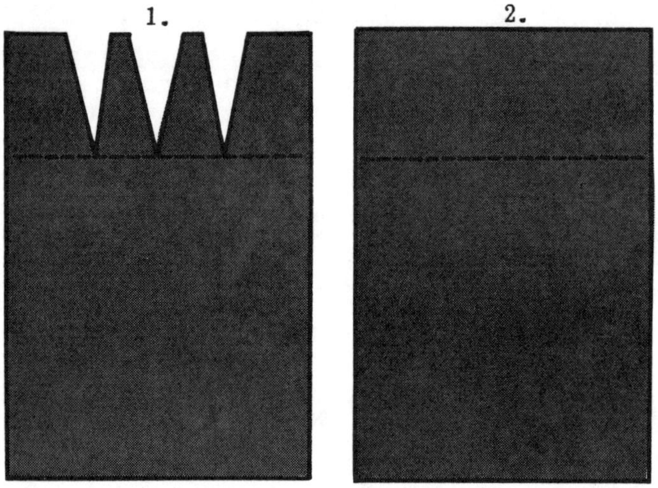

Each lead sheet will be hung from the top edge of the bucket, each opposite the other. Make certain that the cathode hugs the bucket's sides, ie perfectly vertical. This insures against the aluminum piece accidentally touching a cathode and causing a short.

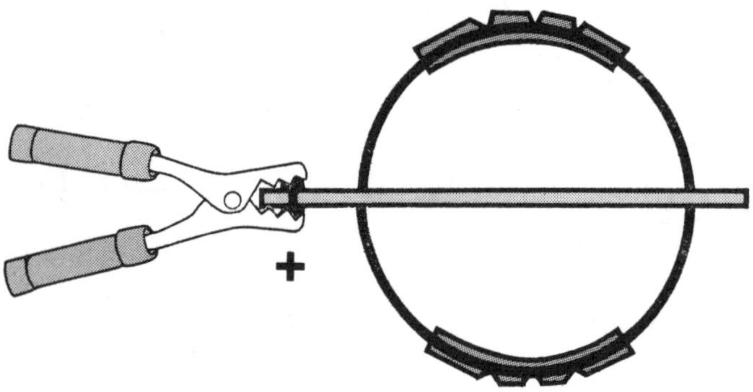

If the anodizing tank is a round bucket, one of two methods may be used to attach the lead sheet to the bucket:

1. The cathode has wedges cut out to facilitate molding it around the compound curve. The dotted line represents where the lead folds over the bucket's top.

2. The rectangular cathode is a little more difficult to mold around the bucket's compound curve, but it can be done with some effort. Just work it into place.

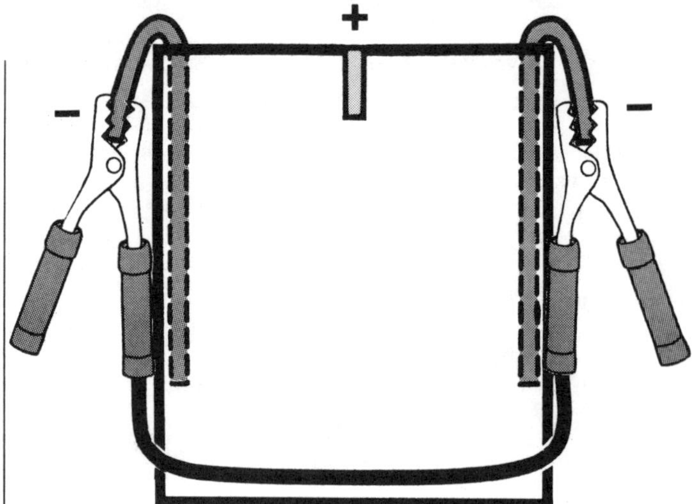

The lead cathodes must cling to the tank's sides and not float up/inward. If the aluminum makes contact with the lead cathode(s), arcing may occur, or, short out your power supply. Attaching the clamps is discussed on page 66, but is also illustrated here for further clarification.

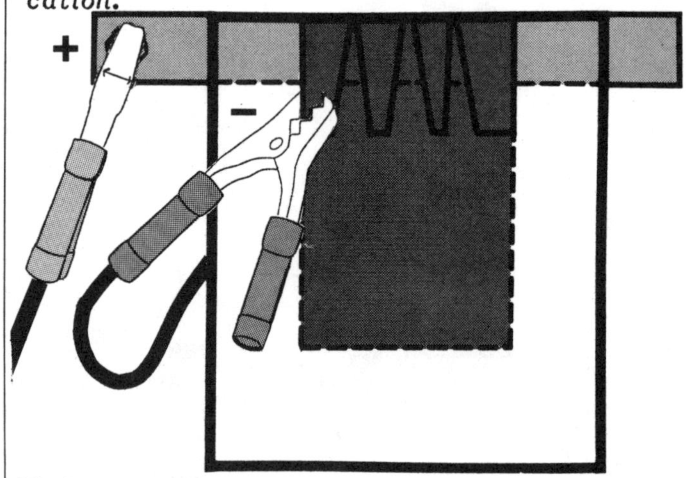

If the anodizing tank is a rectangular form, the lead sheets may be bent over the sides of the tank. This should provide a secure method for hanging the cathodes in the electrolyte.

The two lead cathodes must be connected in series. You need an insulated, copper wire with car battery clips/clamps on each end. These clamps, wires, etc may be purchased at your local electronics store or the automotive section of a hardware store. The clamps may be referred to as clips, and visa versa.

Clamps are rated for amperage, be sure to buy correctly rated clamps. The clamps have plastic tubing color codings, black for negative (cathodes) and red for positive (bus bar).

clamp connection wire ratings:
5 amp power supply = 18 gauge wire
10 amp power supply = 12 gauge wire
15 amp power supply = 12 gauge wire

ATTACHING WIRE TO THE CATHODES:

Measure the distance between the two lead cathodes. Your wire should be a couple of inches longer than that measurement. Strip off about 1 inch of insulation from each end of the black coated wire. Remove one black rubber tubing/handle from each clamp. Slide the tubing over your wire. (Later you will replace the tubing over the handle.) Lay the exposed copper wire into the clamp's groove/handle. Fold over the tabs with pliers, insuring a tight grip/connection with the copper wire. Slide the black tubing back over the handle. One clamp is now attached to the wire. Repeat the same procedure on the other end.

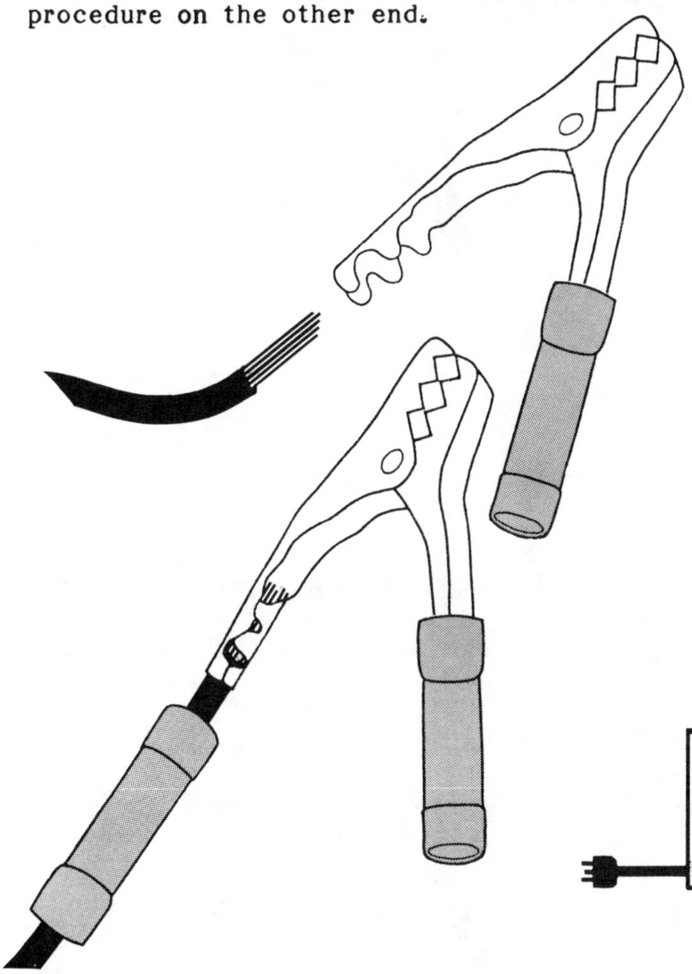

The clamps must be attached to clean lead, so, scrape an area on each cathode clean with your knife blade. Clamp on the clamps.

To prevent a lead sulfate reaction (see page 70), spray the connection area where the clamp and lead meet with battery terminal spray paint, or use vaseline. Be certain to do this after cleaning, but not before you clamp onto the lead. Otherwise, the protective coating will act as an insulator and prevent the current flow.

POWER SUPPLY REQUIREMENTS:

To anodize aluminum, a rectifier* or power supply, changing A.C. to D.C., with at least 12-20 volts and 10-15 amps current capacity, is required. The 10-15 amp capacity is the amount of current drawn (current density) by the size of the aluminum being anodized. In other words, your 10-15 amp power supply should be capable of anodizing a square foot of aluminum.

* Rectify/Rectifier means to change A.C. to D.C.

As far as D.C. power supplies, it seems like you have three ways to go:

1. If you have a friend who can get one for you for free, or build one for you for cheap, you are all set.

-or-

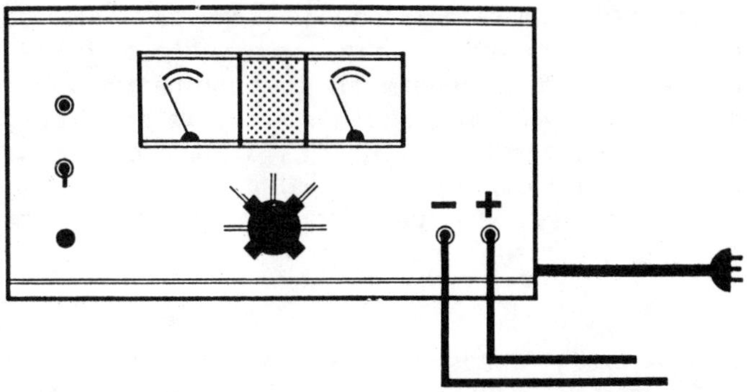

2. You can spend lots of money ($200.00-$500.00+). If you've been doing plating or electroforming and have the more expensive power supply, fine, use it.

-or-

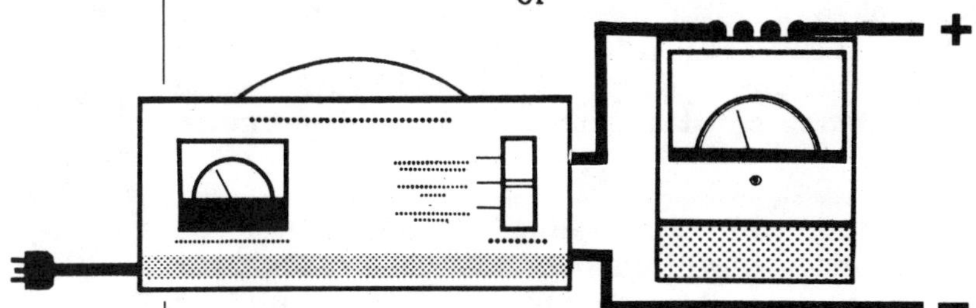

3. You can spend a small amount of money ($25.00-$50.00) and purchase an inexpensive 5-10 amp, 12 volt battery charger. Doesn't that sound better? The inexpensive battery chargers I've tested and used are all capable of producing at least 10-15 amps per square foot. Battery chargers are rated at 4, 6, or 10 amps, take care not to overwhelm your charger with too many square inches of aluminum in the electrolyte. Understand that if you have an old 2-4 amp, 12 volt battery charger, your scale is limited (think small).

This is an illustration of where the wires and clamps connect between the battery charger/ power supply, and, the anode and cathodes.

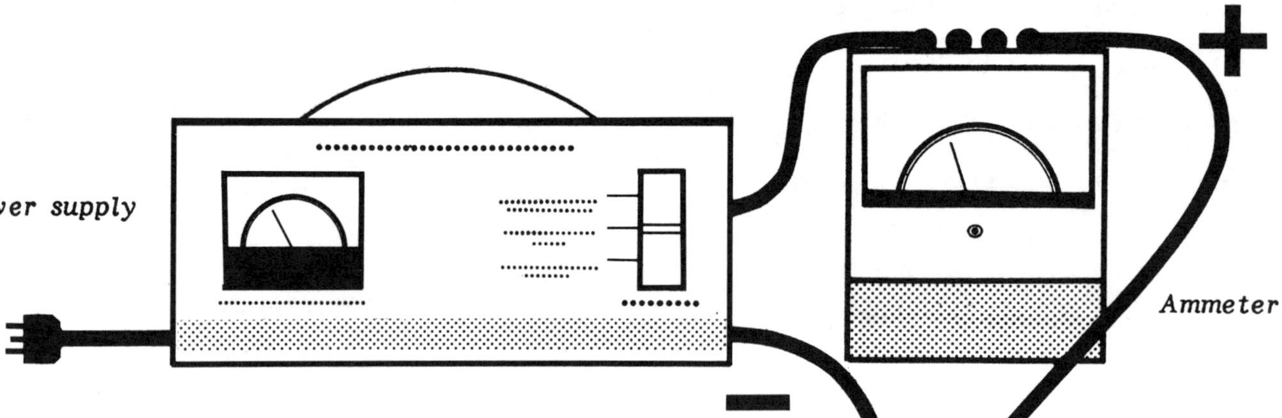

The power supply and the anodizing tank are on two different levels ONLY to illustrate the connections as clearly and as large as possible. Normally they are on the same table together (or level with each other). But, sometimes the power supply is on a shelf, above the anodizing tank, as illustrated. This may allow more room on your anodizing table -and- position the power supply's switches and knobs at a more convenient height.

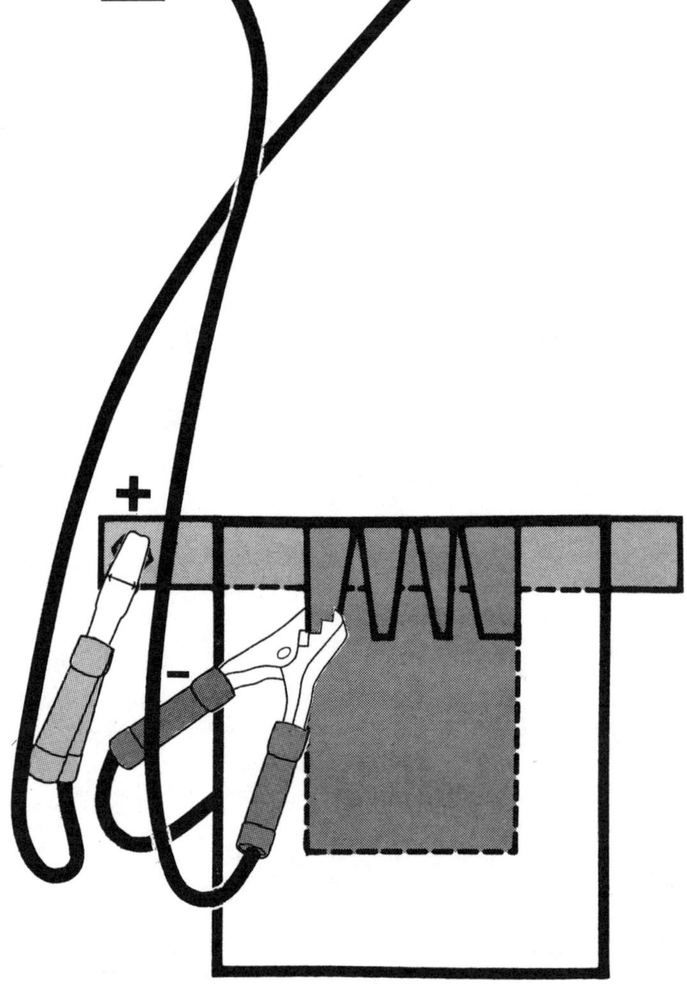

The battery charger I use in my studio is a Schumacher Brand, Model #SE 12-50. It cost about $35.00-$40.00 on sale at a local discount store. It has served me faithfully for many years.

Yes, because of some grant money, I bought the $250.00 power supply. It doesn't do any more than the Schumacher. In fact the more expensive power supply hums incessantly while it's running and the Schumacher makes no noise at all.

No matter what, to do aluminum anodizing, you will need a power supply that can produce 12-20 DC volts. The amperage capacity should be at minimum 5 amps. Better yet, I recommend the 10 amperage capacity power supplies. The 10 amp source will allow you to put many pieces in the anodizing tank at one time without burning out your power supply.

Clamp the positive (+) or red color coded clamp to the bus bar. Small clamps can be screwed in with the connection screw, while large clamps can sandwich/clamp both ends of the connection screw and the bus bar. See photo on page 103 for large clamp connection from the Schumacher battery charger shown on this page.

See page 71 for information on stirring and cooling the electrolyte.

ILLUSTRATED STEP-BY-STEP
FILLING THE 5 GALLON TANKS:

1. DEGREASER:
Fill a bucket with 4 gallons of tap water. Add the recommended amount of degreaser per gallon. Stir with a plastic rod.

The degreaser is added to the water to reduce foaming /bubbling.

2. WATER RINSE:
Fill another bucket with 4 gallons of tap water.

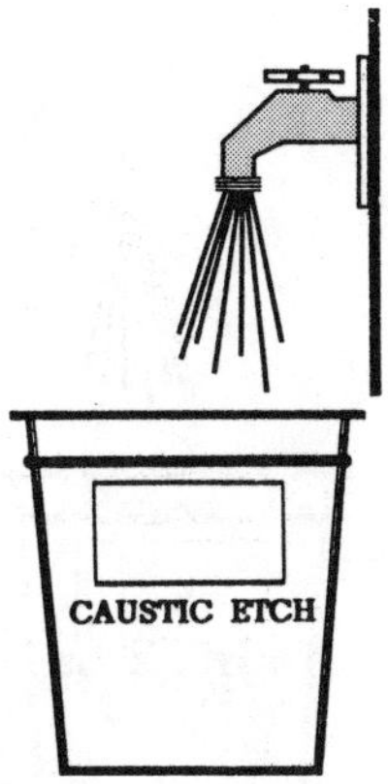

3. CAUSTIC ETCH:
Fill a bucket with about 2 gallons of tap water.

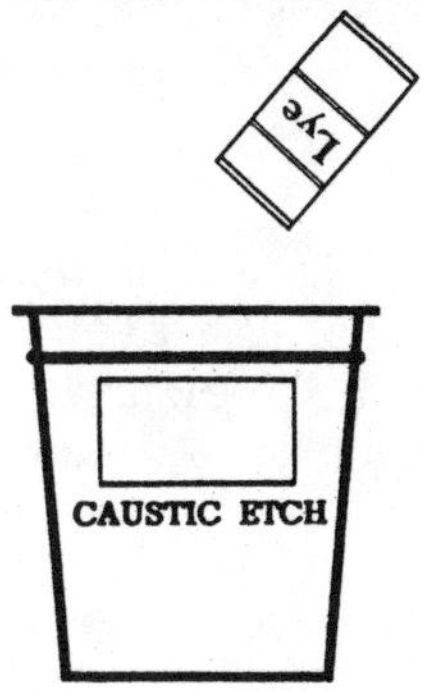

Add 4-6 ounces per gallon of sodium hydroxide (lye) slowly to the water and stir with a plastic rod. I use Red Devil brand. Ounces per gallon is based on the final volume (4 gallons).

Sodium hydroxide fumes are not safe to breath. Work under a fume hood or be certain to exhaust fumes away from you. Do it outdoors with the wind blowing away from you. Do not allow solution to splash on you or your clothing.

Add more tap water to bring up the level once the lye has been dissolved. Cover the container with a lid.

4. WATER RINSE:
Fill another bucket with 4 gallons of tap water.

5. NITRIC ETCH (SMUT REMOVER):
Fill a bucket with 4 gallons of tap water.

Add 13 fluid ounces of nitric acid per gallon of water (a 10% solution). Fluid ounces per gallon is based on final volume (4 gallons). Add the nitric acid slowly. Pour the acid down the side of the measuring container.

Pour the acid slowly down the side of the bucket.

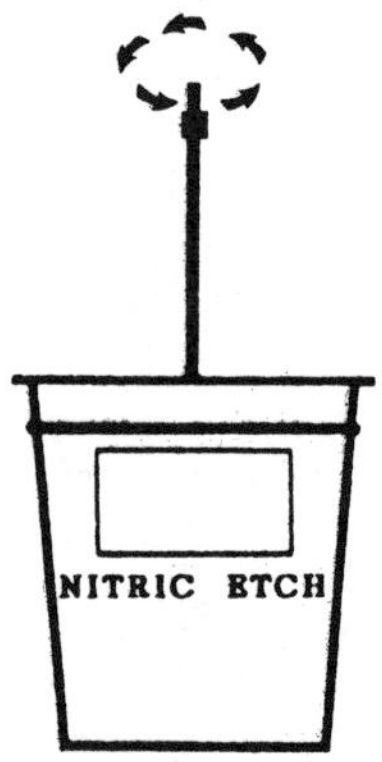

Stir slowly with a plastic rod. Be careful not to spill or splash the solution. Cover the contain with a lid.

6. WATER RINSE:
Fill another bucket with 4 gallons of tap water.

NO, this is not a misprint. This second water rinse is designed to eliminate nitric acid drag-out. Make it up.

7. WATER RINSE:
Fill still another bucket with 4 gallons of tap water.

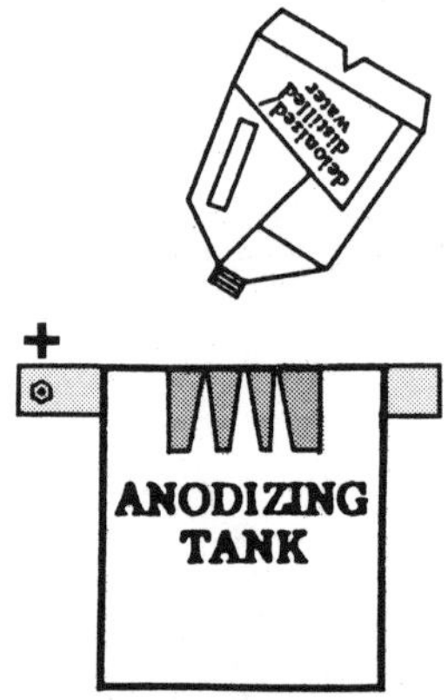

8. ANODIZING TANK:
Fill the anodizing tank with 4 gallons of distilled/deionized water. Be careful not to pour the water onto the bus bar.

Add 22 fluid ounces of sulfuric acid per gallon of water (a 15% solution). Fluid ounces per gallon is based on the final volume (4 gallons). Add the sulfuric acid slowly.

Pour the acid down the side of the measuring container.

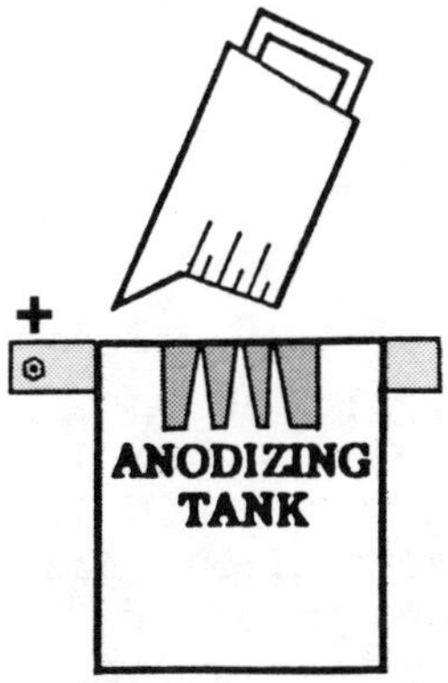

Pour the acid slowly down the side of the anodizing tank. Be careful not to spill the acid on the bus bar.

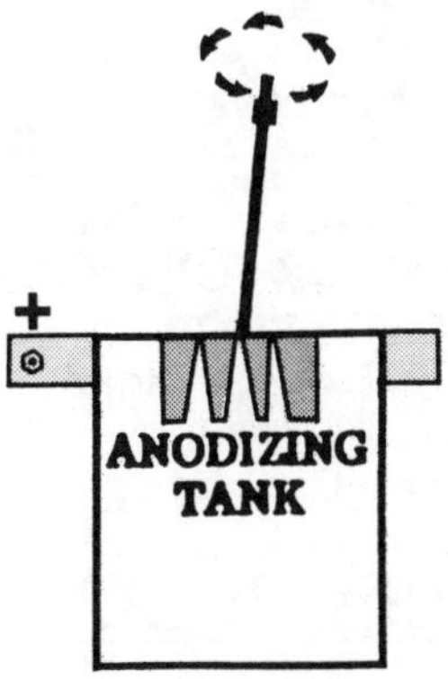

Stir slowly with a plastic rod. Be careful not to spill or splash the electrolyte.

9. WATER RINSE:
Fill another bucket with 4 gallons of tap water.

10. NEUTRALIZER (BAKING SODA) RINSE:
Fill a bucket with about 3 gallons of tap water.

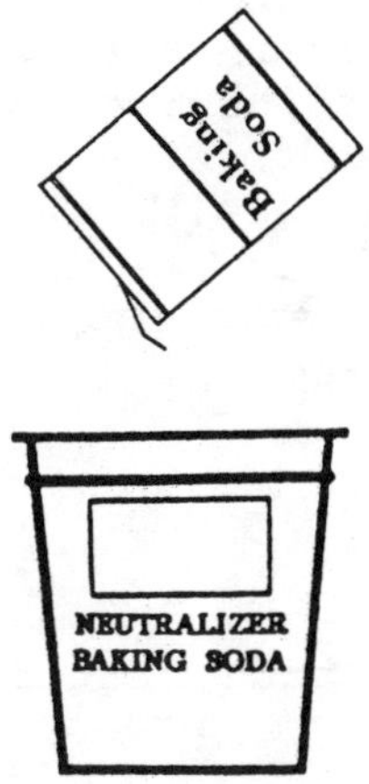

Add 1/2 lb of baking soda per gallon of water, based on final volume (4 gallons).

Stir with a plastic rod to dissolve the baking soda. Since the baking soda settles out over time, you will have to stir prior to each use.

Add more tap water to bring up to the 4 gallon level.

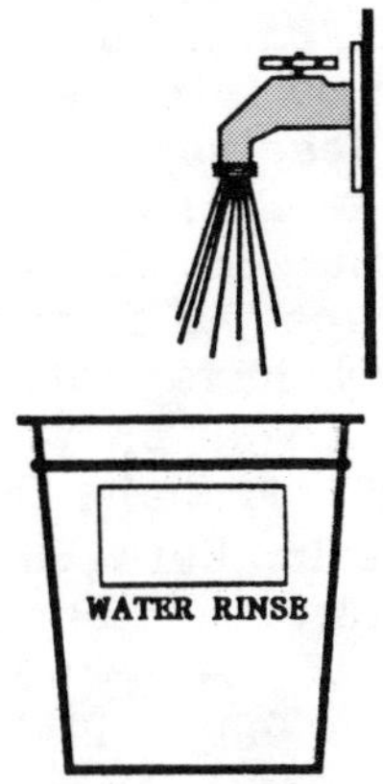

11. WATER RINSE:
Fill another bucket with 4 gallons of tap water.

NO, this is not another misprint. This water rinse is designed to reduce sulfuric acid drag-out. Fill the bucket.

WATER RINSE:
Fill still another bucket with 4 gallons of tap water.

Asides & Other Thoughts

LEAD SULFATE/BAKING SODA

A mixture of baking soda and water will serve as a cleaner/neutralizer. During anodizing, unless you cover the D.C. power supply negative clamp/lead connection area with battery terminal paint, or vaseline, the lead sulfate that develops on the lead surface and the clamp can create current flow problems. This coating will reduce the current flow. You may have noticed the same build-up on your car's battery terminals. This is the reaction between the lead and sulfuric acid and extends on to the battery cable clamp. Copper plated iron clamps will last longer between cleanings than non-plated clamps. Check your local hardware or auto parts store for sprays to coat either your D.C. power clamp or the battery terminals in your car (or both).

To clean the clamp, fill a plastic container with enough water to immerse the end of the clamp. Add 5-6 tablespoons of baking soda to about 2 cups of water. Stir until dissolved. Turn off and unplug the power supply (for obvious reasons) before immersing the clamp into the solution. Remove when clean. Rinse and re-attach the clamp to the lead cathode.

FLOATING AQUARIUM THERMOMETERS

Floating aquarium thermometers work well for monitoring the electrolyte's temperature. Be careful not to let the thermometer run into the aluminum pieces.

CIRCULATORS

Large anodizing shops often use compressed air to stir or circulate the anodizing and dye solutions. A large air compressor with the appropriate filter to eliminate oil from contaminating the baths is required.

PREVENTION OF ANODIZING SPRAY MIST/VAPORS

I am aware of two products that can be added to the anodizing tank to reduce the spray mists/vapors from the electrolyte.

Breatheasy, PMS from:
NORTHEAST CHEMICALS, INC.
240 Mary Street
Hackensack
New Jersey 07601
201-342-8774

Ekaline F Liquid from:
SANDOZ CHEMICALS CORPORATION
4000 Monroe Road
Charlotte
North Carolina
28205
704-372-0210

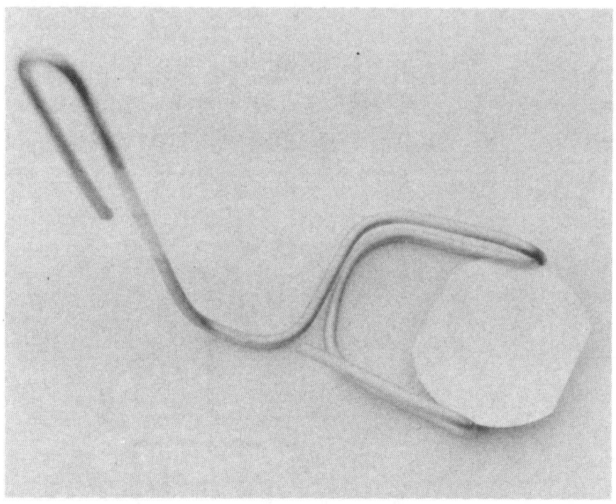

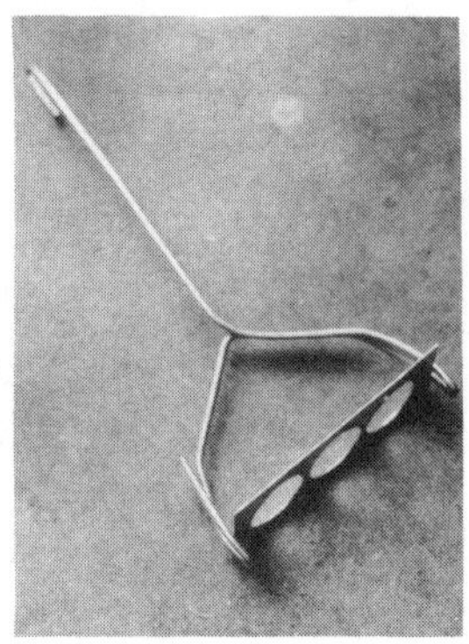

Two examples of aluminum wire, home made racks that support aluminum sheet. Remember to bend the loop inward, creating a springy effect, providing more complete support and current flow. The smaller of the two racks was suggested by Jan Peters.

ELECTROLYTE TEMPERATURE CONTROL

During anodizing, the overall reaction creates heat. As mentioned earlier, the optimum temperature for anodizing ranges is 68-72 degrees F. (60-80 degrees F is the total acceptable range.) A floating aquarium thermometer will let you monitor the electrolyte's temperature.

If you only anodize a few small pieces, or single pieces, then the heat generated may not create a problem.

If the temperature raises above 72 degrees F. you may wish to lower it. Two suggesting for lowering the temperature follow.

1. David Tisdale has an excellent suggestion for controlling the electrolyte's temperature. He suggests that you install a small immersion pump into a bucket of ice water, with polyethylene tubes recirculating the ice water through the anodizing tank. (No illustration.)

2. Another way to insure consistent electrolyte temperature throughout is to circulate air or to stir the solution. A small PVC pipe structure with holes drilled at regular intervals into the pipe on the tank's floor and hooked to an aquarium air pump, is an inexpensive and simple bubbler.

This drawing illustrates the bubbler. I purchased some white opaque PVC tubing (3/8" OD, 1/4" ID), some elbows, T-connections, glue and clear aquarium tubing from my local hobby store. The PVC pipe was heat formed, glued together into the illustrated shape. The holes were drilled every 3/4" with a #56 drill bit at the angle shown in the drawing.

Drilled hole *Drilled hole*

Then an aquarium air pump was connected to the unit and it bubbles away merrily.

An alternative to the T-connection shown is to purchase a non-corrosive aquarium gang valve.

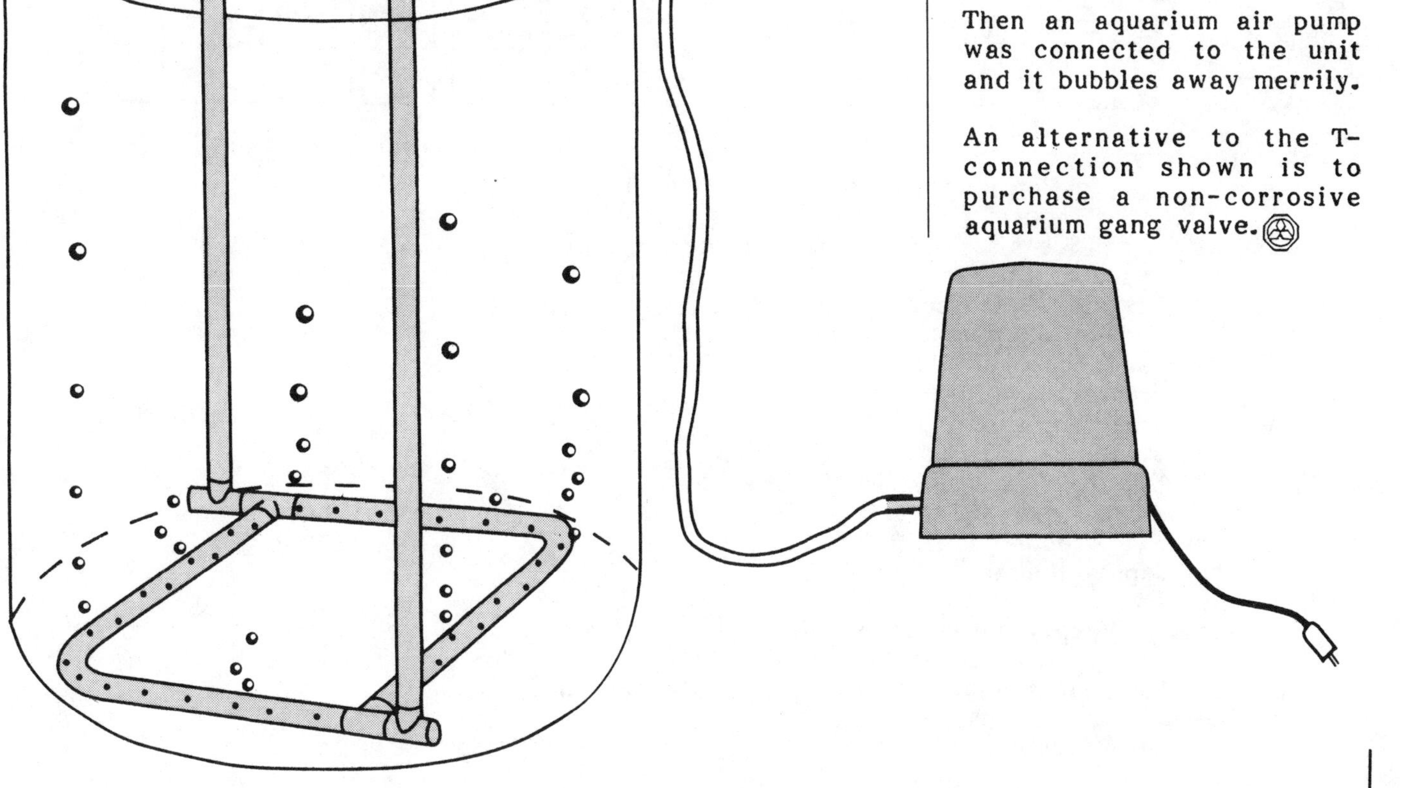

Dye Bath Set-up

The equipment and supplies you will need and where to buy them.

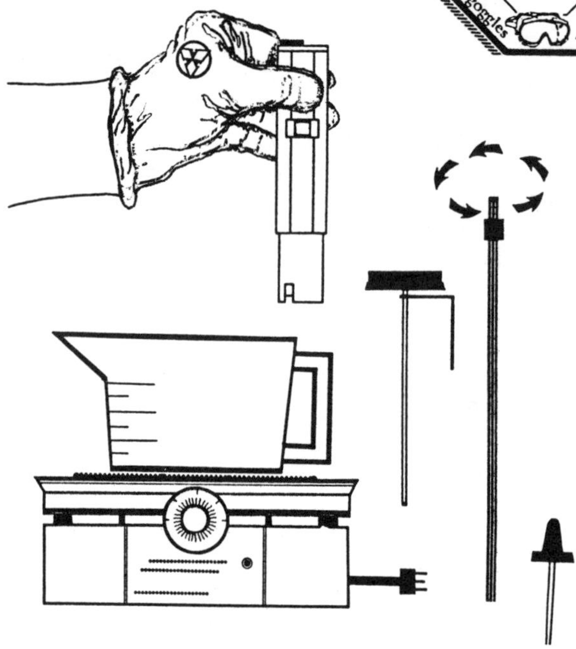

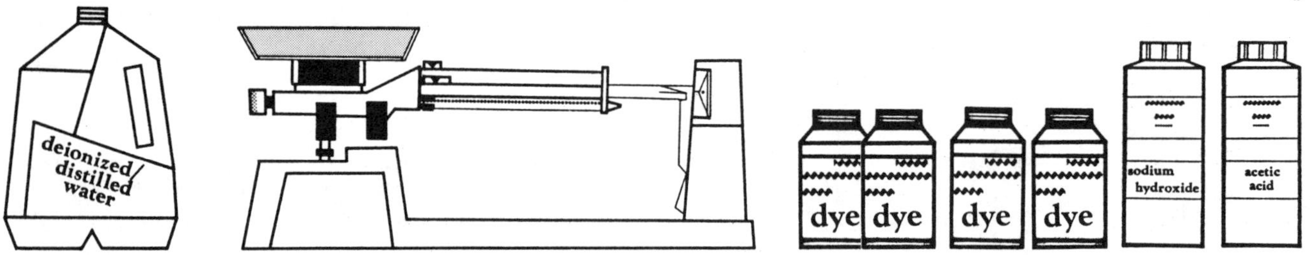

Only the major items are illustrated.

Grocery Store:
- Plastic Containers with Tight Fitting Lids
  - (1 quart, 1 to 2 gallon, etc capacity)
  - (See also - Possible Containers Chapter)
- Pyrex or Stainless Steel Measuring Cup
- Distilled/Deionized Water and candy thermometer
- Bamboo Skewers/Aluminum Wire

Hardware, Discount Store:
- Electric Hot Plate

Chemical Supply House
- Triple Beam Balance
- Sodium Hydroxide (50% Solution)
- Acetic Acid (Glacial)
- Eye Dropper

Dye Supply House:
- Dyes

Plastics Supply House/Hobby Store:
- Plastic Rod for Stirring, add black electrician's tape to designate dry end.

See pH Testing Chapter:
- pH Meter

Obviously many of the above items may be purchased at other locations. This list provides a start.

There are three types of dyes for aluminum:

A. Dyes specifically designed for aluminum
B. Household dyes
C. Fiber reactive dyes

A. DYES SPECIFICALLY DESIGNED FOR ALUMINUM:

These dyes are the best. They have the best light fastness, which means they will not fade as quickly as other dyes. Three important basic facts you should become know about the dyes are:

1. dye concentration -- grams/liter
2. optimum solution pH
3. lightfastness

} see appendix

DYE CONCENTRATION

The concentration of grams per liter is the amount of dye you will be mixing with the distilled/deionized water. The dye is usually packaged in plastic bottles with screw top lids. When mixing the dye be certain to wear rubber gloves, goggles and a respirator. The dye powder is very dangerous to your lungs and skin. The dye can be absorbed into your system quite easily through skin absorption. Be careful. When working with the gram scale, set the scale weights to correct grams per liter weight as per dye manufacturer's chart. For example: If you are mixing Sandoz Bordeaux R, the chart says 2 grams per liter of water. If you have a 1 gallon container for the dyebath you will need 8 grams of dye powder. Remember 1 liter is equal to 0.9463 quarts or 1 gallon equals 3.7853 liters. Sure! Let's make it easy -- 1 gallon equals 3.8 liters. Multiply 3.8 X 2 and the amount of dye in grams would be 7.6 for your 1 gallon dye bath. One dye dealer has calculated concentrations in both grams per liter and grams per gallon for ease of dye mixing -- EAST WEST DyeCom.

Although each color of dye has a specific amount of dye powder in grams to be mixed with the distilled/deionized water, consider cutting the amount in half if you wish a more pastel tone. Another way to arrive at lighter values of color is to reduce the immersion time in the dye. As dyes get older, more exhausted, you may really like the more subtle color. By making variations on the theme so much more visual excitement is possible. Live!

OPTIMUM SOLUTION pH

Each dye has a optimum solution pH. The manufacturers of the dye have developed specific pH ratings for each color they sell. The color of the dye on the chip chart (see color section) is the color you will get if you adhere to the pH rating. However, if you wish some variation of that color, you may adjust the pH to provide your own special results. Experiment!

ADJUSTING pH

The pH of a dyebath is the measure of its acidity, an important factor to monitor. A pH of 7 is neutral. pH values below 7 define the solution as an acid, while values above 7 make the solution alkaline. Acetic acid (glacial) will lower the pH, while sodium hydroxide will raise it. For the size dye baths we would be using, eye dropper sized quantities would be used to add the acetic acid or sodium hydroxide for pH adjustment.

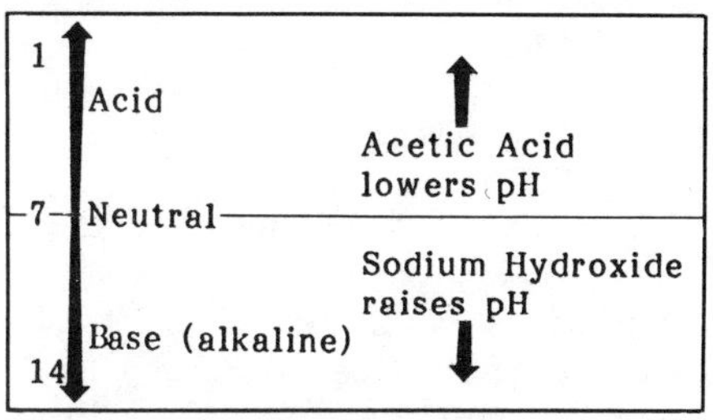

The pH of a dyebath can be measured in two different ways: **COLORIMETRIC OR ELECTROMETRIC.**

COLORIMETRIC -- This method requires the use of pH papers that measure the pH to about plus or minus 0.2 units through a color change in the paper as it is immersed into the dye bath. The only drawback with this method is that often the dye obscures the paper color. Therefore I recommend using pH paper to test the distilled or deionized water prior to adding the dye. The average cost of a box of pH papers is \$15 to \$20.

ELECTROMETRIC -- Instead of paper, a battery operated pH meter is used. The degree of accuracy is plus or minus 0.01. The pH meter device has a probe which is lowered into the dye and allowed to remain in the dye for about one minute. During that time the meter will finally adjust to the pH of that given dye. pH meters, if purchased from a chemistry supply, can cost as much as \$200 and up. Garden shops often stock \$25-\$30 pH metering devices that are used to measure soil pH. I bought one to test out and found the results, when check against a more expensive pH meter, were inaccurate. I shopped around and finally found an inexpensive one for about \$40. I'd recommend that nifty pH meter, shown in the Testing section.

LIGHTFASTNESS

Lightfastness is the measure of how well the dyed aluminum will resist fading. The higher the number, the better the degree of lightfastness. The numbers go from 1 through 8. In jewelry there is not too much to worry about fading, unless the pieces are worn for long extended periods of time in direct sunlight.

Lightfastness is influenced by: thickness of oxide layer, length of time spent in dyebath, degree of sealing, etc.

B. HOUSEHOLD DYES:

Household dyes such as Rit brand work fairly well. They do seem, however to have less lightfastness, and therefore may fade sooner than dyes made specifically for aluminum anodizing. Rit dye is sold in both powder and liquid form. I usually mix the entire amount of powder or liquid with the distilled/deionized water. Follow normal dyeing safety precautions.

C. FIBER REACTIVE DYES:

The fiber reactive dyes provide an excellent variation on the colors that are available from Sandoz, C.H. Patrick, and U.S. Specialty Color Corporation. Check out your local weaving/spinning store and see what lovely colors are available. I use similar amounts of dye, in grams per liter/gallon, as the major dye manufacturer's chart recommends. For example: If a particular shade of red uses 2 grams per liter, I use 2 grams per liter of the red fiber reactive dye. Don't forget your rubber gloves, goggles, and respirator when preparing the dyebaths. Safety counts.

When making up each new dye bath, do the following: No matter what type of dye you are mixing, powder or liquid, be sure to mix with only distilled/deionized water, never tap water. Tap water generally contains chemicals that could contaminate your dyes.

All dyes must be completely dissolved in the water. If not, un-mixed dye granules may spot your piece. This is a nice effect, that is, if you like it. Most anodizers want consistent color without spots. When mixing powdered dyes, mix the dye with a small amount of heated distilled/deionized water in a Pyrex or stainless steel bowl. The temperature of the water should be around 140 degrees F. Stir the dye thoroughly to insure that it is completely dissolved. Pour this solution into the dye bath container. This container should have the rest of the water necessary for this dye bath and that water should also be at 140 degrees F. Stir the mixture thoroughly.

Dye tanks must be made of either stainless steel, plastic, fiberglass, or some other inert substance. Do not use copper or steel. These metals will cause strange reactions and greatly affect your results.

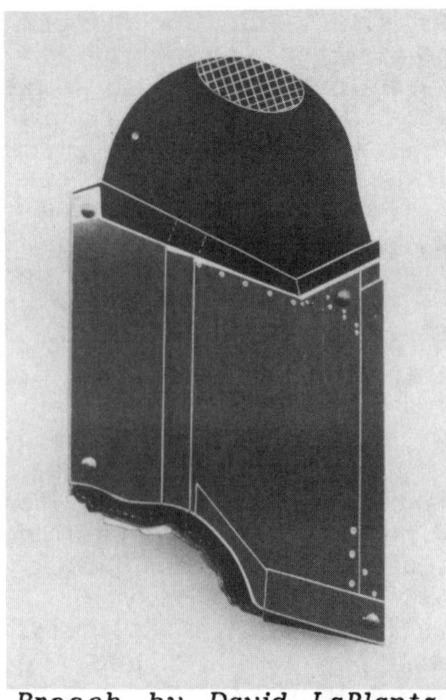

Brooch by David LaPlantz, fabricated, painted aluminum, engraved, actual size, 1983.

ILLUSTRATED STEP-BY-STEP

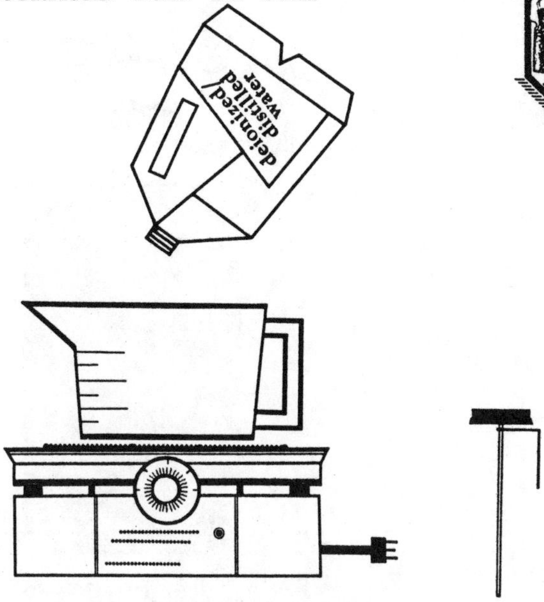

1. Heat distilled/deionized water to 140 degrees F.

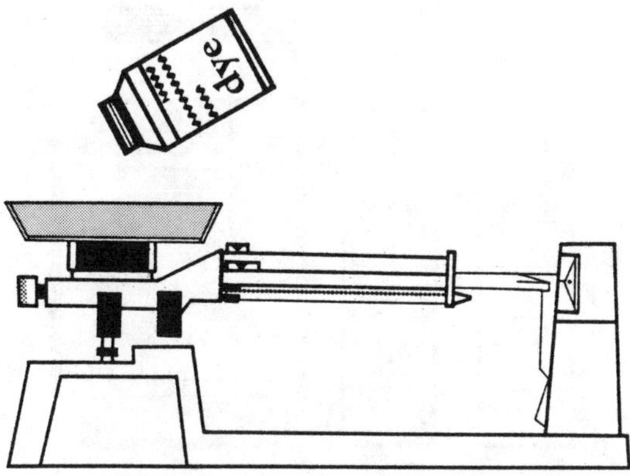

2. Weigh the dye to the required amount, as per dye manufacturer's chart, pages 195-197. Spoon the dye onto the scale's stainless steel tray.

3. Add the dye, from the scale tray, slowly to the hot water. Tap the tray so that all the dye powder transfers.

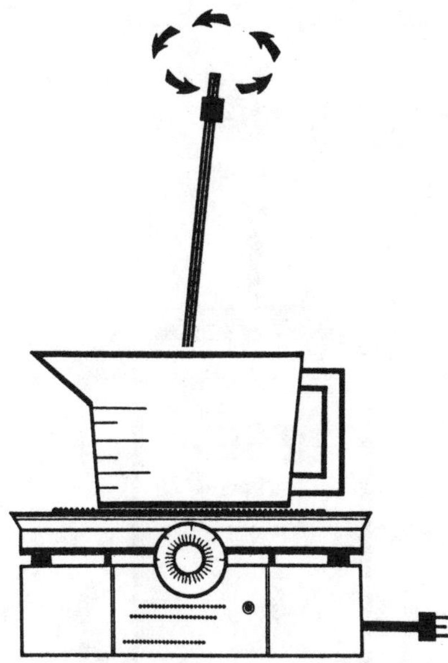

4. Stir with a plastic rod until the dye is totally dissolved. Be careful not to splash the dye.

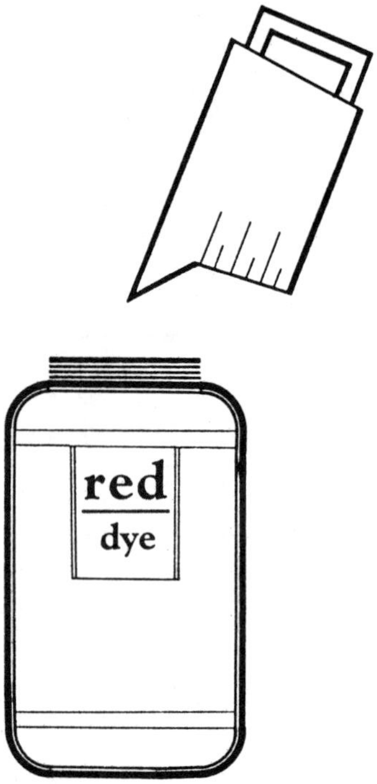

5. Pour the dye into the dye bath container, holding the additional 140 degree F. water. Label dye container.

6. Stir the entire mixture well. Be certain the dye is totally mixed.

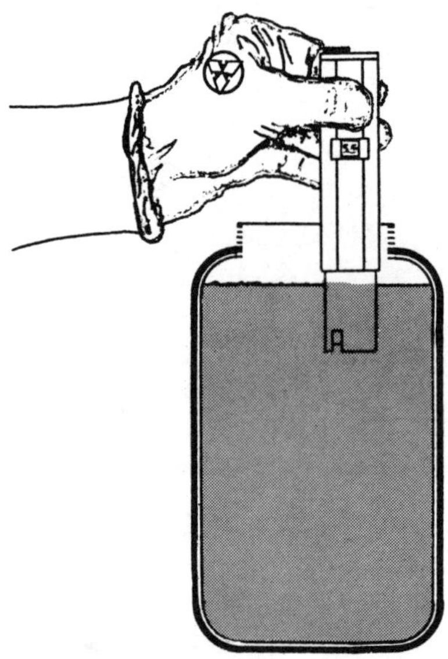

7. Check the pH -- see Testing section for how to check the pH. For the most accurate reading, immerse the pH meter into the dye for a period of 1 minute. Check the dye manufacturer's chart for optimum pH value. If the pH is correct, the dye is ready to be used.

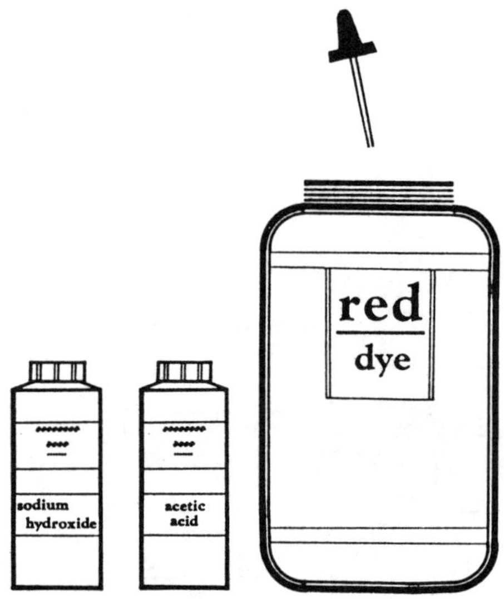

8. If the pH is off, it must be corrected. If the pH is high, acetic acid will lower it. If the pH is low, sodium hydroxide will raise it. To adjust the pH, use the eye dropper to slowly added whichever chemical is required.

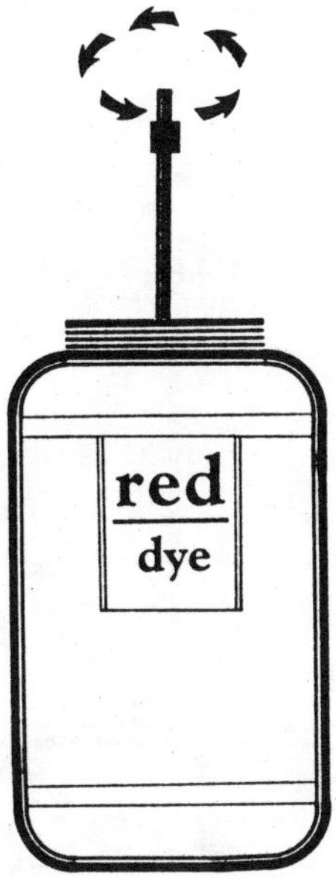

9. Stir with a plastic rod.

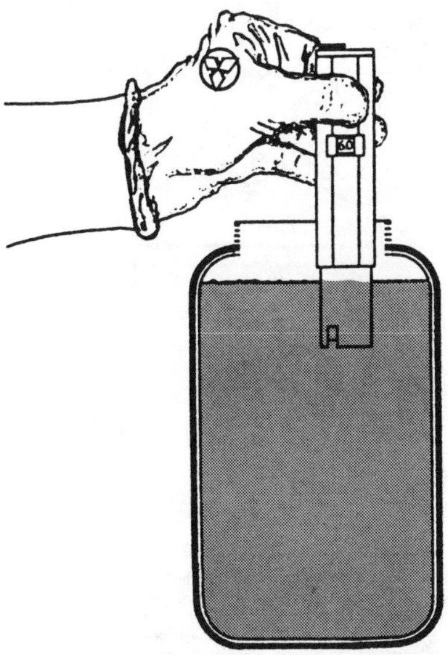

10. Check the pH once again.

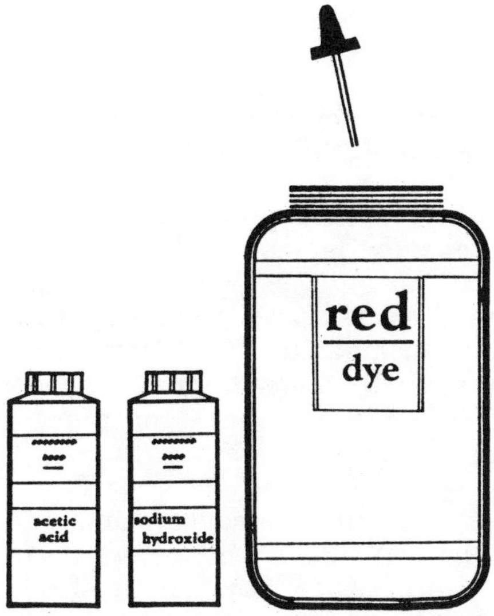

11. Add more pH adjusting chemical, if required. Stir with plastic rod.

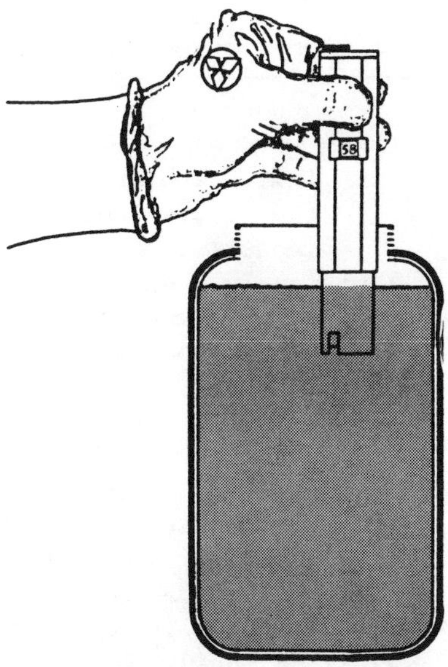

12. Check the pH once again. If the pH value is correct, the dye is ready to be used. If not, repeat the above.

Sealing Set-up

MAKING UP THE SEALER

Sealing is the plugging, or closing off, of the porous oxide layer.

The simplest and most economic method is to use boiling distilled/deionized water. There isn't much to say about this method other than, fill a stainless steel container with water and heat it.

The following demonstration will show you how to mix powdered nickel acetate sealing salts. Nickel acetate may produce nickel rash/itch and is a known carcinogen, therefore it should be used with the utmost care. Liquid sealing solutions are also available.

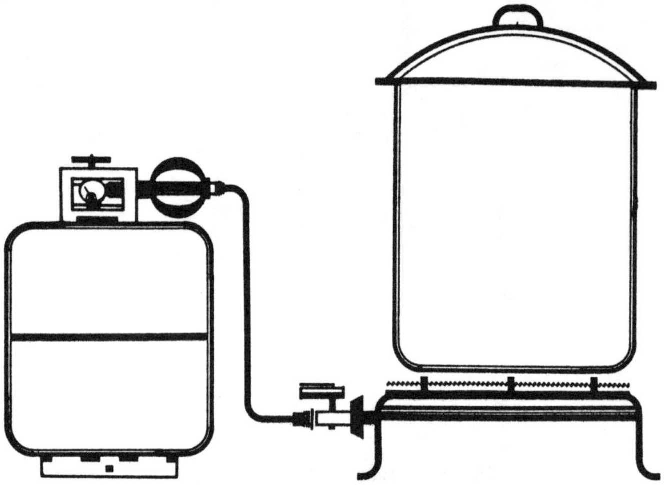

Only the major items are illustrated.

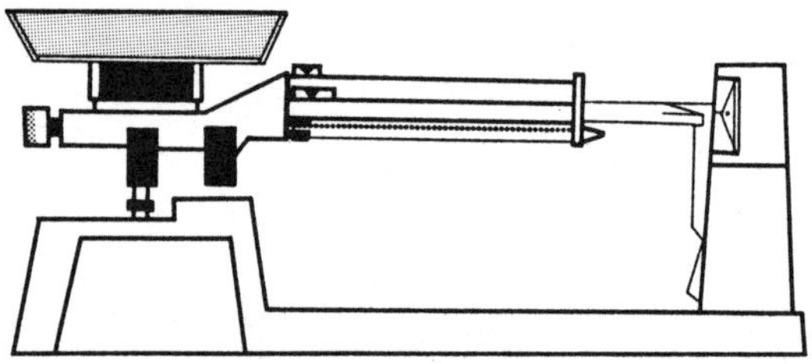

The equipment and supplies you will need and where to buy them.

Grocery Store:
- Distilled/Deionized Water
- Candy Thermometer
- Bamboo Skewers/Aluminum Wire

Dye Supply House:
- Sealing Salt (Nickel Acetate)

Restaurant Supply House:
- Stainless Steel Stock Pot/Lid

Jewelry Supply House:
- Triple Beam Balance
- Spark Lighter

Hardware, Discount Store:
- Electric Hot Plate

Propane Gas Dealer:
- Propane Tank, Burner, etc

Plastics Supply House/Hobby House:
- Plastic Rod for Stirring, add black electrician's tape to designate dry end.

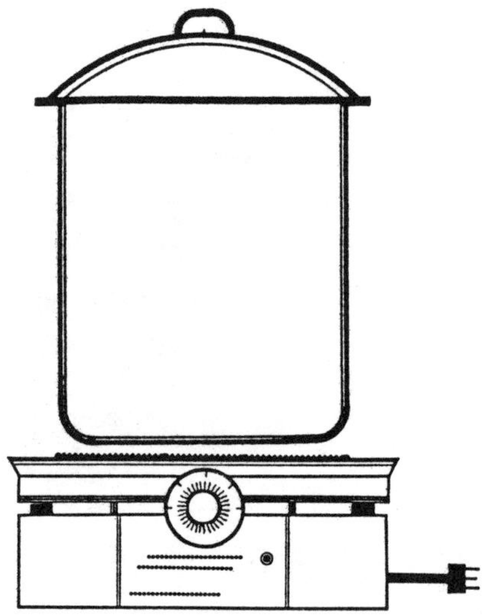

Obviously many of the above items may be purchased at other locations. This list provides a start.

Natural gas is an alternative to propane. Here is a hot water heater burner. If used with a metal grate and bricks for support, it makes a simple heating device.

A Restaurant/Kitchen Supply stainless steel stock pot and lid filled with distilled/deionized water sealing solution. Notice the notch in lid for candy thermometer probe.

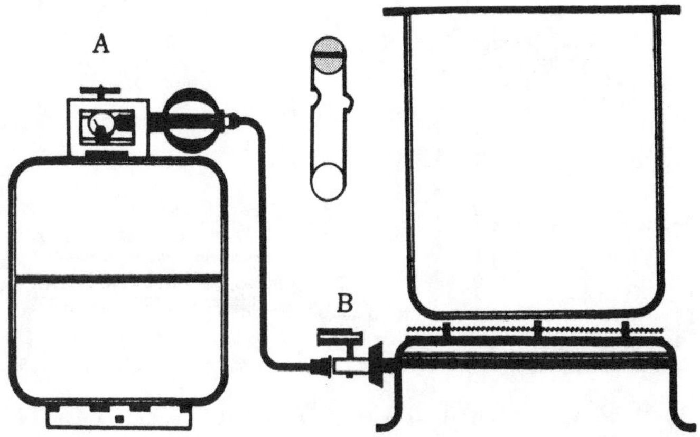

1. Open main propane (tank) valve (A). Open gas burner valve (B), ignite flame with spark lighter and adjust height of gas flame. Place stainless steel container on burner. Add distilled/deionized water. Fill container to about 1/3 capacity.

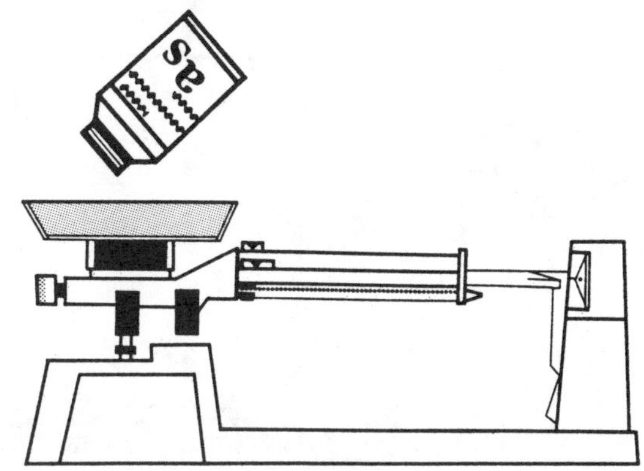

2. Weigh out the required amount of Sealing Salt AS, according to manufacturer's instructions. (Generally, 6 grams of nickel acetate [sealing salt AS] per liter [23 grams per gallon], based on final volume [1 gallon]).

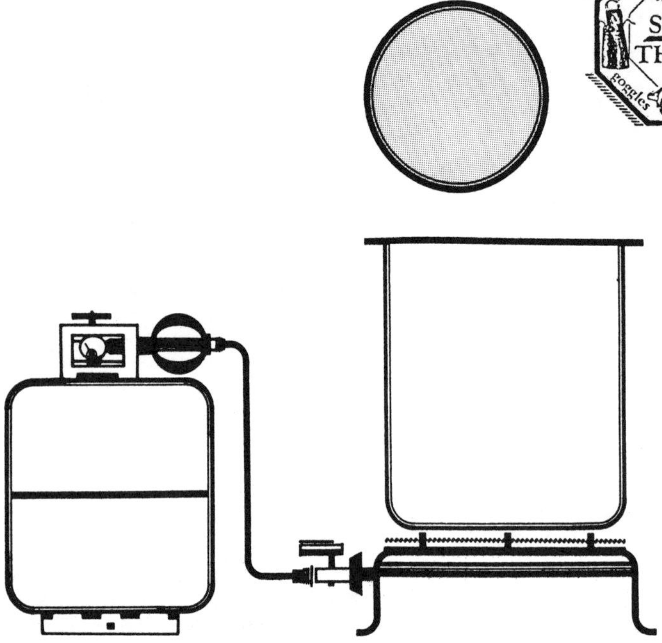

3. Use the triple beam balance tray to transfer the Sealing Salt AS to the water in the stainless container. Add the sealing powder slowly and allow it to dissolve into the boiling water.

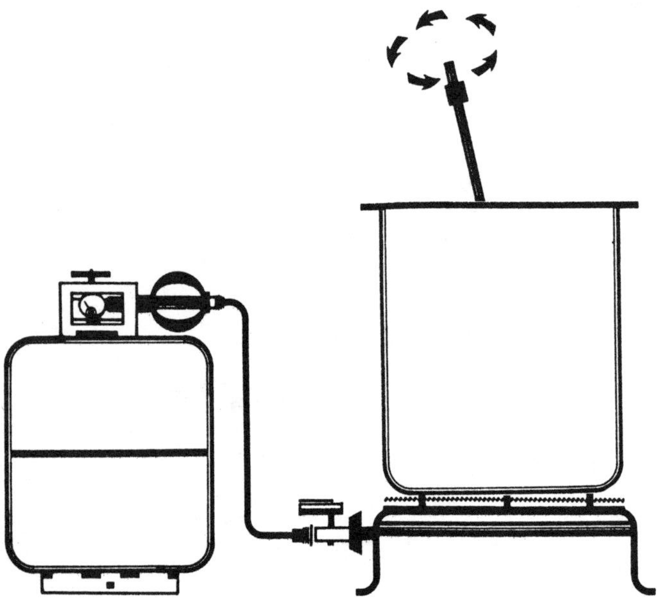

4. Stir slowly with the plastic rod. Dissolve all the Sealing Salt. Be careful not to spill or splash the solution on you or your clothing.

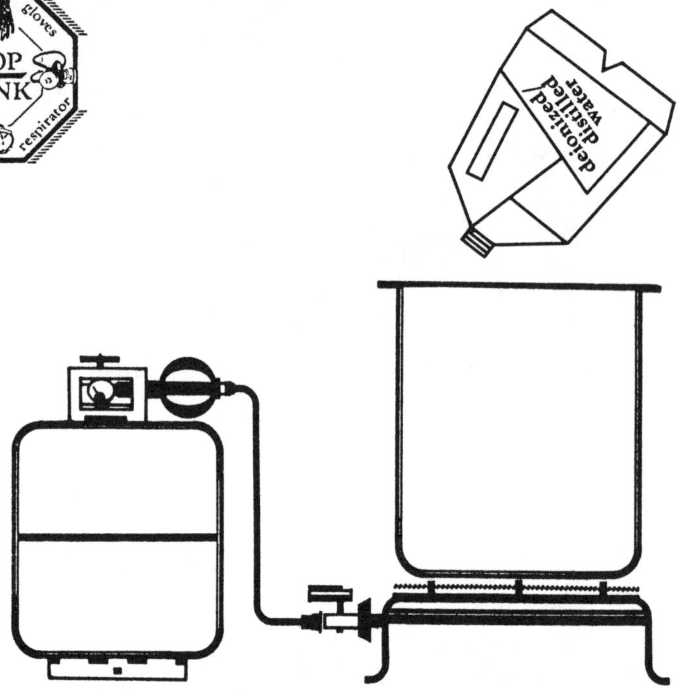

5. Add the remainder of the required distilled/deionized water to top off the sealer. I suggest the liquid level be about 1- 11/2 inches below the top of the container. Stir slowly with the plastic rod.

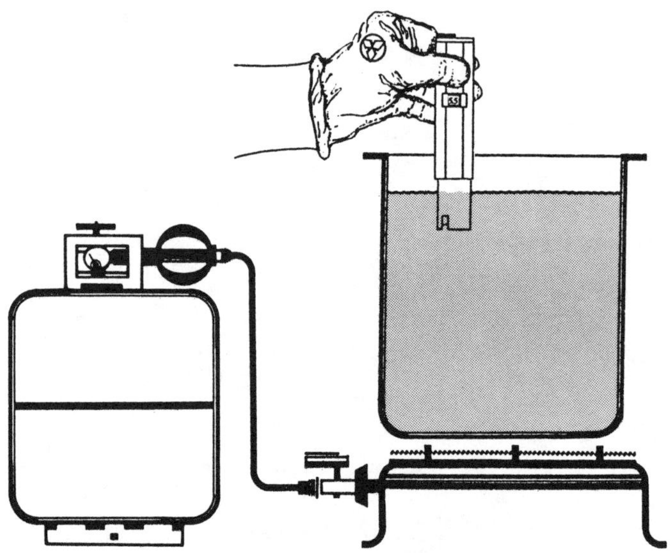

6. Check the pH. A pH of 5.5 is optimum. If the pH is off, lower with acetic acid or raise with ammonia. Do you recall the eye dropper method for adjusting the dye bath pH? Use your eye dropper again and stir. Regular pH checking, control, and correction is paramount to successful sealing.

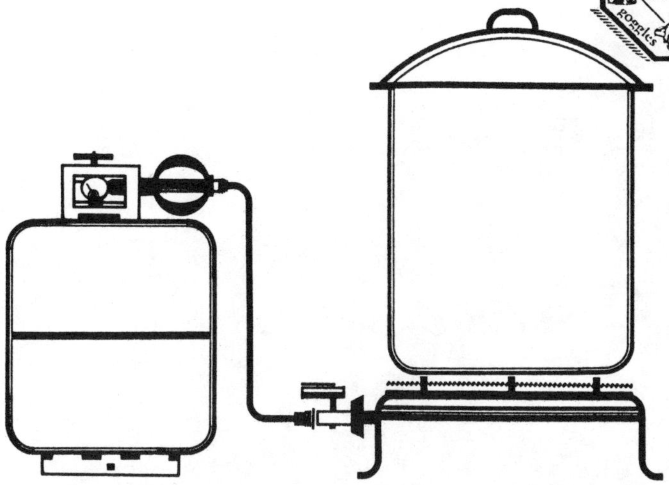

7. Place the lid on the sealer and bring it up to the proper sealing temperature of 208-212 degrees F. Use an inexpensive candy thermometer to monitor the exact sealing temperature. If you experiment with the valve on the gas burner, you will discover the exact setting that will hold the sealer at a given temperature. This knowledge will save you **money** on gas and water!

8. Fill a bucket with 4 gallons of tap water for the rinse that follows the sealing.

When you purchase a candy thermometer, a stainless hanging device is attached. Due to the design, this hanging device forces the probe in towards the center of the sealing container. This may cause inadvertent contact and damage to your work. Therefore I recommend removing the hanging device. I use two methods of holding the candy thermometer in the sealing bath:

An alternative: an electric hot plate will also heat the sealer. Notice the thermometers in stock pots and the aluminum foil used as a lid.

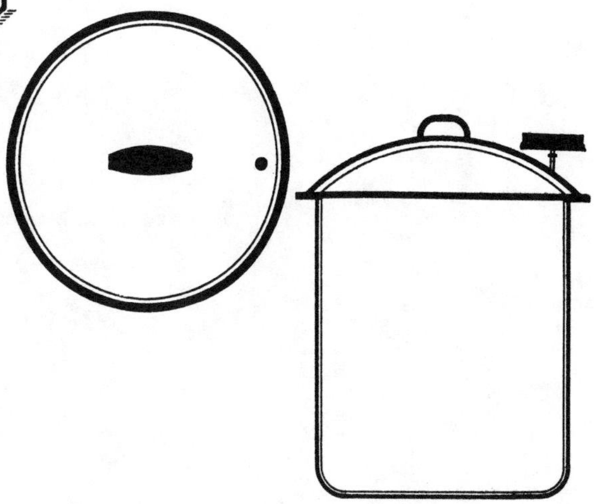

1. Drill a hole in the container's lid, slightly larger in diameter than the candy thermometer's probe. Locate this hole near the edge of the stainless lid, but in alignment so the thermometer will penetrate the sealer when the lid is in position.

Do not drill the hole in the middle of lid because the thermometer probe might come in contact with and damage your pieces.

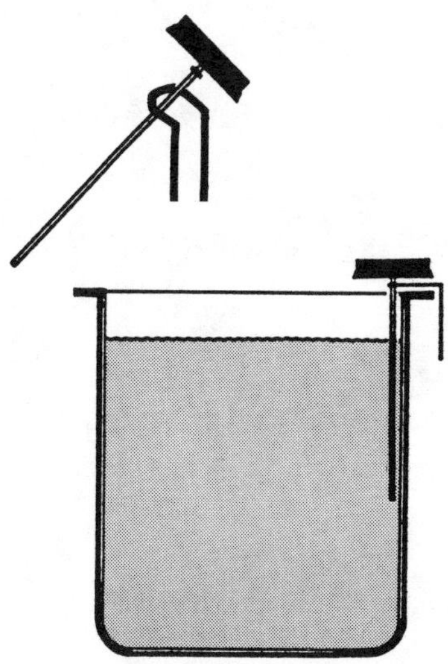

2. Bend a 6 inch length of heavy gauge (12) aluminum wire into this shape.

The "legs" of the wire will hang over the outside of the stainless container, while the probe slips through the "loop" inside, by and penetrating the sealing bath. An easy and simple solution!

SURFACE TREATMENTS

Pre-treatments
Alternative Surfaces
Etching

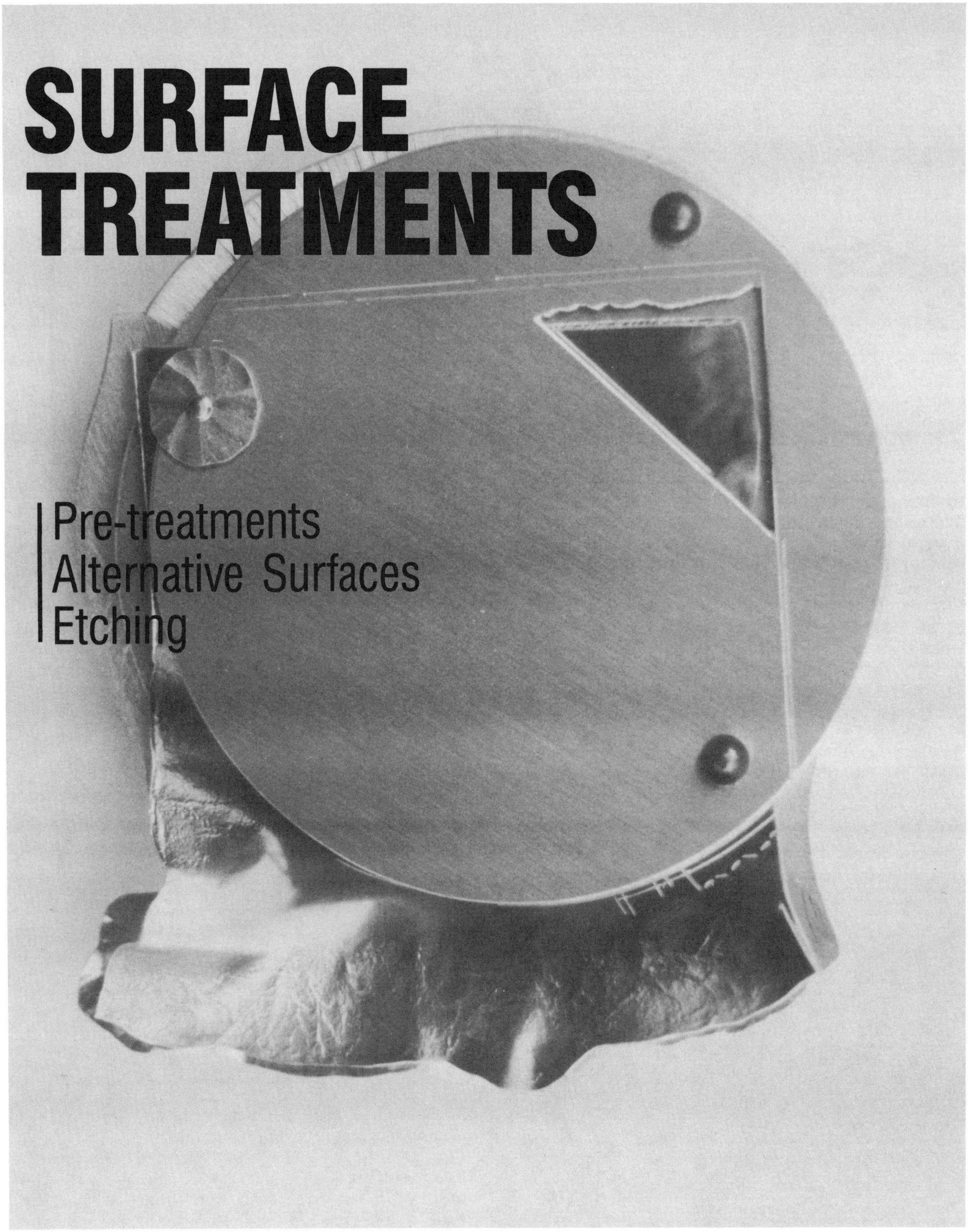

Brooch by David LaPlantz, fabricated, painted and anodized aluminum, engraved, 2 3/4" X 2", 1986.

Pre-treatments

The surface finish of the aluminum prior to anodizing is critical. If you desire bright, reflective colors--you must high polish the aluminum. A high polish is not required. Many commercially produced aluminum products have a non-reflective, brushed finish. These matt finishes do not show damage as easily as the highly polished surfaces.

This chapter attempts to show you several methods of matt finishing metal. If you high polish the aluminum first, then add matt finishes to certain areas, then, after the aluminum is anodized and dyed, there will be a range of color value differences.

If you are a hard core jeweler or sculptor who adores smooth, shiny, extremely reflective surfaces, the following will offer you some unusual tangents.

Large, more expensive, standing sand/beadblaster with vacuum cleaner attached.

Smaller, less expensive, sand/beadblaster, without vacuum cleaner.

MATT FINISHES ON ALUMINUM

SANDBLASTING

Sandblasting is a real favorite of mine for producing beautiful matt surfaces. In sandblasting, various grades of sand or other media, such as silicon carbide are used. The problem with sand is the eventual breakdown and the danger from the silica in the sand. Silica is the major health hazard in the disease called silicosis. Of course, since most health hazards are cumulative, you've got to spend many years exposing yourself to the particular problem to be in real danger. As I see it, though, there is no need to be foolish about your health. Besides, why risk mixing chemicals inside a wonderful test-tube, like the human body?

BEADBLASTING

Beadblasting is the same as sandblasting, but with finer surface results. You also do not have to contend with the danger of silicosis.

HEALTH HAZARDS RELATED TO SAND/BEADBLASTERS

Sand or beadblasters should be respected for a couple of reasons:

1. The first is for health hazards. Be very careful not to stick your head into the box, especially if the access door hinges upward, immediately after ceasing work. Yes, you guessed it, tiny particles of sand or glass will fall onto your head and into your eyes. From my own rude experience, I can tell you how painful sand in the eye can be!

2. Know the setting of your air pressure valve. Be certain to know how high the valve is set versus the thickness or thinness of the object. Also be aware of how close the orifice of the gun is to the object. I've seen some fine pieces of art warped/destroyed by too high a pressure setting or a thin object being too close to the end of the gun. Another tip: To insure constant and consistent air pressure, thus producing the desired results, be certain to hold the gun on an angle instead of shooting straight onto the object. The straight-on blasting produces sand or glass which is bouncing off the work while slowing down the effectiveness of new incoming media.

WET OR DRY SILICON CARBIDE SANDING PAPERS

Wet or dry silicon carbide sanding papers are another excellent source of alternative finishes. I'd recommend experimenting with all grades. Remember that the lower numbers, such as 100 and 220, are rougher finishing papers. 320 sanding paper is a medium grit. And the higher numbers, such as 400 and 600, denote finer finishing papers. For super fine finishing, ask at your local automotive store for 800, 1200, or 1500 grit papers. These papers are wet or dry silicon carbide papers produced by 3M for color sanding cars.

I work quite a bit with wet sanding and have had great results. By working wet, the dust from the metal and the silicon carbide is reduced (a safety suggestion and design additive). The surface sanding is much more regular and consistent when sanding wet, which creates more visually interesting surfaces. These surfaces will be enhanced by the anodizing and dyeing techniques.

Be sure to play with directional sanding. Sanding in one direction, then covering that area with masking tape, and sanding at 90 degrees to the covered area, will create a metal surface that appears to be bent, but in reality is totally flat!

Who said that you have to sand in straight lines or directions, anyway? Consider sanding in curved or curving directions. Have you ever tried sanding with a very coarse paper, such as 100, and then going back over the same area with 600 in the opposite direction? And so it goes on forever! Keep playing with possibilities that are out of the norm.

In order to sand surfaces totally flat some jewelers use plastic sanding sticks or master finishers, as some dealers call them. I use my plastic sanding stick daily, but of late, have moved onto another tool when I want a softer surface supporting the sanding paper. I wrap sanding paper around my hard felt stick with the felt side supporting the sanding paper. This provides a less harsh finish. It works. Another time to chalk up those thoughts of pushing our tools and ideas further than their intended purposes. As long as we are not hurting ourselves, or the tools, who cares? A new method or manner of thinking may be born!

A round metal rod of iron or brass, about 2 to 3 inches in length, 1/8 inch in diameter, with a saw cut 1 1/4 inch from one end of the rod will create a nifty tool for your flex shaft. Cut a strip of wet or dry silicon sanding paper 3 to 4 inches in length by 1 1/4 inch wide. Wrap the sanding paper counter clockwise around the rod by first inserting the end of the paper in the slot. Chuck this device in your flex shaft handpiece and see what magic you can do!?

FILES

Files are another special way to matt finish the aluminum. I do not mean to use the files in a normal way, but to use them to scrape and to literally gouge the surface. A file, used the wrong way can render quite an interesting and different surface finish. Instead of using the file to create a smoother finish prior to sanding and polishing, why not use it to create an expressionistic set of textures, much the way Jackson Pollack used color to excite his paintings? I often use the very end of the file or the tang. Have you tried laying the file on its side and then dragging it across the metal's surface? The following suggestion will make some of you cringe, but here goes -- using a heavy rasp or thicker file, try slapping the surface of the aluminum. The surface texture is different. If you are careful and use files of at least 1/4 inch in thickness, damage to files should be non-existent. Play with the possibilities of mis-using the file in all sorts of kinky ways -- you might be surprised at what you will end up with?!

POCKETKNIVES

Pocketknives lend another interesting signature of their own. One day I was thinking of how I used to whittle on scrap wood when I was a kid. Near me, at my bench, was a pocketknife and some round aluminum rods about 1/4 inch in diameter. One thing led to another and soon I was whittling on the aluminum like it was a piece of wooden dowel. The results were quite gratifying and exciting! The surface of the aluminum rod began to look like carved wood. The only difference was the shimmering silver color. I then could imagine the surface transformed by anodizing and being dyed any color I could create!

Often getting ideas comes from seeing one thing, but really seeing something totally different in our mind's translation!

HACKSAW BLADES

Hacksaw blades are sawing devices -- period. Or are they? One fortuitous day, during a demonstration, when I was trying to be spontaneous, I spied the hacksaw on the wall. Taking it down from its hanger, I viciously attacked the aluminum's surface. I actually shocked myself as much as I did the group. The results were intriguing. If you play with this possibility, you too may be pleasantly surprised.

GRINDING MACHINES

Grinding machines provide a multitude of finishes. Sculptors have, for years, been using portable grinding machines to create circular patterns on their aluminum surfaces. Instead, I'm recommending using a stationary grinder. As you're using the grinder, consider the position of your aluminum to the grinding wheel. I must caution you about using soft metals such as aluminum on your grinding wheel-- unless you are prepared to clean your wheel -- frequently. The aluminum will clog the wheel and cause problems when it comes time to work other harder metals. I suggest that you keep a separate wheel just for aluminum.

If your grinding machine comes with a grinding wheel on one side and a wire brush on the other, you have a double threat tool. Carefully position your metal against the wire brush and you should be quite pleased with the results. The wire brush creates a finish that looks like blown sand or is leather-like in appearance.

With experimentation, you will discover many variations on the theme, and after anodizing and dyeing, you may shout for joy.

Electric grinder/wire brush combination machine. The wire brush on the right side of the machine is a dandy way to create texture on the aluminum surface prior to anodizing.

Here is a sample of what the wire brush will do to the surface of the aluminum. The wire brush textured area is on the right side. Immediately to the left is the scotch-bright texture and the remainder is the stock rolled finish as you see it when the aluminum is purchased at the hardware store or lumberyard.

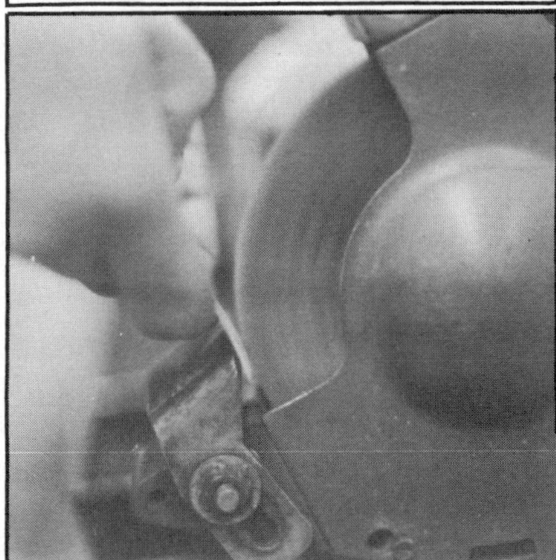

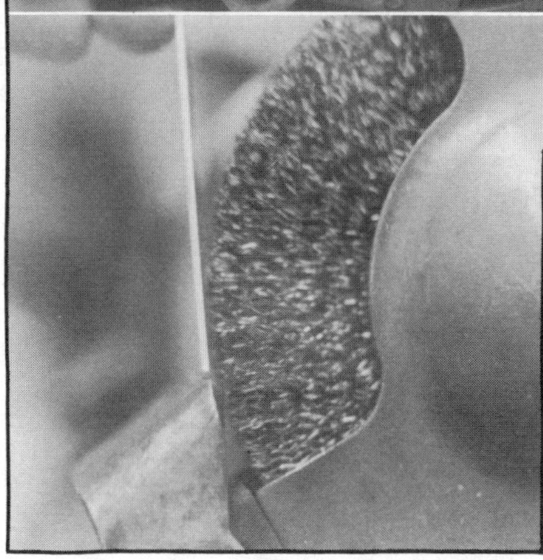

Notice the angle at which the aluminum meets the grinding wheel. Take care to hold the metal securely. The wire brush is turning toward you and will want to pull the aluminum away from you.

Full eye and face protection must always be worn when you operating power equipment, especially like the grinder on this page.

MISCELLANEOUS OTHER TIPS

These tips include grinding burs, and other flexible shaft attachments, careful mis-use of drill bits, either in drill press or flex shaft, and the old stand by the jewelers saw blade! The saw blade you say? Yes, the saw blade. I have had great fun beadblasting round surfaces of aluminum tubing and then carefully going back onto the matt surface with the saw blade, creating very shiny lines. When the tubes are anodized, then dyed with a single color, I achieve a dual colored effect due to the difference in surface value--matt to shiny. What a difference a little playing around can accomplish.

For hammer effects, how about using a ball peen hammer on polished, matt sanded, filed, beadblasted, etc. surfaces? A cross peen forging hammer or a damaged hammer head (customized?) is another alternative.

Have you considered using dull faced chasing or stamping tools on a polished aluminum surface? This will give you a series of matt indentations on a bright surface. The reverse of this would be high polished chasing or stamping tools on matt surfaced aluminum.

A DIGRESSION

The world of art and life can generally be divided into the following groups:

THE DULL ROARS AND THE LOW ROARS: This group usually includes those who talk a lot, but never seem to muster the courage or energy to do more than blow hot air.

THE MIMICS, COPYCATS, AND WAGONEERS: This group includes those who find it more creative to use someone else's ideas and personal imagery instead of looking inside themselves for stimulation. The wagoneers are those who find joy in the ride presented as others discover or rediscover techniques or other methods of expression. A perfect example of a wagoneer is the person who flirts with every new technique or breakthrough in a particular process. This is manifested in the so-called new work seen within weeks after major articles or new books are published. Instead of developing one's own personal direction technique-wise, there is a constant search to be like the "hit" of that season. The only problem with this person is that they always have to be on the lookout for the next big hit on the horizon. They never develop their own personal style and stymie themselves as artists.

THE LEADERS OF THE PACK: Certainly a smaller group of true individuals who work hard to know who they are as complete individuals. They worry less about the hit of the season, but plod onward, playing with possibilities within their imaginations. The techniques they employ are important only as a means to getting to the art. Technique is important, but not the technique that seems to be in vogue or the upcoming hit pick. Instead they are interested in that technique they have committed their lives to artistically, in the search for excellence. The leaders of the pack impress me, due to the courage they show yearly in the search for excellence within a particular technique, regardless of how significant that particular process might be at the time. The stick-to-it-tive traits constantly displayed warm my heart and give me the courage to continue with my own work.

THE TRULY ORIGINAL: I know what you are thinking -- original is a much bantered word and of course nothing is really original in the world, blah, blah, blah. There are some truly original people and we all recognize them and their work, when we see it. The truly original are certainly ahead of the leaders of the pack, but since their numbers are smaller, we just don't see them as often or as clearly as we should. Be cautious -- the truly original are visible -- we must sharpen our eyes to focus in clearly on them.

Alternative Surfaces

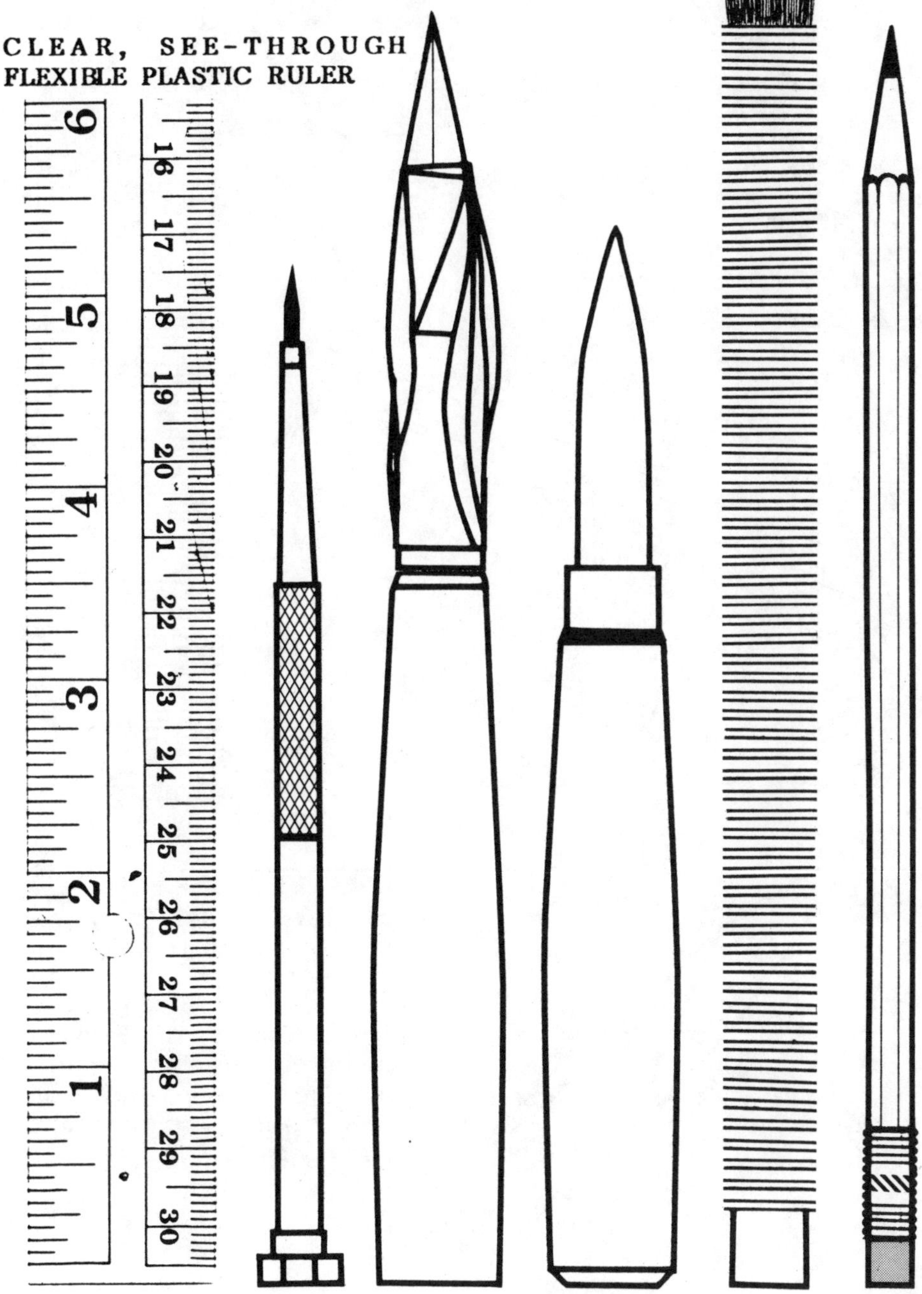

Are you tired of using the same old tools in the same old, boring ways? This portion of the workshop has been included to encourage you to consider alternative tools and techniques to create some highly personal alternative surface finishes on your aluminum.

Of course, there are the **normal** finishes: buffing/polishing, sanding (wet or dry), bead or sandblasting, etching, etc. that produce a specific finish on the aluminum, but why stick with the tired, tried and true? Where is your sense of adventure?

Imagine if you pre-treated your aluminum with any of the above mentioned **normal** finishes you could:

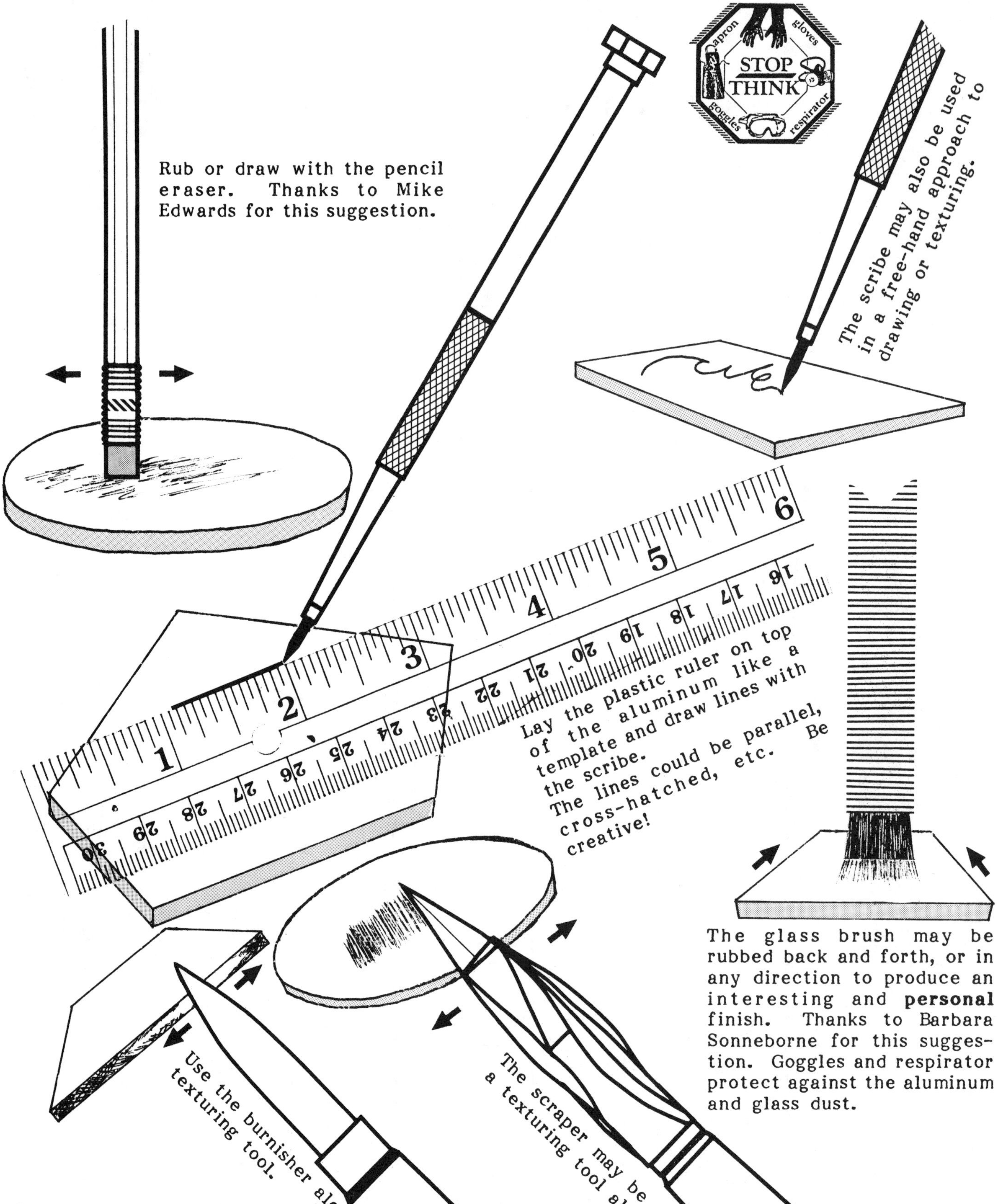

Rub or draw with the pencil eraser. Thanks to Mike Edwards for this suggestion.

The scribe may also be used in a free-hand approach to drawing or texturing.

Lay the plastic ruler on top of the aluminum like a template and draw lines with the scribe. The lines could be parallel, cross-hatched, etc. Be creative!

The glass brush may be rubbed back and forth, or in any direction to produce an interesting and **personal** finish. Thanks to Barbara Sonneborne for this suggestion. Goggles and respirator protect against the aluminum and glass dust.

Use the burnisher alone as a texturing tool.

The scraper may be used as a texturing tool alone.

Combine the scraper with the smoothing out effect of the burnisher.

Etching

There are two types of etching: overall surface etching and decorative design etching. The basic caustic etching solution we have already made up will create an overall surface etch.

Decorative design etching allows your creativity to spring forth. One etching solution is made from:

2 ounces hydrochloric acid (muriatic acid)
8 ounces distilled/deionized water
1/2 teaspoon copper sulphate crystals

Calculate the general volume of etching solution required as per the overall size of the piece(s) you expect to produce. This can be accomplished by visualizing the piece(s) in a given volume or container of acid. For example: If you work flat, how many pieces would you etch at any one time, and how many square inches of container do you need? Continue this concept if you work three-dimensionally into cubic inches/volume. Occasional etching requires only a small amount of etching solution.

Increase the above formula proportionally to fit your volume requirements.

The following demonstration will show you how to mix the above etching solution and to etch your aluminum piece(s).

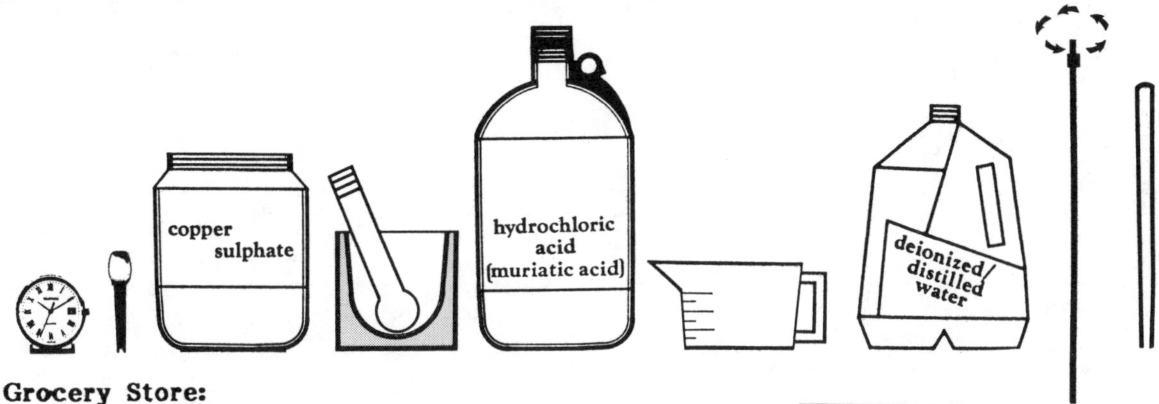

The equipment and supplies you will need and where to buy them.

Grocery Store:
- Plastic Containers/Lids
- 1/2 Measuring Teaspoon
- Plastic, Pyrex or Stainless Steel Measuring Cup
- Distilled/Deionized Water

Chemical Supply House:
- Mortar and Pestle

Farm/Feed Store:
- Copper Sulphate (Blue Stone) (also known as Blue Vitriol)

Plastics Supply House/Hobby House:
- Plastic Rod for Stirring with black electrician's tape designating the dry end

Jewelry Supply House:
- Copper/Bamboo Tongs

Water Company:
- Tap Water/Spray Rinse

Discount House:
- Timer(clock)

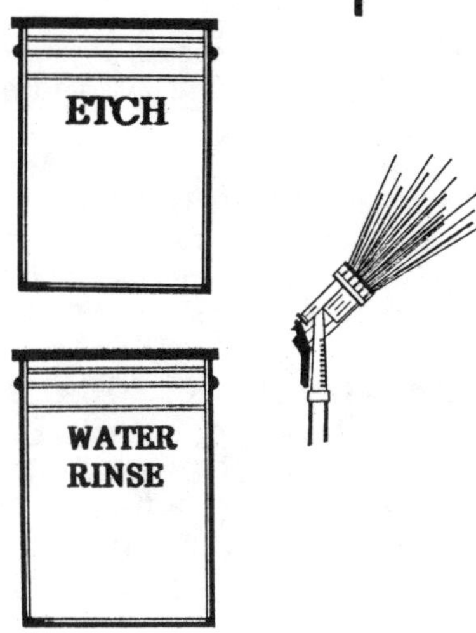

Obviously many of the above items may be purchased at other locations. This list provides a start.

ILLUSTRATED STEP-BY-STEP

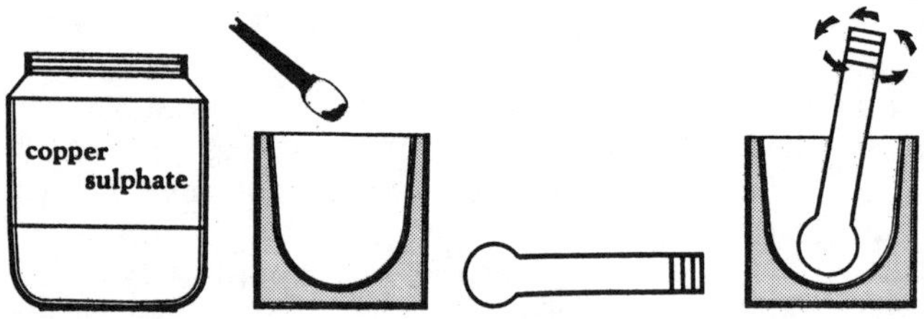

1. Using a measuring spoon, add the required amount of copper sulphate crystals to the mortar. Using the pestle, grind the copper sulphate into a fine powder.

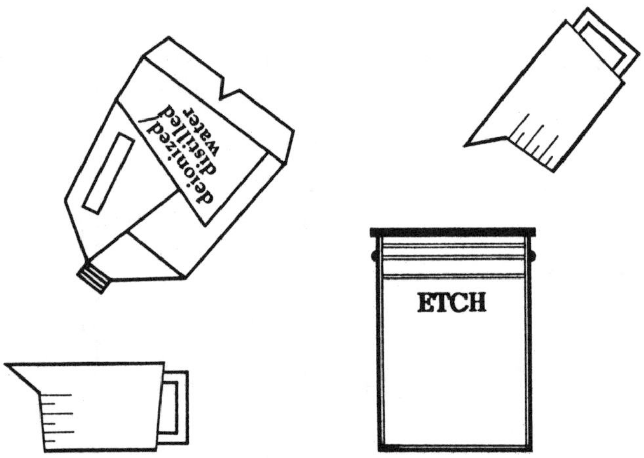

2. Measure the required amount of water into the measuring bowl and fill the etching bucket/ container.

Make sure the proportion of chemicals used is written on the bucket along with the word **ETCH** so you know what this bucket contains. This is a safety factor.

ALL CONTAINERS IN YOUR STUDIO SHOULD ALWAYS BE MARKED. YOU CAN'T TAKE FOR GRANTED THAT YOU WILL REMEMBER TWO WEEKS FROM NOW WHAT SOLUTION IS IN EACH UNMARKED CONTAINER.

3. Add the required amount of hydrochloric (muriatic) acid to the measuring bowl. Pour the acid slowly down the bowl's spout/side.

The fumes are not safe to breath. Work under a fume hood or be certain to exhaust fumes away from you. You may also work outdoors with the wind blowing away from you. Do not allow solution to splash on you or your clothing.

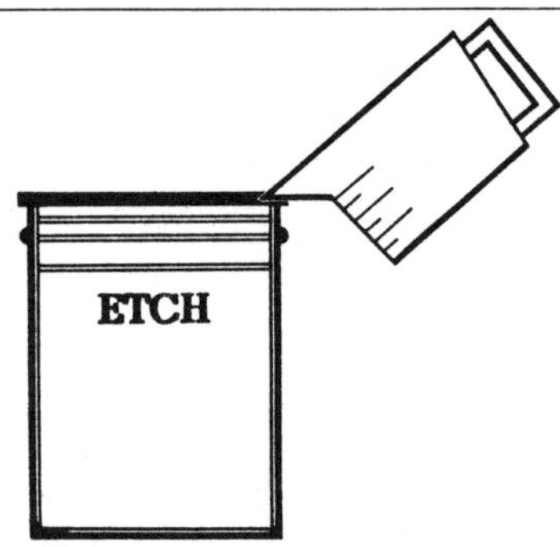

SLOWLY add the acid to the water by pouring it down the side of the bucket. Always add acid to water. Tip: Remember A & W (like root beer).

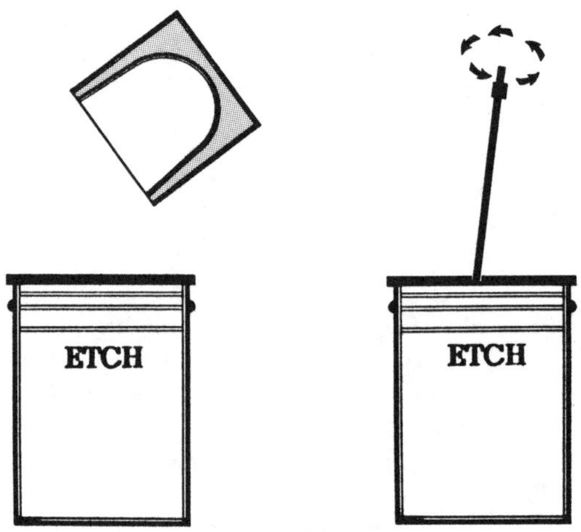

4. Pour the copper sulphate powder from the mortar slowly into the bucket. Stir with the plastic rod until all copper sulphate is dissolved. Cover bucket with lid.

5. Before you apply a resist, you must make sure the aluminum is chemically clean/degreased. Let's try a test: Run some tap water over the aluminum's surface. Does any water remain on the surface? I'll bet the water remains in only a few areas. This water break test will tell you if you need to degrease. A chemically clean surface will show water totally covering everything. Oil films resist the resists. (Oil films come from our fingers, etc.)

Now you can apply the appropriate resist to your aluminum as per your design. Consult the various resists Visiting Artist Christina DePaul discusses in her demonstration.

I like to use fingernail polish. It is easy to apply, since almost all fingernail polishes come with a small applicator brush. Use colored polish -- it's easier to see than clear. Buy the "on sale," yucky close-out colors no one else wants to buy from your discount stores. Those prices can't be beat.

OTHER RESISTS AND THEIR SOLVENTS:

fingernail polish	acetone or alcohol
asphaltum	acetone or turpentine
paint (brush or spray)	paint stripper
electroplater's tape	your fingernail
electrical tape	your fingernail
masking tape	your fingernail
contact or shelf paper	your fingernail
hard ground/soft ground	acetone
wax	boiling water
press-off type/ graphic tapes	your fingernail
rubber cement	rub it off
peelable lacquers	rub it off

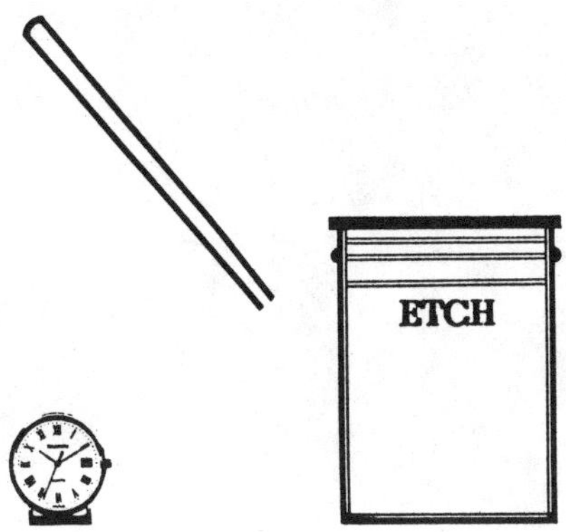

6. Once the resist has dried, use your tongs to pick up and place the piece(s) in the etching solution. Keep track of the etching time so you can repeat the results.

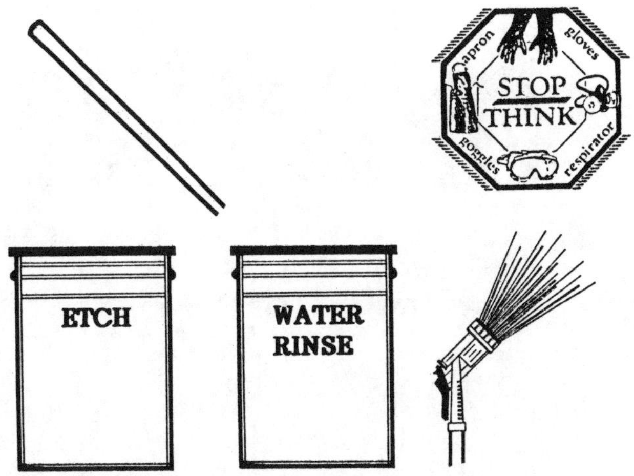

7. Remove the etched piece(s) with tongs and rinse thoroughly in a rinse container or use a spray rinse.

8. Using the appropriate solvent, remove the resist. Rinse piece(s) in water. Dry your metal. You are done -- or are you?

After the resist has been removed from the aluminum, notice a dark color in the relief areas. Do you like this? If so, proceed to rack and anodize. The dark areas will remain dark throughout anodizing, dyeing, and sealing. This dark color is smut, similar to that you will experience when the aluminum comes out of the caustic etch. If you do not want this dark color, use the nitric dip (smut remover -- see the Anodizing chapter) to remove the discoloration. Rinse.

9. Now, examine your work. If you don't like it, consider re-applying more resist and etching a second time. The initial etch may only begin to show your total image. Give your work some time to tell you what it could/should be. Etched surfaces that are etched to different levels will provide a "sparkle" that single etchings will never create. Think about it.

OTHER ETCHING SOLUTIONS

Archer Etchant -- ferric chloride.
(available at Radio Shack)
Caustic Soda

Try various dilutions to see what works best for you. Remember to follow safety suggestions on bottle. Good luck.

When using ferric chloride be sure to place the metal in the etch -- face down. Agitation of etch will improve results. Be careful of fumes and the heat generated by the etching action.

RACKING

Racks and Jigs

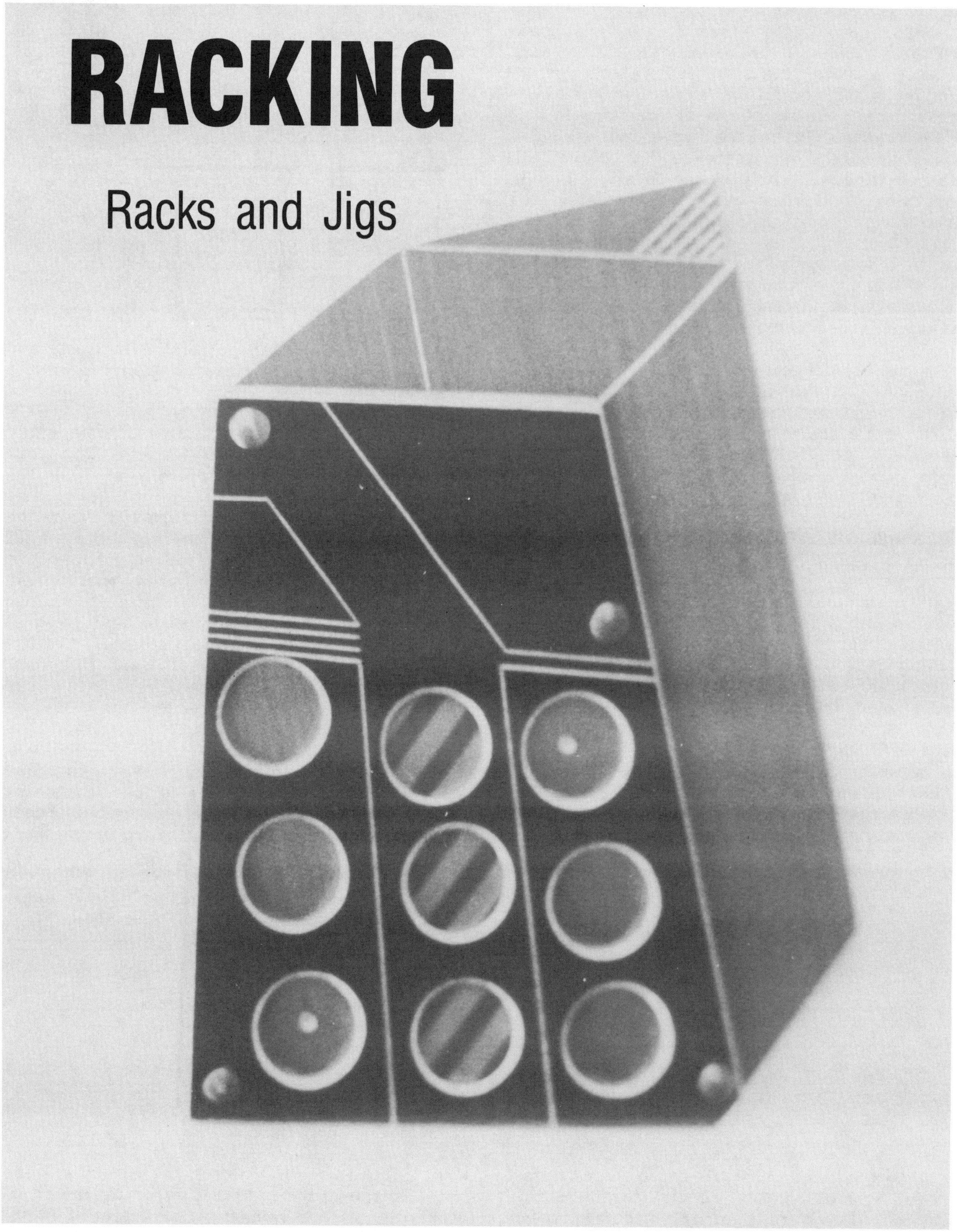

Brooch by David LaPlantz, fabricated, anodized aluminum, engraved, graphic arts tape, 2" X 1 3/8", 1983-4.

Racks and Jigs

The art of racking is almost as important as the art you're racking.

In order to physically handle the aluminum easily -and- to assure the proper connection for the current flow, we must rack each piece separately. In other words, we must either make a rack, or use a commercial rack, that will serve both as the handle and the electrical connection for the piece. Once racked, greasy or oily fingers should never touch the aluminum's surface.

It is important to note that this rack should have the following qualities:

1. Be made of titanium or aluminum (aluminum of an alloy close to or the same as the work piece).
2. Be sturdy enough to both support the work piece while attached to bus bar,

-thus-

3. insuring positive contact to provide current flow.

RACKS AND JIGS

Most commercial racks hold through tension, while handmade racks attach through a hole in the aluminum.

The art of racking involves the briefest possible connection between the rack and the aluminum. Wherever the rack comes in contact with the aluminum, the piece won't dye -- think tie-dye.

When visiting commercial anodizers, it's always a treat to watch them examine my work. I can see their eyes scrutinizing every square inch, trying to figure out how the aluminum was racked. Good racking will almost defy discovering of where the connection point(s) were. I buy many commercially anodized products to see how well they are anodized and to locate the rack marks.

HOMEMADE RACKS

Simple racking works best, when an aluminum rack wire and a small hole (drill # 56-64) by design/function (rivet hole) in the piece, have a firm connection.

Commercial knitting needles are perfect for making rack wires. I use #1 or #2 knitting needles (about 1 to 2 millimeters). This size is appropriate for jewelry scale pieces. As your scale increases, or the number of works to be anodized grows, you will need to modify your rack wires or possibly purchase commercial racks. More on this later.

Let's change a knitting needle into a rack wire:

1. Cut off the knob end of the knitting needle.

2. Most knitting needles are anodized (insulated). Therefore, to conduct electricity, the oxide layer must be removed. Immerse the needle into the caustic etch to strip off the anodized/dyed coating. Leave it in the caustic until all color has been removed, about 1-3+ minutes. If the caustic is new and clear, the bath's color will

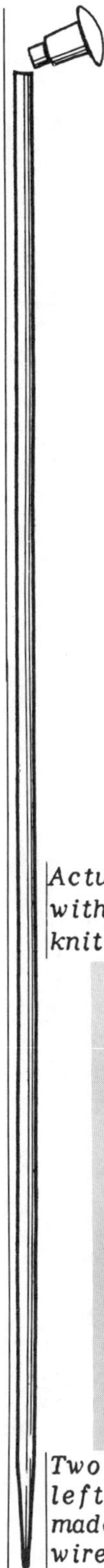

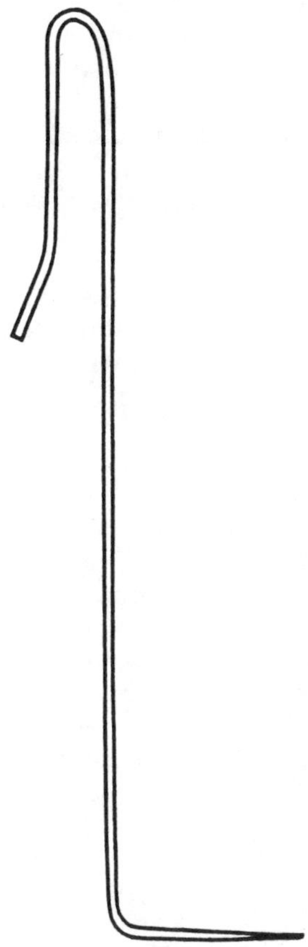

Actual size knitting needle with knob end cut off. A knitting needle rack wire.

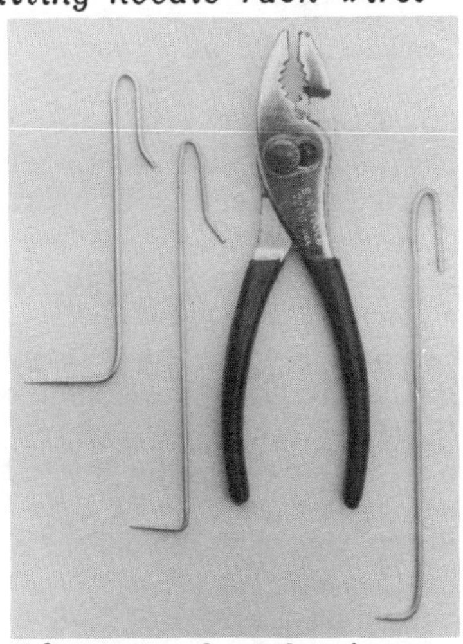

Two functional rack wires on left, with the pliers that made them. The other rack wire lacks functional taper.

change to the needle's color during the stripping process.

Some knitting needles are covered with a coating that is impervious to caustic. Sand or beadblast to remove. Regular sanding also removes the coating, use 220 wet-o-dry silicon carbide paper. Once the anodized coating is removed, test for continuity, if you have a multimeter. See Testing chapter.

3. Remove the needle from the caustic bath carefully. Avoid splashing that will harm your skin and clothing.

4. Rinse well in clean water.

5. Depending on the aluminum alloy, when the needle comes out of the caustic it will be a light gray to a dark black color. This color is smut and is an indicator of the metals that make up that particular alloy. I'll bet smut is a word you have always wanted to use in polite company -- right?

6. To remove the smut, immerse the wire in the nitric acid. The nitric acts as a cleaner and de-oxidizer. It will change the wire to a beautiful silver color.

7. Remove the needle from the nitric, rinse in clear water, and dry.

8. The knitting needle comes with a tapered end. The tapered end is probably too big to fit into the rack hole in your piece. File taper the end of the needle by filing around the wire, thus reducing the cross-section diameter, but still retaining the round shape. Each rack wire will be filed to fit snugly into a hole on the aluminum.

BENDING THE RACK WIRE:

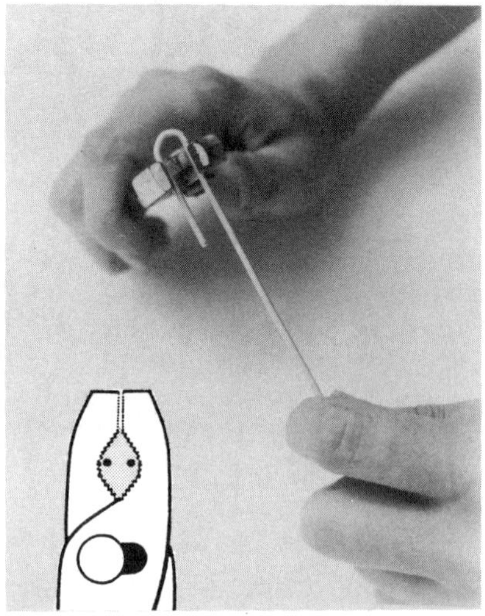

When I say snugly, I MEAN SNUGLY, AS IN -- ONCE CONNECTED TO THE PIECE, THE RACK WIRE AND WORK PIECE SHOULD **STAY FIRMLY CONNECTED** TO EACH OTHER.

Using regular household pliers, bend the wire so it fits around the bus bar. The rack wire hook should fit snugly over the bus bar. I insure a tight connection between my rack wire hook and bus bar by always using a spring clamp. Bull Dog clamps from a stationary store also work.

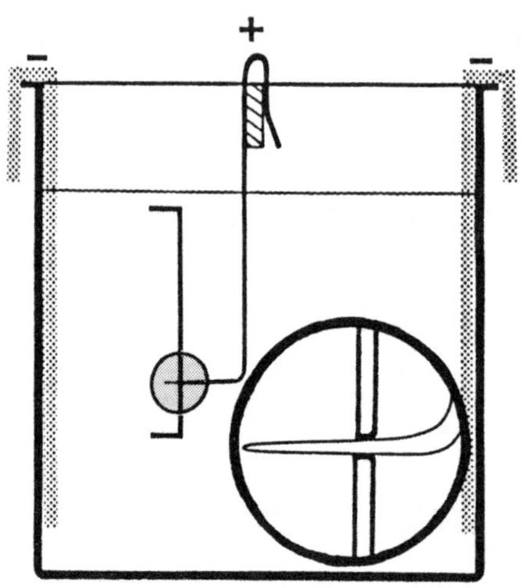

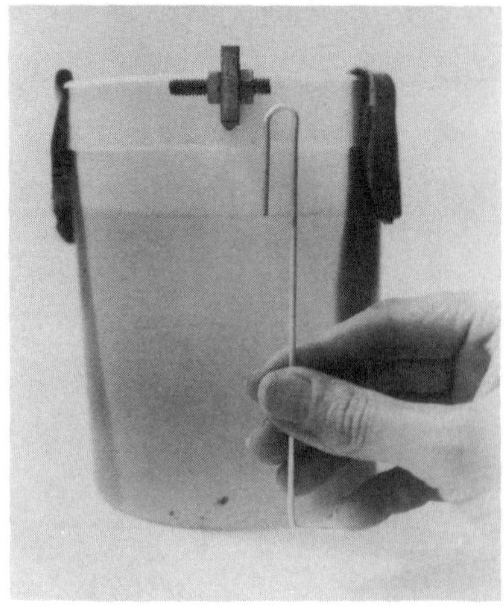

When forming the rack wire, use the open, super serrated part of the plier's jaws--they help you keep the rack wires parallel, giving a better connection to the bus bar.

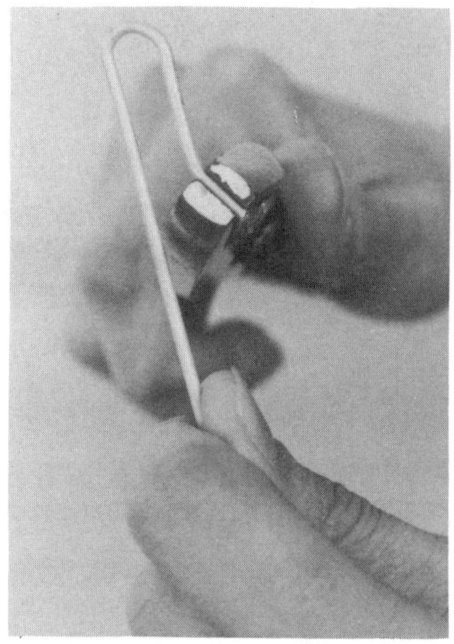

In order to facilitate removing the rack wire hook from the bus bar, a small bend/lift is added to the end of the wire -- see photo.

Inset: An excellent example of functional taper for firm connection with the aluminum. Overall diagram: How the rack wire and aluminum piece hang in the anodizing tank. Notice how the rack wire hook firmly connects to the bus bar.

9. Bend the rack wire at 90 degrees, about 1 1/2 inches from the tapered point. This length allows for the shortening caused by repeated filings and strippings during subsequent rackings. You'll see how this occurs as we move along.

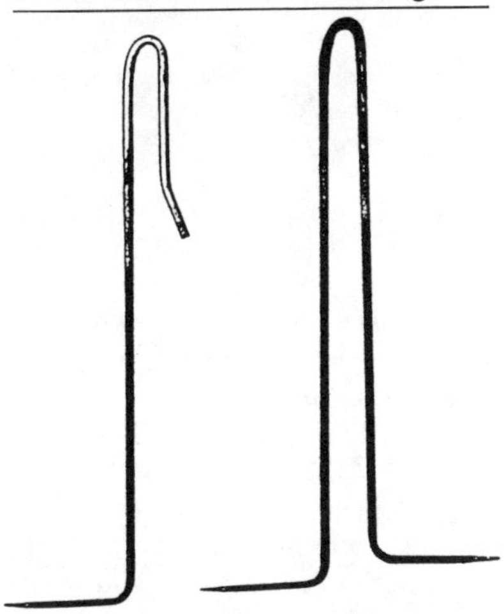

Single and double rack wire.

10. Look at the illustration of an average rack wire. This is the shape you want to make from your knitting needle. The size and shape of your copper bus bar dictates what the rack's hook will look like.

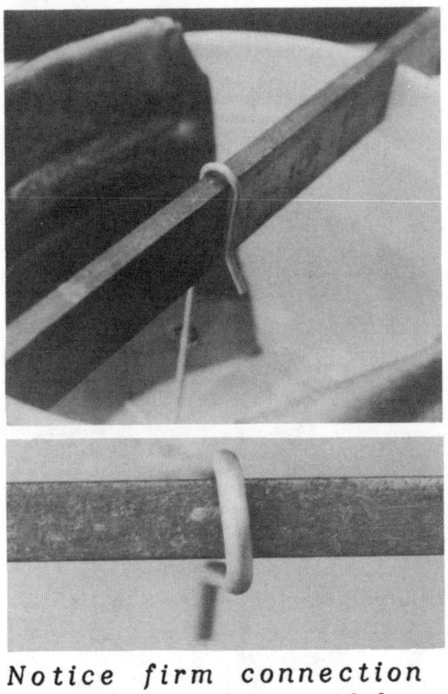

Notice firm connection between rack hook and bus bar.

Remember:

1. The rack wire is your handle throughout multiple immersions in many different liquid solutions.
2. A successful connection will eliminate frustrations caused by poor current flow, creating a poorly anodized surface,
3. or the fun of searching for your piece in the bottom of some solution tank.

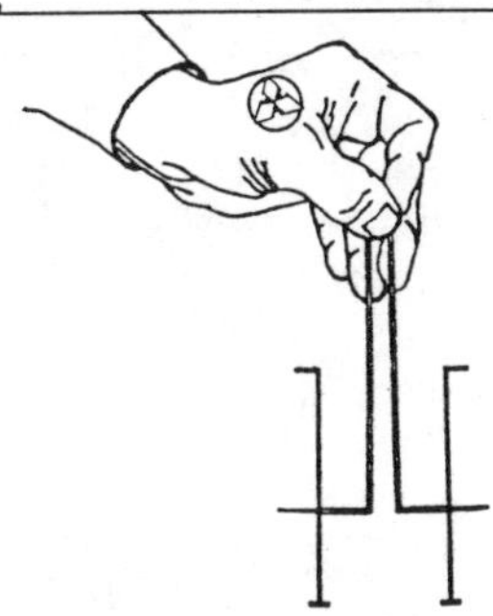

Double rack wire in action.

THE CONNECTION

Now that the rack wire has been prepared, it's time to make the first connection.

The logical relationship between the tapered end size and the hole in the work piece is hard to define. See illustration on page 94.

The tapered end on my rack wires usually extends past the aluminum piece by about 1/4" to 3/8". This gives a firmer connection. Racking with the single rack wire, is not always easy, take some practice, but WORKS!

I use the aluminum knitting needles for racks because they are harder than most of the stock I anodize. When the tapered end is forced into the rack hole, the hole slightly enlarges, insuring a firm connection. If the rack wire is too soft, as it penetrates the rack hole, the wire will decrease in diameter. This will cause a loose connection.

The rack hole can be a rivet hole. When the rack wire is forced through the hole, its size increases. So, drop back a bit and make the hole smaller than you need.

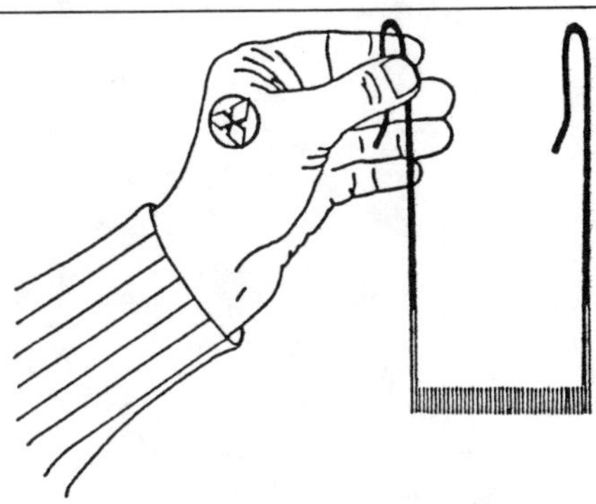

Double hook for anodizing a heavy coil.

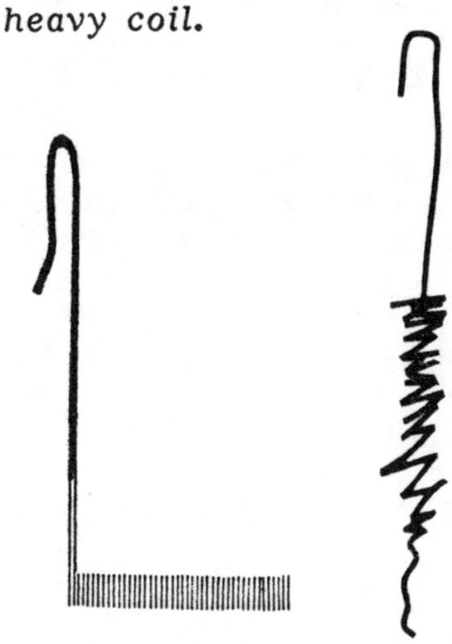

Single hook for a coil.
Single hook for wire squiggle.

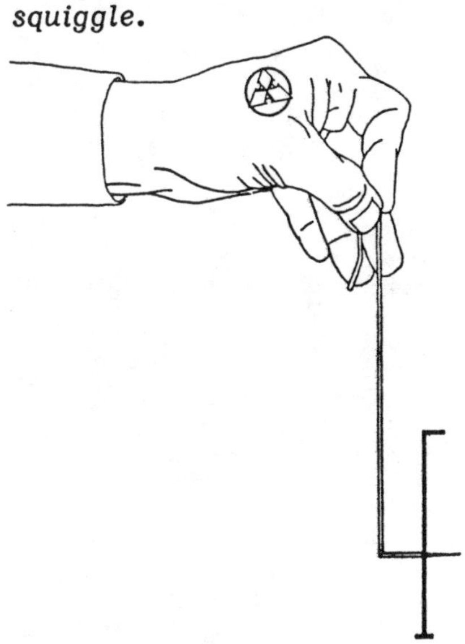

Side view of single rack wire and aluminum piece.

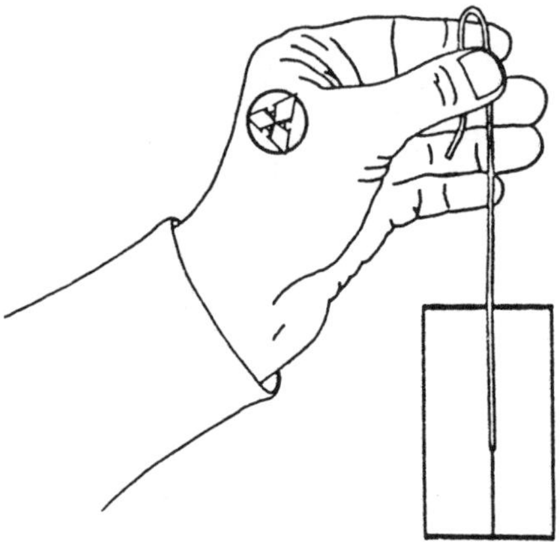

Single rack wire and aluminum piece, front view.

If you want to anodize a bracelet, for example, using a single rack wire -and- you want your connection as invisible as possible, design a small hole into the piece. In doing a piece where the rack hole is an integral part of the image, make certain that the rack wire's point is pointing away from the inside. Why? This insures that the entire face of the piece, including the interior of the hole, can be dyed. How? Because the rack wire is tapered. That means; inside where the fattest part of the tapered rack wire came in contact with the work piece, there will be a non-dyed area. But on the outside, the dye slips in between the skinny part of the rack wire and the hole.

The rectangular, horizontal plate has holes drilled and threaded to match the machine bolts. Another excellent example of maximum support for the aluminum pieces along with good current flow potential.

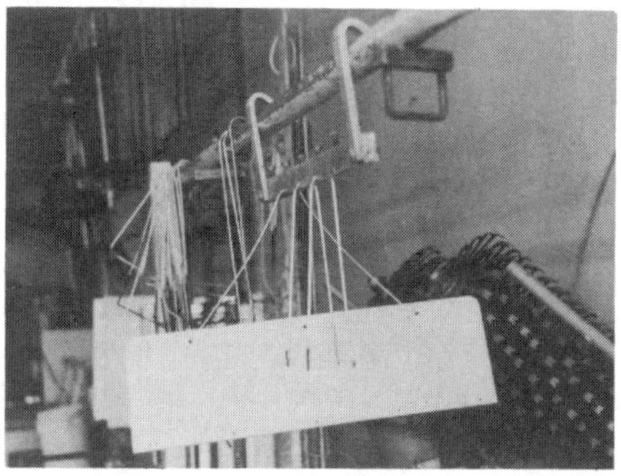

Another view of the same rack shows how the wire is sprung and then slipped through the outside holes of the aluminum pieces.

Here is a great detail photo of another commercial titanium rack. Notice how the springy fingers are pulled together, slipped into a hole on the back of the aluminum piece and then allowed to spread under tension. Thus, good current flow is maintained along with maximum support.

This is a custom designed rack. The aluminum wires that support the pieces are slipped through holes and held in place with hex-head machine bolts.

Notice the gripping action of the machine bolt.

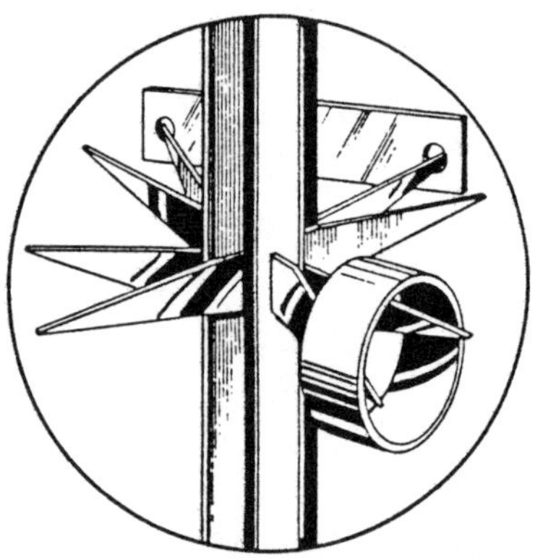

A commercial rack supporting typical aluminum shapes.

COMMERCIAL RACKS

A commercial rack is comprised of the following:

1. Hook with Machine Bolt and Nut.

2. Spline.

3. Rack----Tapered Tip Disc.

 ----Wire Pin Disc.

 ----Notch Tip Finger Disc.

In the standard commercial rack system the spline and discs are made entirely of aluminum or titanium. The hook is usually made of aluminum or bronze. The aluminum racks do not last as long as titanium because after each anodizing they must be stripped in the caustic etch. Eventually the caustic etching reduces the aluminum to a thickness, or shall we say thinness, not suitable for continued anodizing.

Discs may be purchased in aluminum or titanium. The splines, however, come either in titanium or titanium clad over an aluminum body, which is cheaper to purchase. Titanium clad splines are often guaranteed against defects for up to five years worth of normal usage.

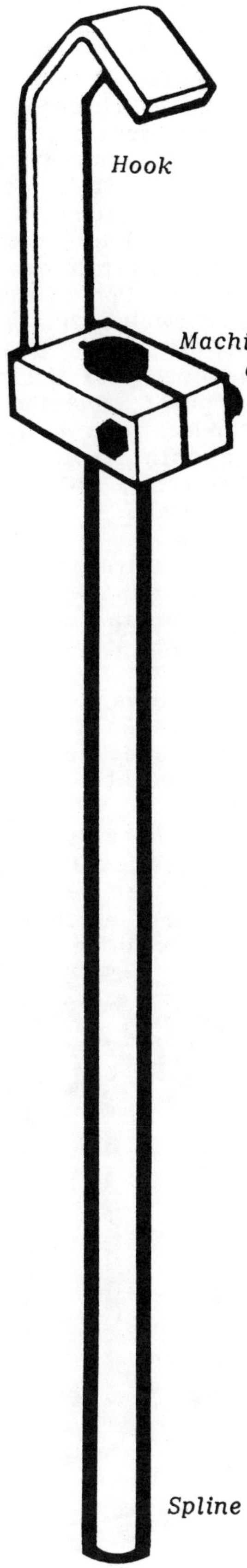

Tapered Tip Disc

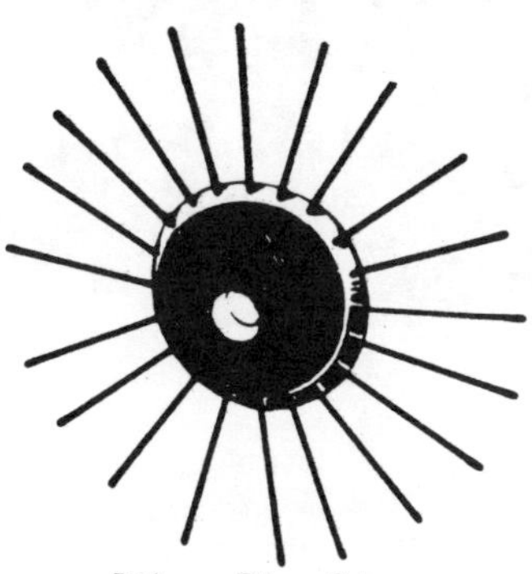

Wire Pin Disc

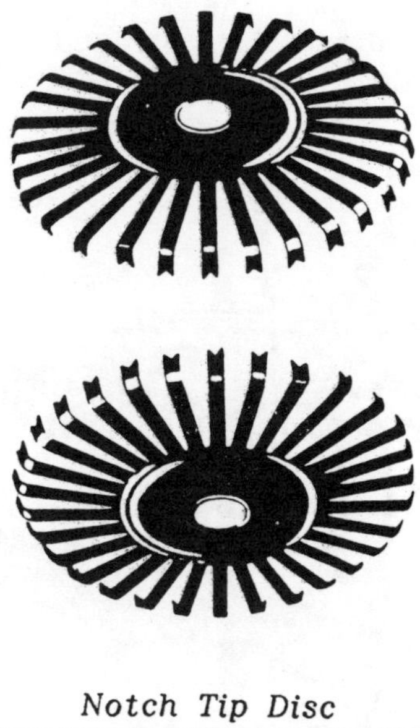

Notch Tip Disc

Why use only aluminum and titanium for racks, and not other metals, like copper? This explanation may amuse you: Sulfuric acid electrolyte is, in essence, a super pickle, because of the D.C. current that's pushing through the solution. When copper is used as a rack, the reaction between the electrolyte, D.C. current, and the copper itself causes a dissolution or dissolving action. Bye, bye copper if it is left in too long! A horror story: I once had the "brilliant idea" of riveting two pieces of aluminum together with silver rivets. The plan was to anodize the bracelet and after it was dyed, the sterling silver would still be white! Don't I wish!? Luckily, I checked the tank after about twelve minutes. The silver rivets were about eighty per cent dissolved and the rivet holes had enlarged. I then realized that the electrolyte with the silver wire and the "push" from the D.C. current was, in effect, a super pickle. The action of the electrolyte literally dissolved both rivets and eroded away the aluminum around the holes, with great emphasis. I hope you learn from my mistake.

Titanium racks are not affected by the anodizing and do not require stripping. The color of the titanium will change, due to the current's passage in the electrolyte, but will remain otherwise unaffected.

Clips, tips and various other clamps are also available. There are racks to accommodate almost any shape imaginable! Check out the Supply Sources chapter. You will be amazed at what you can buy!

If multiples of the same shape are anodized clear, or dyed the same color, only one type of disc rack would be attached to the spline. In other cases, where many different shaped objects are to be anodized clear or dyed the same color, different disc racks would be attached to the spline.

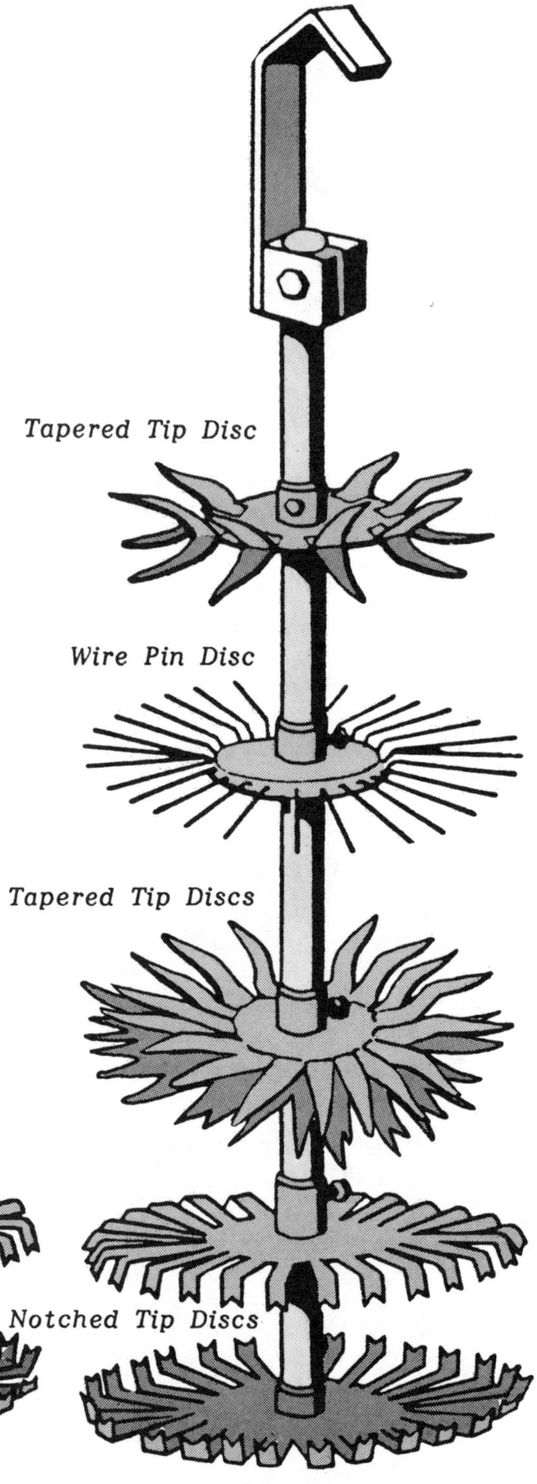

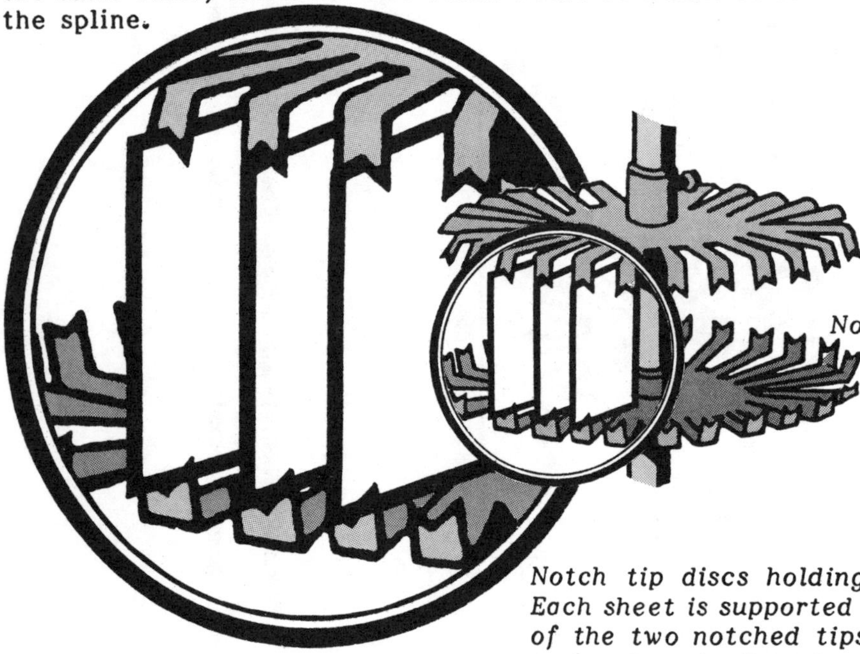

Notch tip discs holding pieces for anodizing. Each sheet is supported between the "V" groove of the two notched tips. Each tip has enough spring to grip the aluminum sheet tightly.

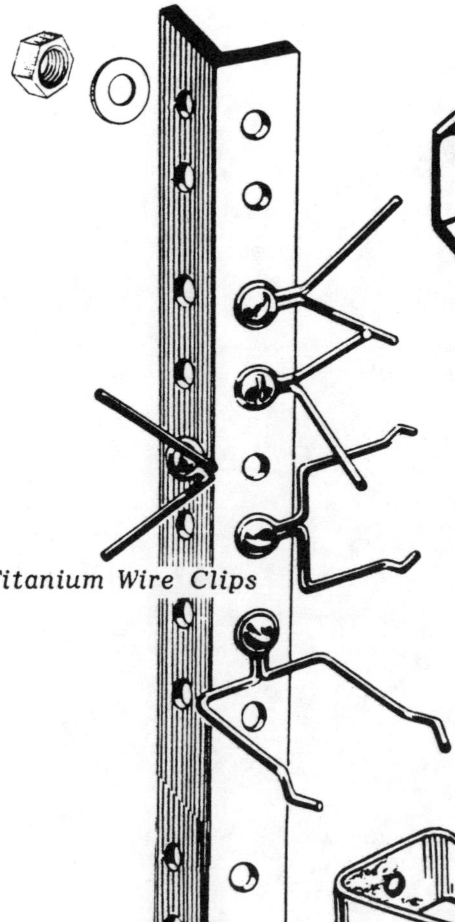

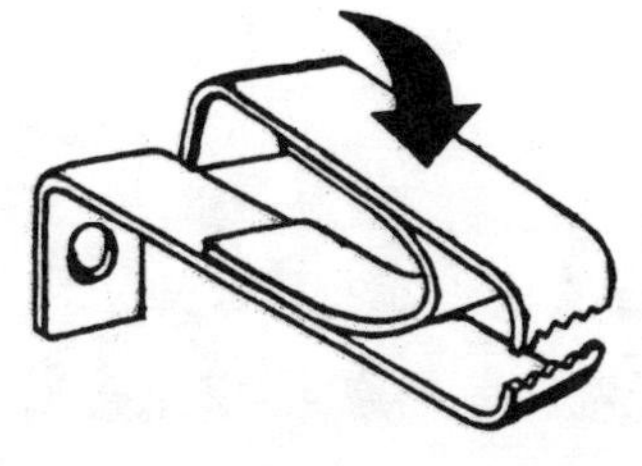

Titanium Grip Clip

ALL ITEMS ILLUSTRATED ARE AVAILABLE
FROM: SERVI-SURE CORP.
3551 N. Sheffield Ave.
Chicago, Illinois 60657
312-528-7575

Construction angles, wire clips, general purpose clips, grip clips, hex-head nuts, bolts and washers allow you "do it yourself" racking. I use these component parts to rack my work with superior results. They are so easy to work with and can be used either vertically or horizontally.
The grip clip is especially useful when you can't attach your piece any other way.

Titanium Wire Clips

GENERAL PURPOSE TITANIUM CLIPS

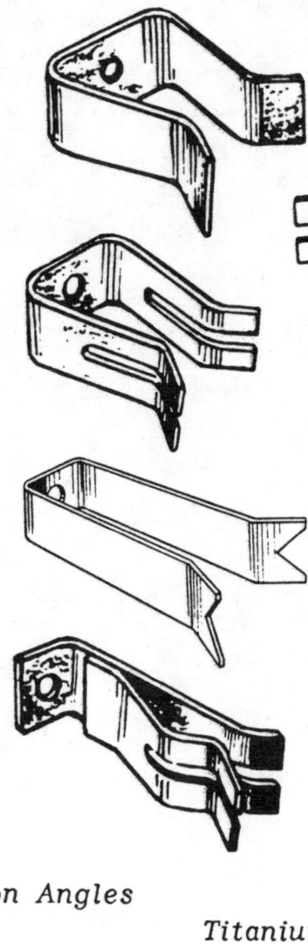

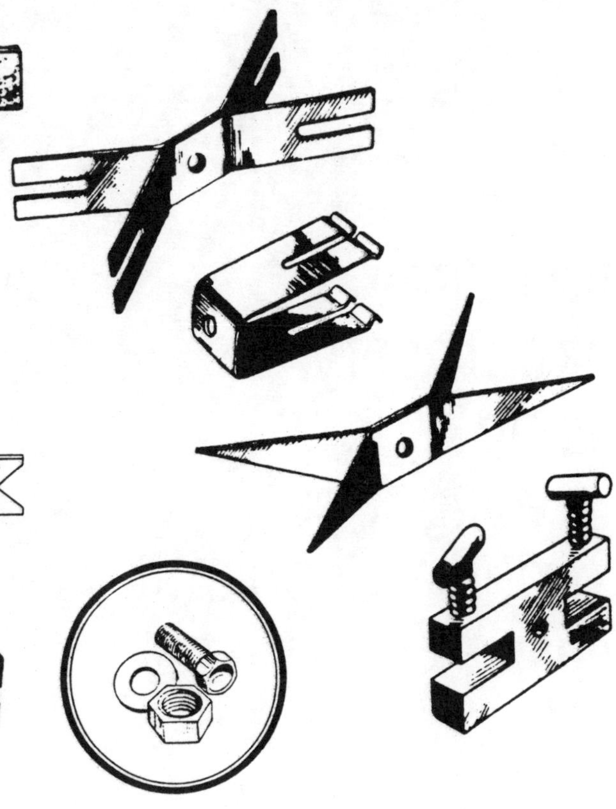

Titanium Construction Angles

Titanium Hex-Head Bolts, Nuts/Washers

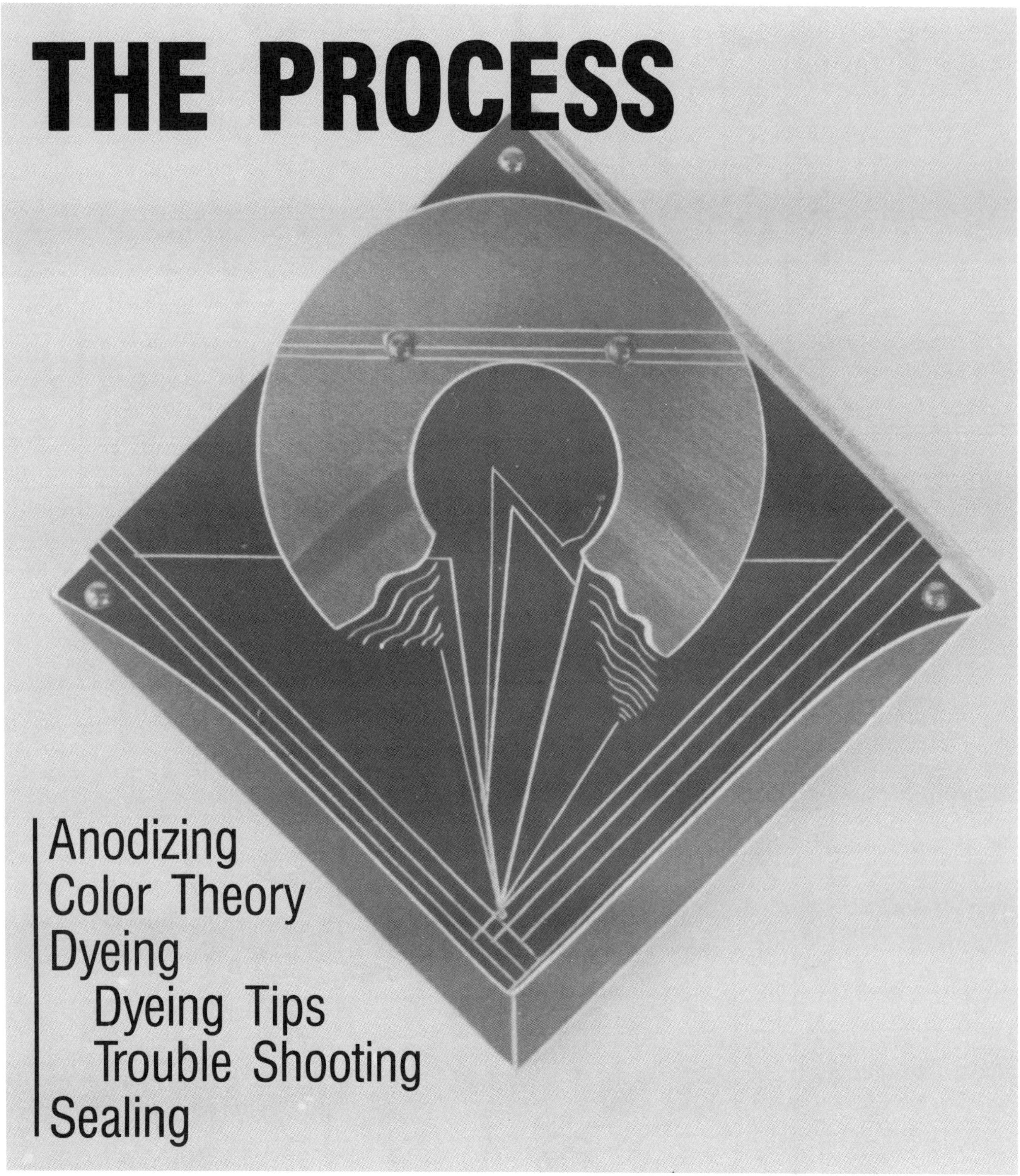

THE PROCESS

Anodizing
Color Theory
Dyeing
  Dyeing Tips
  Trouble Shooting
Sealing

Brooch by David LaPlantz, fabricated, painted aluminum, engraved, 2" X 2", 1983-4.

Anodizing

Okay, let's go through the complete process of anodizing.

Before anodizing: The object to be anodized is completely finished. It has been formed, forged, chased, etc. It already has drilled holes for riveting or design effect or both (or at least one hole to connect the rack wire, I like drill bit numbers 56-64).

1. The rack wire is firmly connected to the object. Henceforth, you will always handle the object to be anodized by the rack hook. The metal must now be completely degreased. Hang the piece in the degreaser from a bamboo skewer or aluminum wire laid across the tank. Hang your piece in the degreaser until it is clean, generally 1-5 minutes. You may also use an ultrasonic cleaner or steam jets, if you have them, to degrease.

2. Rinse off all excess degreaser. Make certain your piece is well rinsed. Proper rinsing is one of the most critical aspects of anodizing. How clean is clean? The type of clean I am referring to is "chemically clean." Do you know what the water break test is? The test: Run some water over the aluminum. If the water sheets smoothly over the metal, it is clean. If it spots, it's dirty. If dirty, degrease and rinse again.

3. If the object to be anodized is to have a matt finish, I will suspend the piece in the caustic bath using a bamboo skewer or aluminum wire to hang the rack wire from for a period of 1-5 minutes.

If the object to be anodized is to retain a polished finish or some other finish that would be harmed by the caustic bath, assuming the piece has been degreased, you proceed directly to the anodizing bath. A polished piece of aluminum will dull slightly during anodizing.

For the sake of demonstration, I will take you through the process with a slight degree of matt finish created by the caustic bath.

Remove the object from the caustic bath, allowing as much caustic solution to drip back into the bath before moving on to the cold water rinse. This reduces the amount of chemical drag-out. Drag-out, the pulling of chemicals past their intended solution, will create imperfections in anodizing and dyeing. If you are working with a hot caustic bath and your piece has larger, flat surfaces, quickly rinse, as hot caustic will cause streaks if left on the work too long. Personally, I always work with a cold caustic. It's safer. Hot caustic will cause a greater reaction and more <u>gas</u>. Depending on your ventilation system, your lungs won't like the gas.

At this point the aluminum has turned a light gray to a dark black color. If the rack wire and object are of different alloys, you may detect different values of gray or black. The purer the aluminum, the lighter the color. Darker colors are caused by the presence of iron, copper, magnesium, manganese, etc. in the alloy. This discoloration is referred to as smut.

4. Dip the piece vigorously in and out of the cold water rinse. To insure a more total rinse, I also move the object in a back and forth motion under the water. Do this with care, however, so as not to unhook the object from the rack wire. A squirt (use a chemistry wash bottle) with clean, cold water should be used to remove excess chemical in holes or other hard to reach areas that might not otherwise be thoroughly rinsed by the dipping activity. A second cold water rinse may be useful as a safety precaution. Let your gut dictate.

5. Suspend the object in the nitric bath from a bamboo skewer or aluminum wire, for 2-3 minutes, or until the piece is entirely white/silver again. The nitric is desmuting or deoxidizing the aluminum. Depending on the strength of the acid, alloy, and the amount of time the piece is in the caustic, you may see the black smut fall off the aluminum immediately when immersed into the bath. This is rather fun to watch! Little pleasures, indeed.

Remove the piece from the nitric bath, again allowing time to drip off excess chemical by hanging over bath.

6. Dip the piece vigorously in and out of the cold water rinse. To insure a more total rinse, I also move the object in a back and forth motion under the water. In anodizing the two biggest problems are inadequate rinsing and inadequate degreasing.

Another alternative for rinsing your work is to purchase an inexpensive shower hose. They have a large coupling to fit onto the end of your faucet, are long enough to maneuver around with ease, and create a healthy spray. Buy one.

7. A second rinse will eliminate drag-out.

8. The piece is now ready to be anodized.

THE ANODIZING BATH:

Since the rack wire was prepared to fit the copper bus bar, all you need to do is to hang the racked object in the anodizing bath. For obvious safety reasons, do the following: Make certain that the power supply is turned off, and put on your rubber gloves to eliminate the chance of electrocuting yourself. I know what you are thinking -- this is such a low voltage, nothing could possibly happen. Think again.

Consider the placement of the piece to be anodized in relation to other pieces already in the bath and to the cathode. You do not want to damage the other pieces when entering the bath. Also, the chances of shorting out your power supply are great if the aluminum piece (anode) is too close to or touching, the lead (cathode). Be careful. I've seen many nice pieces either damaged or fall off the rack wires into the bottom of the anodizing bath. Unless you enjoy refinishing your work or fishing around with your chemically resistent gloves on to retrieve the work from the bottom of the bath, I suggest being prudent about placement in the bath!

As double insurance that the rack wire fit is tight against the bus bar, clamp the rack wire firmly against the bus bar. Use the spring or bulldog clamps here. Turn on the power supply. Set the voltage to 12-14 volts.

The number of objects, or the square inches, will dictate the amount of current flow or amperage. The current density should be 10-15 amps per square foot at a voltage of 10-25 volts D.C. Be careful not to exceed the amperage rating for your power supply.

THE POWER SUPPLY:

Inexpensive battery chargers with ratings of 2-4 amps may only have a 12 volt output. They need only to be plugged into an electrical outlet to operate. It couldn't be easier.

More expensive battery chargers may work like one of the following two examples:

A. -If you have a 6 or 12 volt charger, set the voltage at 12.

-Is there a setting for fast or slow amperage? If so, set it at the fast setting.

-Is there a timer? If so, set the timer for at least 1 hour. This will give you plenty of anodizing time.

This is a Sears battery charger with volt meter, timer, fast and slow amp, 6 and 12 volt settings. Like "A" on this page.

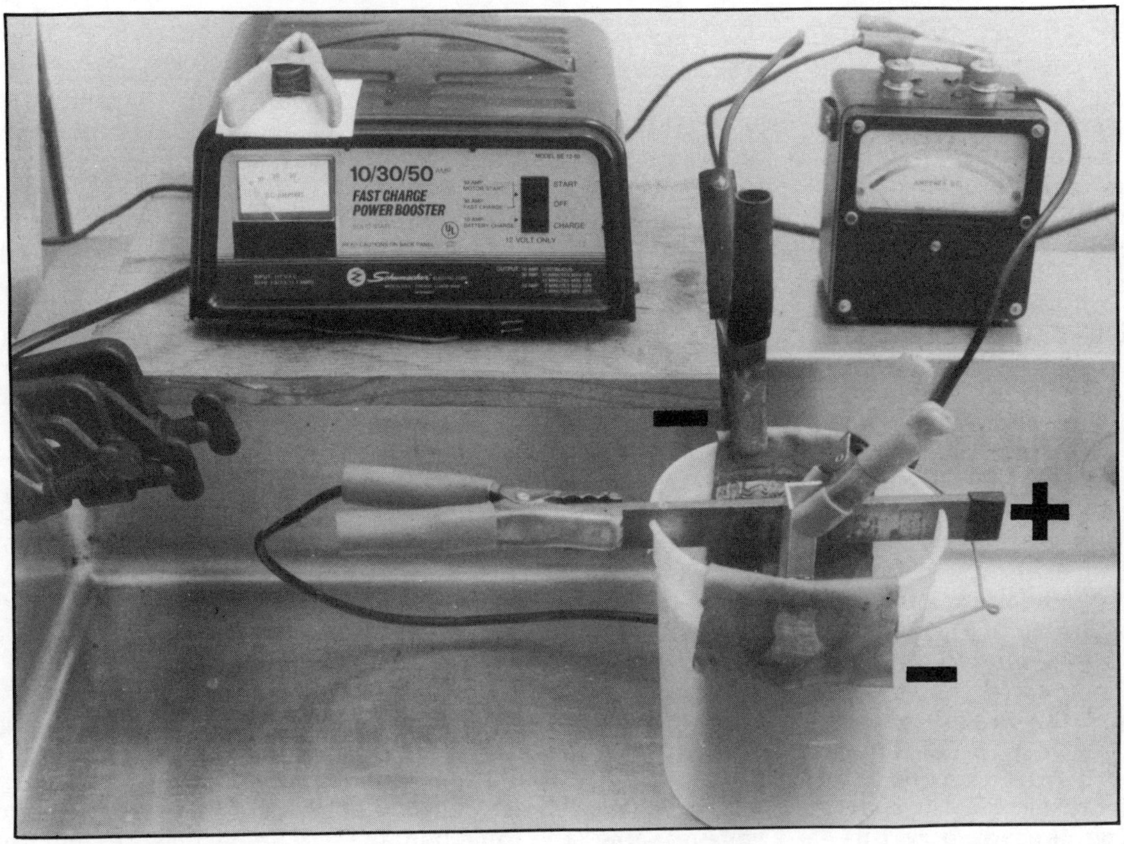

This is a simple anodizing set-up. The battery charger is located in the back left with the ammeter to the right. The ammeter is wired in series with the positive electrode. The 1 quart plastic container in the front contains the sulfuric acid/water electrolyte and has 2 lead cathodes. A 10-12 gauge copper wire (equal to the maximum amperage of the power supply, 10 amps = 12 gauge wire) connects the 2 lead cathodes together. The bus bar is 1/4" thick X 1" wide X 8" long. Black electrician's tape covers the right end of the bus bar. This tape build-up prevents the bus bar from falling into the electrolyte. The positive clamp from the battery charger is attached to one end of the bus bar, while the negative clamp is attached to the lead sheet (negative electrode). An extruded aluminum angle is clamped to the bus bar with a spring clamp. A clamp may be the most obvious method to attach the work piece to the bus bar. To anodize large strips or sheet, simply clamp the aluminum to the bus bar. Clean the bus bar before each anodizing. Clean the bus bar with scotch bright and water. Do this cleaning in the sink away from the electrolyte to minimize contamination.

B. -If you have a 6 or 12 volt charger, set the voltage at 12.

-Is there a multiple setting such as; 10 amps-battery charge, 30 amps-fast charge, and 50 amps-motor start? If so, set the charger at the 10 amps-battery charge setting. The other two settings cannot, according to the directions, be operated for a long period of time or would, as in the fast charge state, be too much for the piece you wish to anodize.

ANODIZING CONDITIONS:

Electrolyte: 15-18% Sulfuric Acid:(15%: 22 oz/gallon)
(16%: 24 oz/gallon)
(17%: 25 oz/gallon)
(18%: 27 oz/gallon)

Voltage: 12-20 volts D.C.

Current Density: 10-15 Amps/Square Foot

Temperature: 60-80 degrees F. Optimum 68-72 degrees F.

Time: 20-40 minutes. Longer times produce thicker films, deeper color if dyed.

Obviously, you will need to experiment to find the best settings and results for your personal needs and the type of battery charger you own. Also keep in mind that you do not want to go over the rated amperage of your battery charger. I was using the bottom of the line Sears 6 and 12 volt charger with a 4 amp capacity and had absolutely no problems. No problems, until I hung too many pieces in the bath and disregarded the amperage meter which was well over the limit of 4 amps! In fact, I ran the charger with a setting of 5 amps for over 12 hours! (I was not anodizing the same pieces for 12 hours, but was putting in and taking out pieces over a 12 hour period.) No wonder it stopped working. Ratings are there for some obvious reasons. Now all I have to do is to respect the rating, watch the meter, and put a few less pieces in the bath. Simple, isn't it? I do hope that you have learned from my error. Snicker, snicker, time will tell.

Some battery chargers have voltage meters, while others have amp meters. I'm not convinced that the voltage meter is that important. The amperage meter will help you to know how many amps per square inch a given piece will draw. Remember the 10-15 amps per square foot? The amp meter will also keep you from going over your battery charger's rated amperage.

When the power supply or battery charger was turned on, you may have noticed some bubbles moving in the bath. If you do not see them, look closely at the area between the piece (anode) and the lead cathode. Do you see the small bubbles? If so, you have a reaction taking place and anodizing is occurring. If you have an inexpensive multimeter, like the one in the Testing chapter, set the control knob on D.C. voltage at the number 50 mark. Carefully touch the red probe to the bus bar or to the rack wire, while the black probe is touching the lead cathode. The meter will immediately read the voltage. The meter should read between 12 and 14+ volts. Now you know by sight and a voltage test that the anodizing is taking place. Periodically, each day, I do this test just to see that everything is working correctly. If you have an inexpensive volt meter (0-15 or 20 volts), wire it to the bus bar and the cathode, remembering the bus bar is the red wire or positive connection, while the cathode is the black wire or negative connection. The meter should read between 12-16 volts. Now you are beginning to know a bit more about electricity. Isn't it fun? Feel the power at your finger tips? Between this and your art -- what more do you need? Such a foolish question, indeed!

DURATION OF ANODIZING:

The amount of time in the anodizing bath will depend on how thick an oxide layer you desire. For jewelry, such as earrings and hairpieces that get less abuse, you may want to go for 15-20 minutes. Brooches, bracelets, belt buckles, and other items that receive more abrasion should go for up to 1 hour. Remember, the longer the anodizing time, the thicker the oxide layer, and the more vibrant the color will be once dyed. On the average, when I anodize with the battery charger, for about an hour, the oxide thickness on the aluminum tests out to be between .8 mils (20.3 microns) and 1.1 mils (27.9 microns). An oxide layer between .5 mils (12.7 microns) and 1.0 mils (25.4 microns) is thick enough for most jewelry applications. A simple conversion would be 1 mil = 25.4 microns = .0254 mm. See Testing chapter for how to measure the oxide thickness.

It is also possible to have the piece too close to the cathode, or to be anodized too long, which will result in surface pitting and too soft an oxide coating. The pitting is not very pleasing, but it is another way to create expressionistic effects. Sometimes doing things the "wrong" way can and will provide you with, if not a better way, at least a different way to communicate visually. If the oxide coating is too soft it will not resist abrasion.

BAD GAS/VAPOR AND VENTILATION:

As the anodizing takes place hydrogen gas is given off as a gas or vapor by the electrolyte. It will make you cough and is not healthy to breath. If you do not have the luxury of a fume hood, simply cover the bath with clear plastic. Food storage bags from the kitchen work well. This cover will keep the vapors from contaminating your shop and lungs. Take care to keep the same side of the plastic down over the bath, since it will become wet.

You need proper ventilation to anodize. I assume by now you have a copy of the ventilation book (Safety chapter) and are following it. Don't play games with your life.

When the anodizing is completed, turn off or unplug your power supply. Remove the clamps holding the rack wire to the bus bar, and take the piece out of the bath. Be certain to gently shake the object over the bath, so as to leave as much sulfuric acid behind and to eliminate potential acid drag-out.

9. Dip the piece vigorously in and out of the cold water rinse. To insure a more total rinse, I also move the object in a back and forth motion under the water. Do this with care, however, so as not to unhook the object from the rack wire. A squirt with clean, cold water should be used to remove excess chemical in holes or other hard to reach areas that might not otherwise be thoroughly rinsed by the dipping activity. A second cold water rinse may useful as a safety precaution. Let your gut dictate.

10. Prior to neutralization in the baking soda bath, be certain to use the plastic rod to stir up or mix the solution. Baking soda has a nasty habit of settling to the bottom of the bath container.

The baking soda bath neutralizes any sulfuric acid electrolyte that may not have been removed in the rinse. Dip your piece into the baking soda bath -or- if the piece has lots of holes or intricate places where the sulfuric acid might still be lodged, hang it from a bamboo skewer or aluminum wire. Let you intuition rule. If you feel the piece should remain in the bath longer--so be it! It is better to totally cleanse the piece now, than to have acid drag-out into the dye bath and ruin both your color on the piece and contaminate the dye. Acid drag-out will cause streaking and other similar blemishes when the aluminum is dyed. In essence, the acid acts as a resist to the dye. The acid drag-out will also effect the dye's pH and influence the overall color. (Each dye has an optimum pH that should be maintained.)

11. Dip the piece vigorously in and out of the cold water rinse. Remember: Thorough rinsing is critical. A double rinse is recommended.

Notes

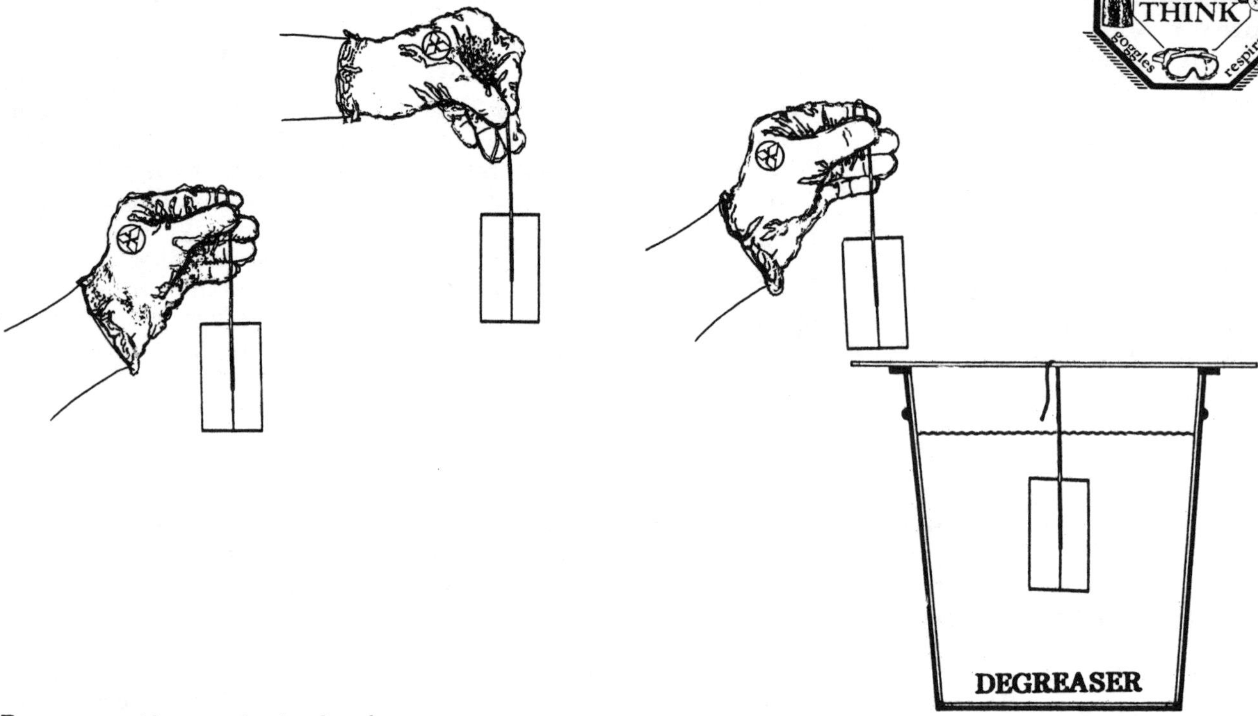

1. Degrease the racked aluminum.

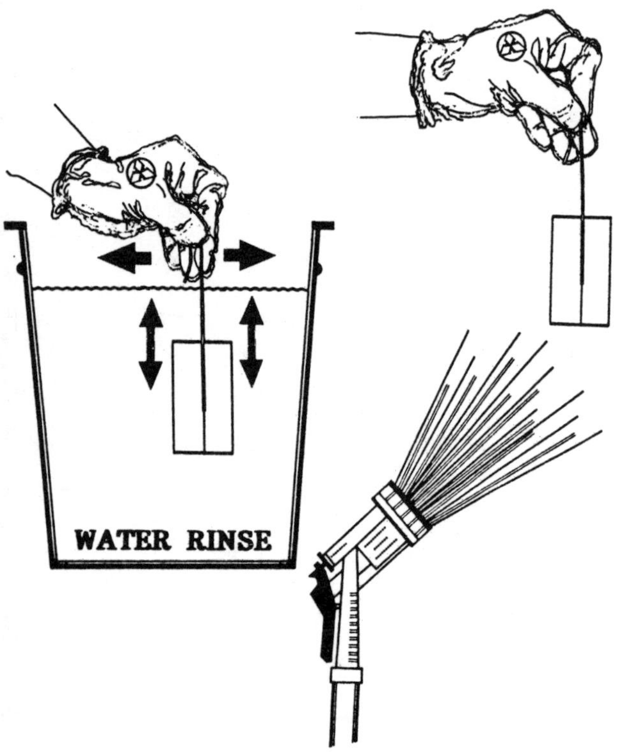

2. Thorough and complete water rinse or spray.

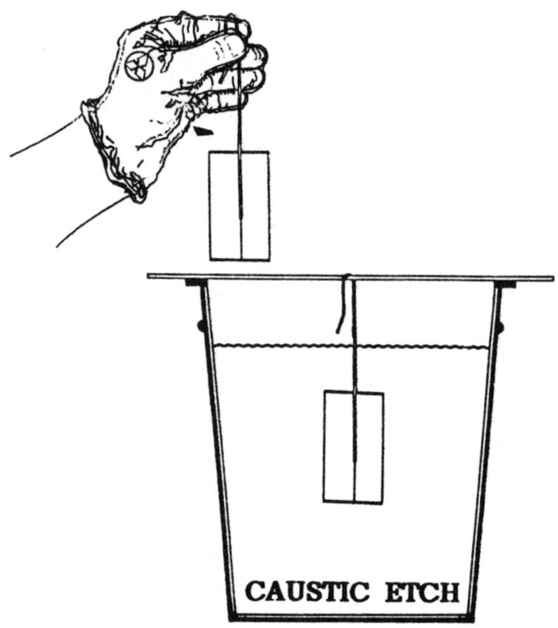

3. Caustic etch.

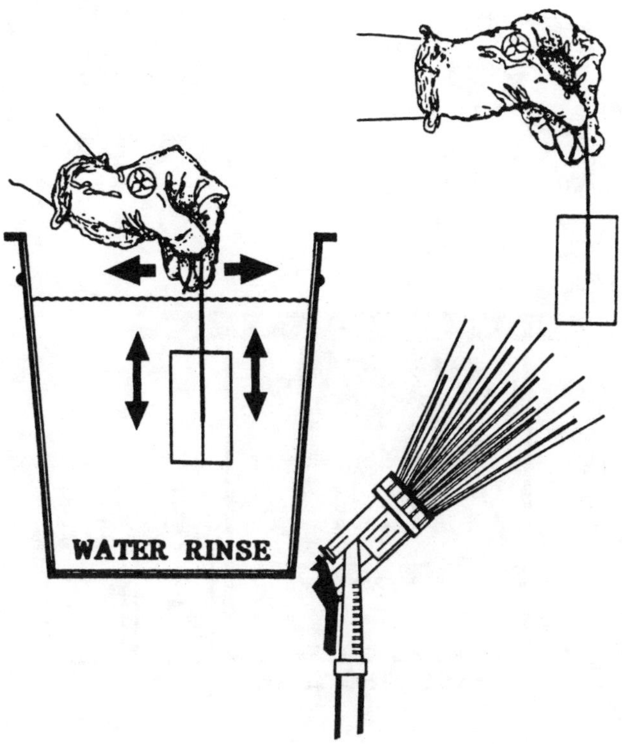

4. Thorough and complete water rinse or spray.

5. Nitric etch.

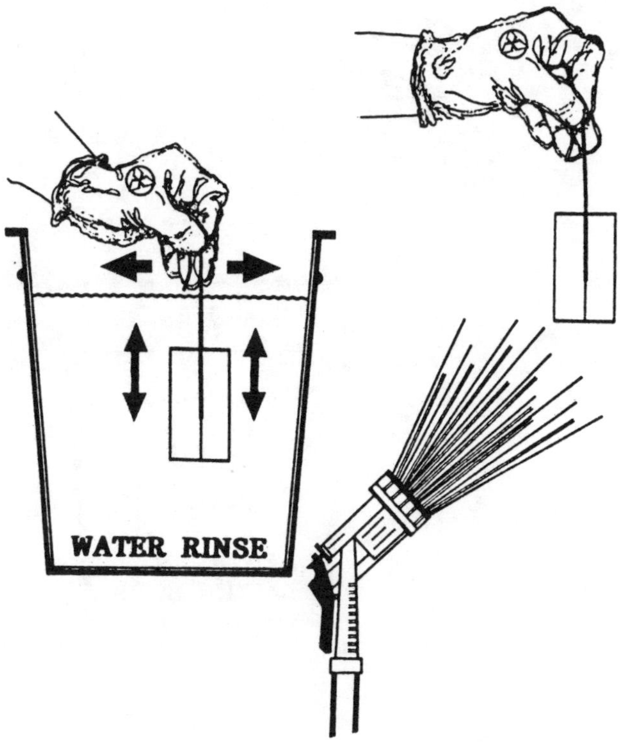

6. Thorough and complete water rinse or spray.

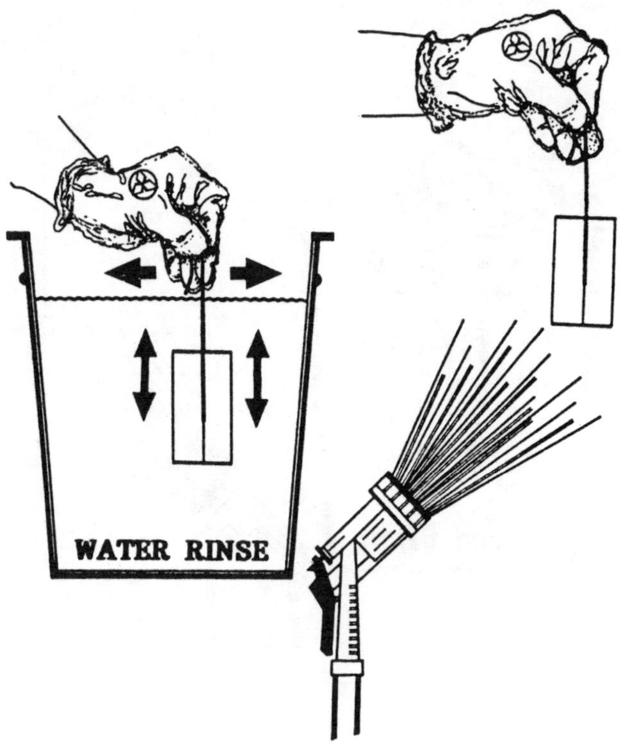

7. Second water rinse.

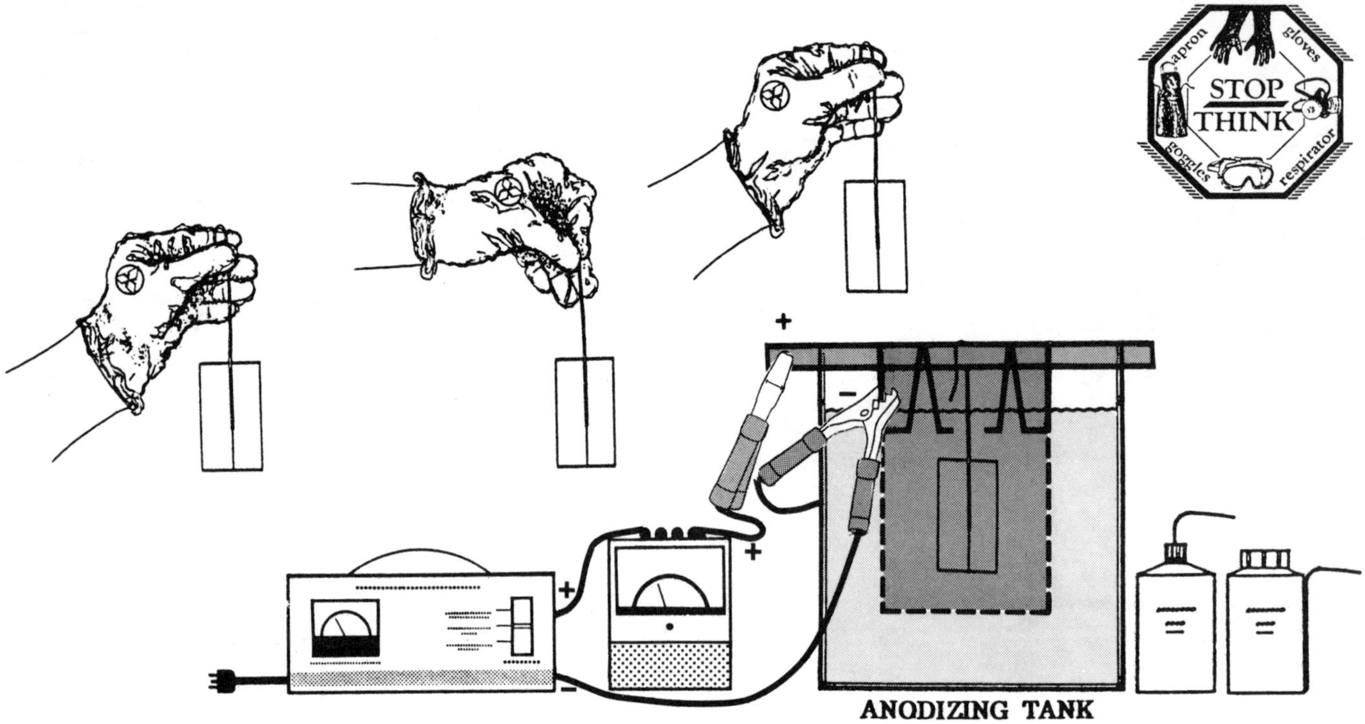

8. Anodizing.

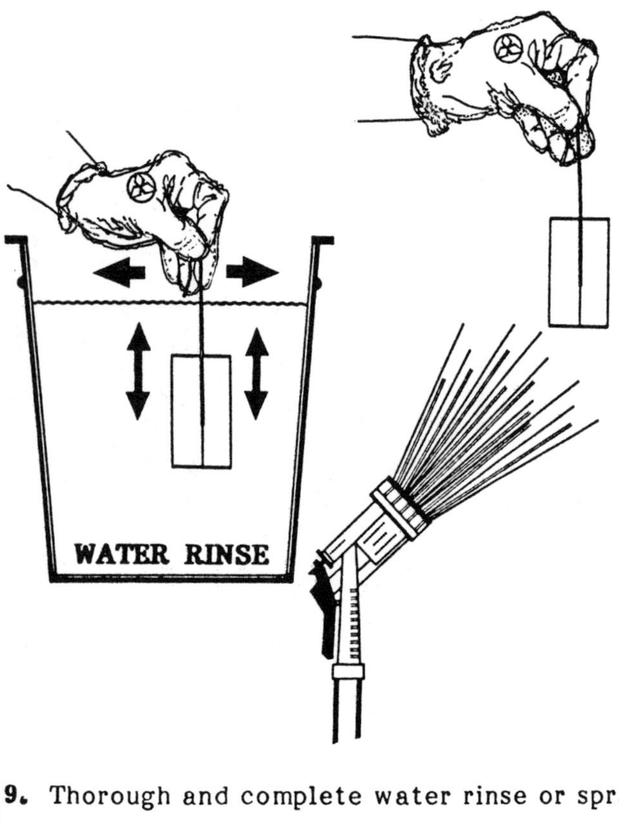

9. Thorough and complete water rinse or spray.

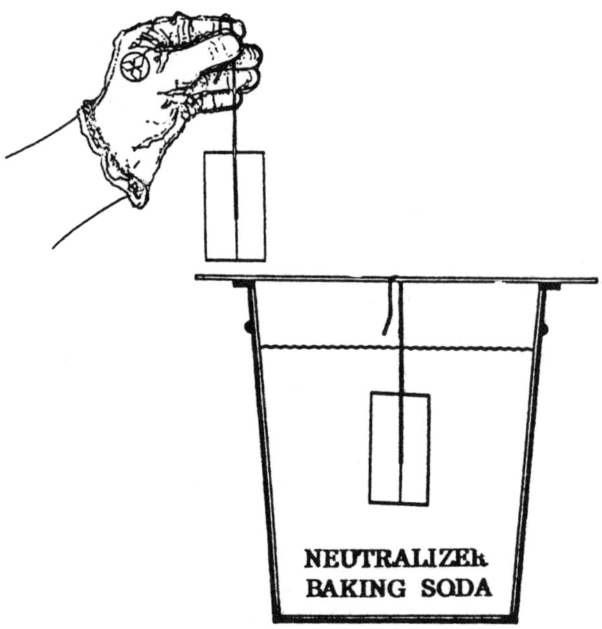

10. Baking soda neutralization.

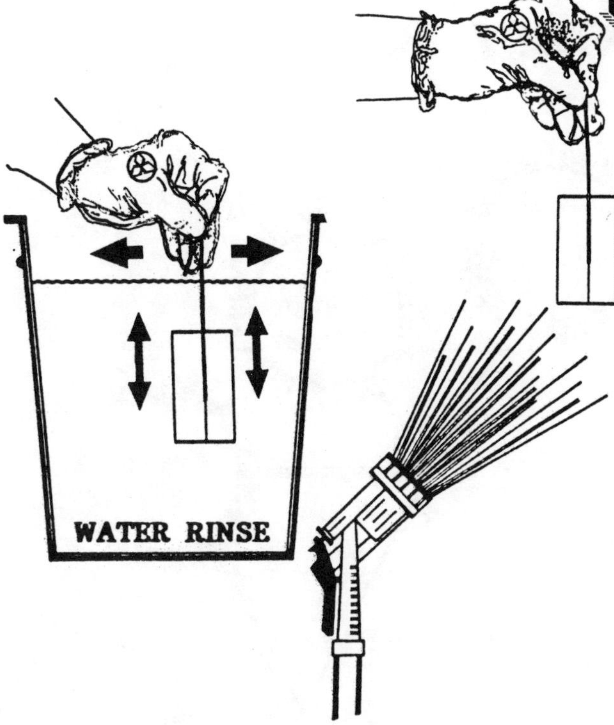

11. Thorough and complete water rinse or spray and second rinse.

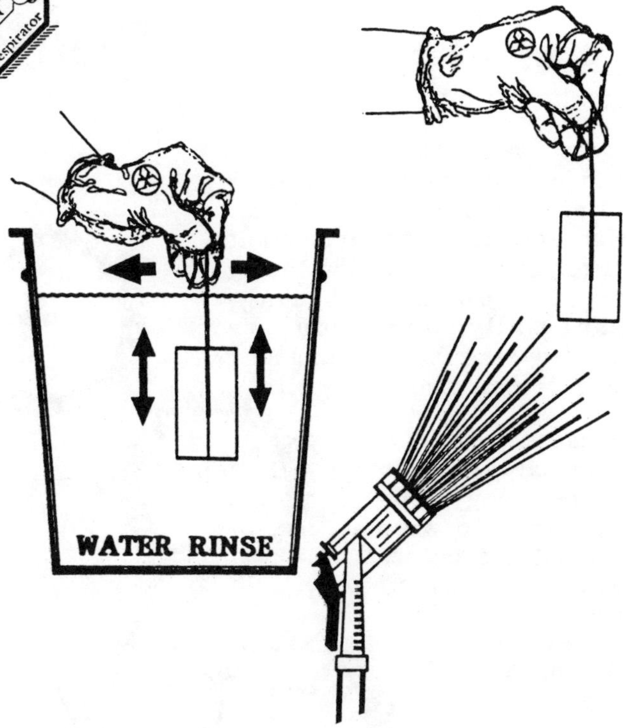

Dyeing time! Now, the real fun starts! Ready? Set? Let's go -- to the Dyeing chapter.

Notes

Color Theory

Dyes are transparent; that means they reveal and alter the color below them. Consequently it's important that you consider color theory when you are dyeing, especially when you are over-dyeing your aluminum. Dye colors also mix well.

Because dyes are transparent, when over-dyeing, you will need to work around the color wheel. Example: red-red/violet-violet, etc. Dyeing across the color wheel produces muted colors -or- you will get brown.

To work in split complements, complements, etc, with dyeing, you will need to resist areas, so that you are either dyeing gray aluminum -or- you can bleach out a previously dyed area.

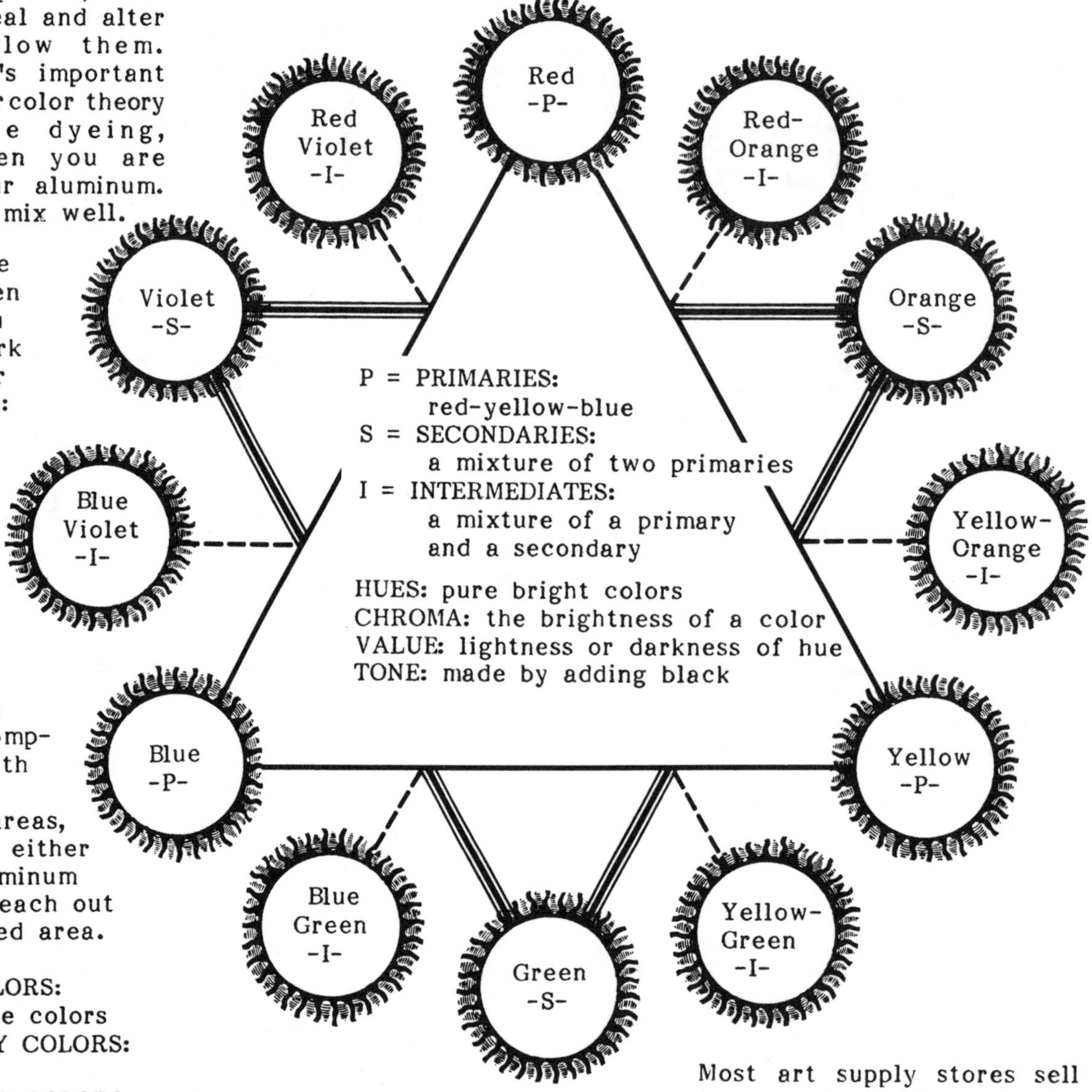

ANALOGOUS COLORS:
side by side colors
COMPLEMENTARY COLORS:
opposites
MONOCHROMATIC COLORS:
one hue with black or gray
TRIADIC COLORS:
three equidistant colors
TERTIARY COLORS:
mixture of three or more colors:
orange & green = citrine
orange & violet = russet
green & violet = olive
SPLIT COMPLEMENTS: a hue and two colors
on either side of its complement
ANALOGOUS COLORS: colors in a row:
yellow & yellow-green
green & green-blue, etc.

Most art supply stores sell inexpensive color wheels. Buy one--they will aid your "eye" for color and help to improve your dyeing techniques.

Personally, I've learned more about color theory by mixing dyes for dyeing aluminum than I ever did in a painting class!

Dyeing

Dyeing time! Now, the real fun starts! Ready? Set? Let's go!

It is critical to dye immediately after anodizing for maximum color potential. If you wait, the anodic oxide layer will begin to close. Thus, the color (dyeing) will not be as effective. Again, is this another hint for a variation on the technique?

Two families of dyes will color anodic films: organic and inorganic. I will be discussing the use of organic dyes (those made specifically for dyeing aluminum). Organic dyes are most often acid dyes. They are soluble in water.

Multiple cold water rinsings are imperative if you wish color consistency in your dyed and anodized aluminum. If the work is not rinsed properly, sulfate ions, or drag-out, from the electrolyte will cause streaks or discolorations. This drag-out also causes a chemical change or lowering of the dye bath's pH. The change in pH causes color variation. In industry, consistency of dye color is paramount.

DYE PARAMETERS

The following four parameters affect dyeing/ color with freshly mixed dye solutions:

1. Dye concentration (grams/liter or gallon)
2. pH of dye
3. Temperature of dye
4. Immersion time in the dye

COLOR MIXING

There are two ways of mixing colors:

1. Overdyeing seems to be the easiest way to mix colors. Simply immerse the aluminum in one color, rinse well, and then immerse it in a second color. This can be repeated for multiple colors.

2. You may also mix two or more dye powders together to create a new color. Mix multiple dye powders the same way as shown in the Dye Bath Set-Up chapter.

For me, it still seems easier and more efficient to mix the dye by multiple immersions. By moving from dye to dye, greater range and richer colors are possible, plus the process itself will be more flexible and enjoyable.

The dyes are transparent. That's why you can overdye and get such rich colors. That also means you must be aware of how the color wheel works, for instance: Dyeing green over red gives black or brown, not green. Don't forget to dip into black dye to modify your colors as you work around the color wheel.

SINGLE COLOR DYEING

Single coloring dyeing is simple. Immerse the piece into the desired dye color, using a bamboo stick/ aluminum wire to hang the rack wire. If you want light pastels, hold the aluminum piece in the dye for a moment, until the desired depth of color is reached. To check the exact shade of color: Lift the piece out of the dye, shaking gently to leave as much dripping dye in dye bath, and rinse in clean, cold water. After the rinse, the color you see is the color you get. (On some colors the sealer will magnify the color.)

For a more thorough rinsing, move the piece back and forth, up and down. Don't forget to use the chemistry wash bottles to spray rinse those intricate areas, overlaps, holes, etc. The through rinse cannot be overemphasized.

OVERDYEING

Double or multiple dyeing requires some forethought. If you have many dye containers, it would be brilliant to line up or organize those colors in one area with your rinse container in the middle. Thus you can move from color to color with a rinse stop in between. Be careful never to carry a dripping dyed piece over another dye bath. Obvious contamination then takes place.

Don't forget to immerse the piece only part way into the new color(s). Be free and bold. Remember, if you do not like your results, you can wipe away the color -- just like that! Simple, isn't it? To wipe out color, simply bleach.

BLEACHING

Bleaching is the removing of unwanted dye/color. You may remove all or part of the color. We can get into the technical way or the easy way -- let's start easy.

Sparex, the regular jeweler's pickle, will bleach out the color, when used at room temperature. Obviously this pickle should not be your normal pickle and should be in a separate container. Use this pickle only for bleaching. You may also use contaminated (with iron) pickle for bleaching, I do and it works great.

A 10%-25%-50%, by volume, nitric acid solution will bleach most dyes in 5 minutes at room temperature.

Some dyes which are difficult to strip/bleach, may require an addition of potassium permanganate to the nitric acid. Use the following formula for this addition:

50 grams per liter potassium permanganate
100 m/liter nitric acid

for 5-15 minutes of bleaching time at room temperature.

Follow all bleaching with a thorough rinsing before re-dyeing.

You may ask, "Is the dyed color this easily removed AFTER the aluminum is sealed?" No -- not quite as easily. The process for removing color on a sealed piece of aluminum is to strip the dye AND the oxide coating off. In other words, starting all over with the caustic etch, rinse, nitric etch, rinse, and re-anodizing, etc. So, it behooves you to decide if the color you have is what you want before you seal. Therefore, think ahead and if need be, strip and re-dye prior to sealing.

COLD VS HOT DYEING

We can either dye cold or hot. The cold dyeing process requires less work and equipment. It is also, to my way of thinking, the best way to achieve the most exhilarating color range. This opinion holds for both single color dyeings and multiple overdyeings. The cold dyes allow a full value range from the lightest, most subtle pastel, to a full, rich, dark shade.

The cold process takes more time however. But time is relative. By using cold dyes you don't have to wait for multiple dyebaths to heat up. Cold dyeing is cheaper on your electrical bill than hot dyeing. Hot dyes do zap color into your aluminum immediately. That saves time, but doesn't allow as large a range of overdyed colors as do the cold dyes.

Hot dyeing, as the name implies, requires a heated dye. The heat may be internal or external.

INTERNAL HEATERS

Depending on your dye container's size, you can use small or large immersion-type heaters. If you are working with small, quart size dye baths, coffee cup immersion heaters will probably suffice. Remember, it may be better to start small and work with inexpensive equipment.

Aquarium immersion heaters will also work. Check your local aquarium store, tell them the volume of liquid and the maximum temperature you wish to heat. There are many types of heaters and I am sure you will have an opportunity to compare and contrast the models available. The aquarium heaters that I've tried had low wattage ratings and heated too slowly for my taste.

If you want to spend big bucks, check out the immersion heaters the chemists use. These heaters are rated for higher wattage, heat up solutions faster, and, of course, cost more.

I have experimented with immersion heaters for water. The results, I feel, were quite slow and disappointing. The heater I used was rated at 200 watts at 110-120 volts. It took 2 hours to bring 1 gallon of dye up to 150 degrees F. Dye manufacturers suggest the optimum dye temperature is 140 degrees F.

I stirred the dye frequently to even out the heat. My immersion heater was only about four inches long. Therefore only a small portion of the dye was heated. External heat sources heat the dye from the bottom and are therefore more efficient. Stirring the dye is also important with external heat sources.

EXTERNAL HEATERS

External heat works the best for me. Why, you say? Inexpensive immersion heaters are slow -or- can only heat up small volumes of dye. I am an action person. I want to see it done NOW!

A good example of an external heater is a combination electric hot plate with magnetic stirring device. The combined heating and stirring will greatly reduce your working time, but is expensive. If you have some good second hand stores, government or state surplus places, or a university close by I'd recommend inquiring when they have their surplus equipment sales. You just might find some real bargains!

After considering how to adequately heat single and/or multiple dyebaths, I've developed a method using a deep fat fryer. See inset for directions.

To use my overall working time to my best advantage, about half way through the anodizing, I start heating my dyes. This way I have all the colors I want to work with ready as soon as the piece comes out of the final cold water rinse. After all, time is money, and if you are involved in doing production art, you must make your time count!

In order to achieve the appropriate dye bath temperature, I use candy thermometers. Simply hang one in your dye bath, using the convenient hanger. Be certain the lid is on the dye bath to retain the heat. Obvious, isn't it? Once the dye baths are heated they retain the heat for upwards to half an hour. This is plenty of working time. Obviously, if you need to work longer you will need a better method of retaining heat in the dye baths.

Frankly with all of the above in mind, you can see why I really enjoy working with cold dyes. Why fool around? The art is what counts.

THE DEEP FAT FRYER METHOD

The best method I have found to heat 1-2 gallons of dye uses an old deep fat fryer/cooker. The particular model that I use is rated at 1150 watts, 110-120 volts. The deep fat fryer will be used as a double boiler.

The deep fat fryer should have a lid to keep in the heat. A stainless steel, enameling screen is used as a trivet, or sub-floor, for the plastic dye container to stand on. Bend two opposite sides of the screen at 90 degrees (1/2" to 1").

Place the screen in the bottom of fryer. Fill the deep fat fryer about 1/3 full of water. Lower the plastic dye container filled with dye into the fryer. Allow dye container to rest on the screen. The fryer's water level should be about 1 inch below the top so as not to spill over. The heating action of the water, plus the water displacement from the dye container, can lead to spilled water and the possible danger of an electrical shock. **Think ahead** and **be careful!**

I have made an assumption here: That your deep fat fryer is wide enough to accommodate the diameter of your dye containers. Make certain that the screen will fit into the fryer's bottom and have enough surface area to fully support the dye container.

My fryer takes 20 minutes to heat 1 gallon of dye to 140 degrees F. It takes 35 minutes to heat 2 gallons to 140 degrees F. Not bad.

DESIGN POSSIBILITIES

Dyeing is the time to be creative!

So far I've made the assumption you wish to immerse the entire piece into the dye. What a silly assumption! Who says all of the piece should be dyed? Perhaps you wish only a portion dyed. Or when mixing colors, you desire a multi-colored approach. Have fun, be creative! I often dip partially into a second or third color, providing a more interesting set of values. If you are shrewd with your racking device, you can carefully maneuver the object into various positions, establishing provocative color patterns without the use of masking out or other design tricks. Think about it. By dyeing the piece, rinsing it well, and handling carefully with clean cotton gloves, you could change the relationship of the rack wire to the object. Thus, adding a different angle of immersion. Consider also the use of pulleys and strings to hang your work at different positions in the dye bath so you don't have to stand there and physically hold onto the rack wire.

To produce an air-brush edge effect on the work, move the piece up and down in the dye. If a straight line of color is the desired effect, simply hold the piece in the dye until correct color is achieved.

AN OUTLINE OF DYEING/DESIGN THOUGHTS

YOU CAN DYE BY:

IMMERSION (these have already been explained)

Full immersion for one solid color.

Dip dyeing -- dipping a part of the piece into the color.

DIRECT APPLICATION -- Use the dyes simmering hot and accept that the colors will be paler than with immersion.

Painting, sponging, using a syringe, eye dropper or straws, spilling dye onto the object.

Spraying, squirting or air brushing onto the piece.

Silk screening -- you can add a thickener and silk screen - I've even seen photographic reproductions done this way.

The dye can be thickened with the following mixture:

Indian corn starch	- 75 parts
Tapioca starch	- 25 parts
Gum tragacanth	- 225 parts
Water	- 1000 parts

It is suggested that you add this to the dye at a 20% thickener, 80% dye ratio, but-- add until you get your desired consistency.

YOU CAN CREATE PATTERN BY:

Resisting -- see Christina DePaul's chapter and the Etching chapter. Other resists include: masking tape, mailing labels (Avery), clear contact paper, press-off letters, graphic tapes, rubber cement/peelable lacquers, etc. I recommend going to a stationary store and see what has adhesive on it, it should make a good resist. Try stars, notebook paper reinforcers, numbers, hearts, etc. A Japanese method of resist dyeing uses the resists to create low dams or walls, then the dye is flowed inside these dams/wells. Perhaps rubber cement or a peelable lacquer would make a good dam for you.

Partial overdyeing -- dipping the piece in fully, then part way, then less part, etc.

When directly applying dye, applying it in a pattern.

Going back in, after the dyeing is done, with paint or a pen and creating an over the dye pattern, drawings, words, etc. Permanent marking felt tip pens may be used to apply color (aluminum must be carefully dried before marking pen color is applied) and will resist overdyeing.

Don't forget dyeing many pieces each a different color, cutting them up and then fabricating them together into a pattern -- like a patchwork quilt.

If you are going to handle the aluminum with your hands, remember to use white cotton (photography) gloves to prevent grease and oil contamination. Buy 2 or 3 pair and have them all on hand, in case you accidentally forget and touch yourself with the gloves on. You can, of course, wash the cotton gloves to clean any body oils off them.

DYEING TIME CAN VARY:

Optimum dyeing time is 5-15 minutes.

Leaving a piece in for the full time will give you a full bodied color

if you have a full strength dye you get an intense full bodied color.

if you have a partial strength dye you get a full bodied pastel.

Quickly dipping or leaving a piece in for a short time will give you a tint, or a pastel. In anodizing, this pastel can be quite vivid -- this might eliminate the need to mix partial strength dyes.

This is the creative portion of anodizing -- go wild.

Further information on dyeing techniques can be found in fabric dyeing books, but perhaps the most creative thing you can do is learn on your own. Without knowing the limitations of fabric dyeing, you may discover new things that can be done with the dyes and the aluminum.

ILLUSTRATED STEP-BY-STEP COLD DYEING

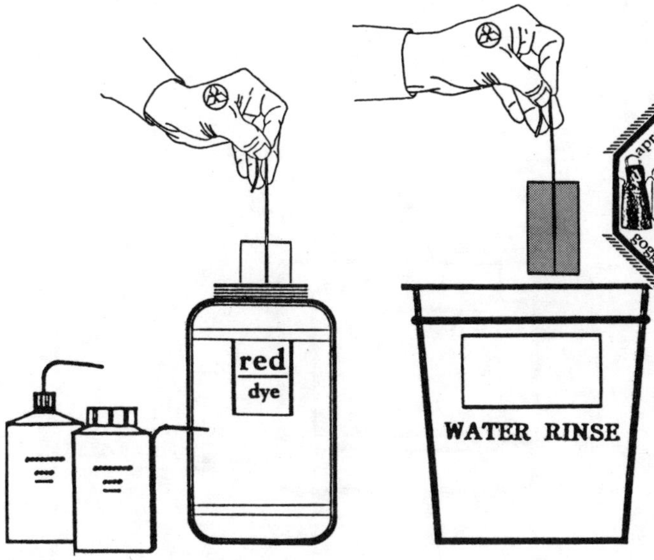

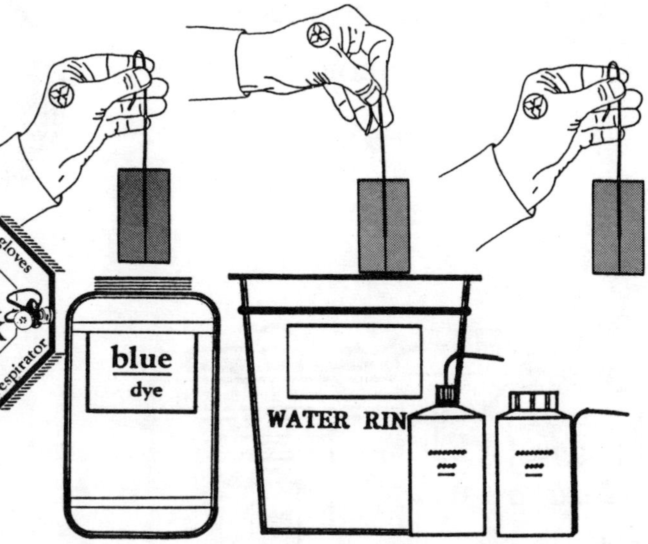

1. Dip your racked and anodized aluminum piece into a dye bath. Leave it in until your desired color depth has been reached.

2. Rinse thoroughly in cold water. Use a chemistry wash bottle to squirt dye out of hard to reach areas, holes, etc. Remember how much fun a squirt gun was?

3. Dip your aluminum piece into another dye bath, if you wish to overdye colors. Leave it in until your desired color depth has been reached.

4. Rinse thoroughly in cold water. Squirt it, if necessary.

The dyeing and rinsing can be repeated until you have the color(s) you wish. Rinse after every dye bath.

Now you are ready to seal.

HOT DYEING MATERIALS:

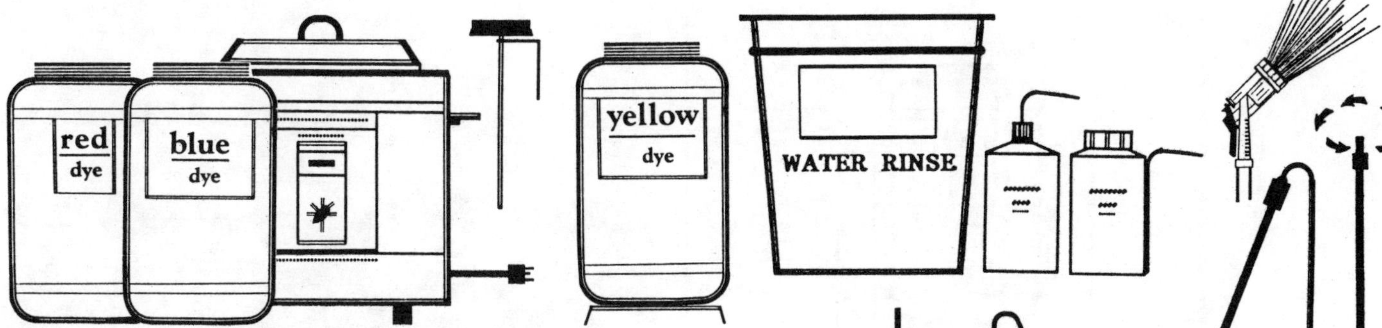

Grocery Store, Restaurant Supply House:
- 1 quart, 1 to 2+ gallon plastic containers with tight fitting lids
- Candy Thermometer
- Coffee Cup Immersion Heater
- 1 - 5 Gallon Plastic Bucket for Rinse Water

Aquarium Stores:
- Aquarium Immersion Heater

Chemistry Supply House:
- Immersion Heater
- Chemistry Wash Bottles

Second Hand Store:
- Used Deep Fat Fryer/Lid, 1000 watts minimum

Jewelry Supply House:
- Stainless Steel Mesh Enameling Trivet, large enough to fully support the dye containers

Water Company:
- Tap Water/Spray Rinse

Obviously many of the above items may be purchased at other locations. This list provides a start.

ILLUSTRATED STEP-BY-STEP
HOT DYEING:
DEEP FAT FRYER METHOD

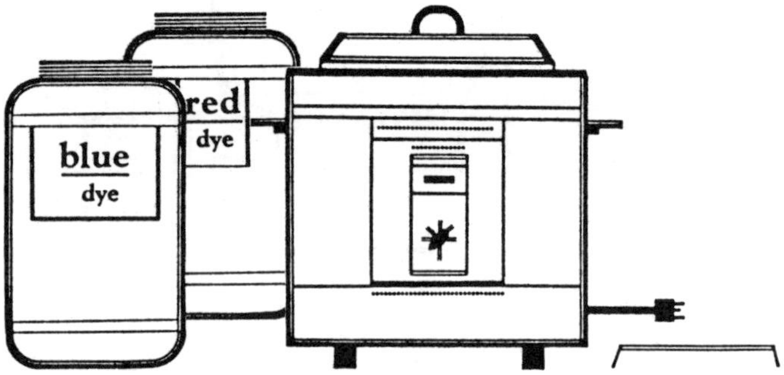

1. Have your dye baths, deep fat fryer and trivet ready.

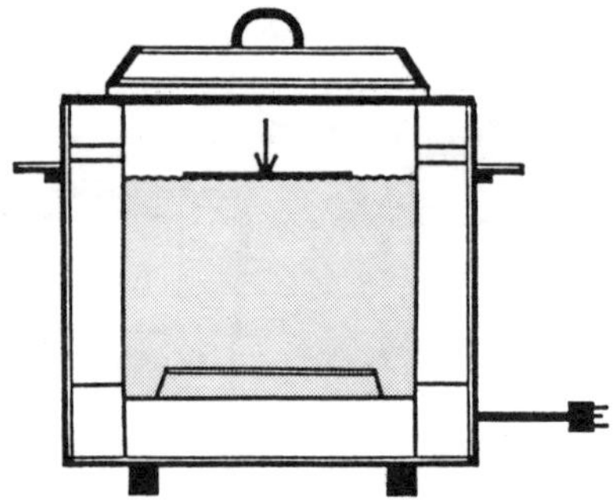

2. Place trivet in bottom of fryer. Fill the fryer with the required amount of tap water. Replace lid to keep in heat. Turn the fryer on "High." Bring water to a rolling boil.

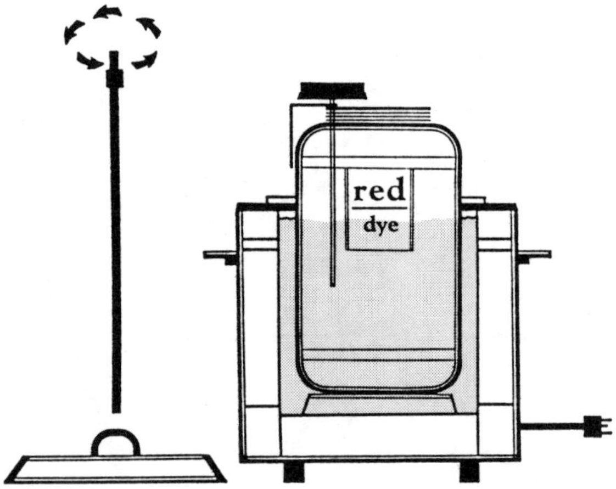

3. Remove lid. Place dye container on trivet in the boiling water. Be sure the dye container lid is on (to retain heat).

Periodically check the dye temperature with a candy thermometer. Stir the dye, with a plastic stirring rod, to even out the heat. At 140 degrees F. the dye is ready to use.

This is the deep fat fryer that I use to heat my dyes. The 2 gallon plastic buckets shown were purchased from a restaurant supply house.

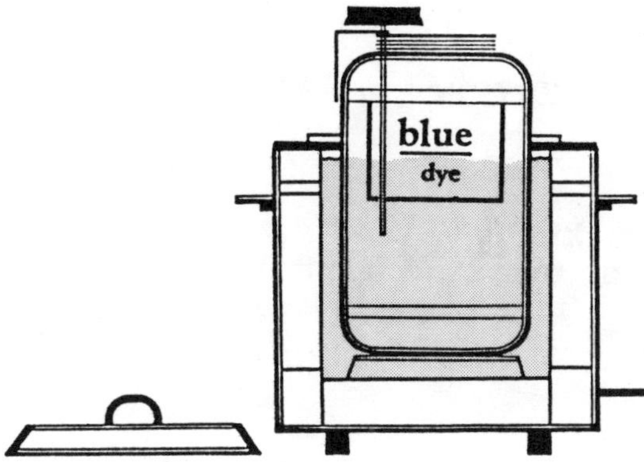

4. If heating more than one dye container, check the water's height in the fryer. Add more tap water, as required. Place the new dye container in the fryer and repeat the above step.

5. Dip your racked and anodized aluminum piece into a dye bath. Leave it in until your desired color depth has been reached.

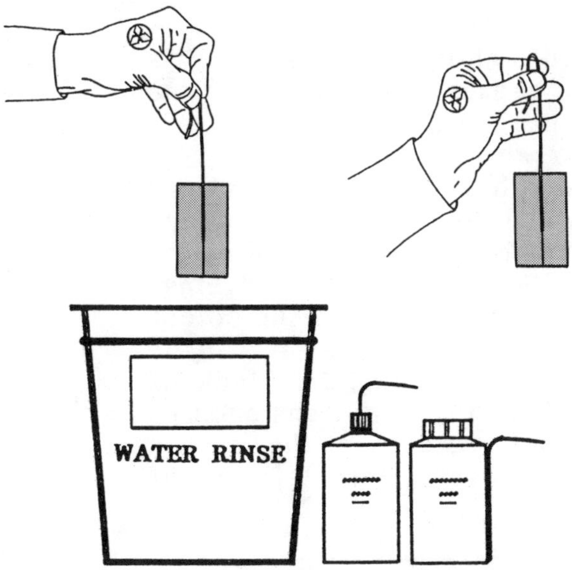

6. Rinse thoroughly in cold water. Use a chemistry wash bottle to squirt dye out of hard to reach areas, holes, etc.

The dyeing and rinsing can be repeated until you have the color(s) you wish. Rinse after every dye bath.

When you are finished with the fryer, turn it "Off."

IMMERSION HEATERS

Coffee cup immersion heater.

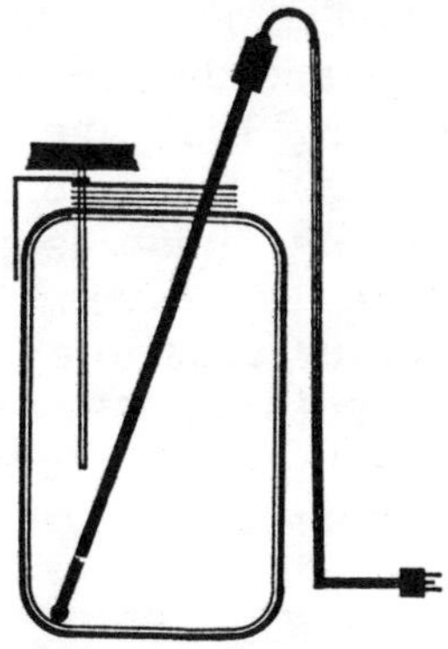

This illustrates how you use an aquarium heater and the larger more expensive immersion heaters. Simply place the immersion heater into the dye container and plug it in. Then proceed as shown above from the last half of Step 3 onwards.

If you do not like the color you have, remember bleaching and re-dyeing.

Sealing is next.

Trouble Shooting

DYEING FAULTS, THEIR CAUSES AND PREVENTION

The following information has been kindly provided and permission given to reprint by:
Sandoz Colors and Chemicals
4000 Monroe Road
Charlotte, North Carolina 28205
704-372-0210

When faults occur in the colour finishing of aluminum leading to rejection of the work, there is a widespread tendency in the industry to blame the dyeing process or even the dyes, without looking further into the cause of the trouble. Experience has demonstrated that very often undetected or undetectable deviations in the pretreatment of the metal are responsible for colour faults.

The following information should help to dispel this misconception by indicating possible causes of colour faults and outlining suitable action for their prevention.

The term "colour faults" is understood here to mean deviations from a quality standard which are visible to the unaided eye and are due to errors in the preliminary surface treatment or dyeing that are not visible until after completion of dyeing. The relationship between the process conditions and the colour of the finish is not considered except in so far as it can be used to prevent or correct colour faults. Similarly lack of colour constancy and physical or colour imperfections such as defects on the metal surface or self-colour in the anodic coating which are detectable prior to dyeing, are left out of account.

The following tables are arranged under three column headings:

1. Description of colour fault

2. Possible cause or causes

3. Preventive or corrective action

A. COLOUR DIFFERENCES FROM BATCH TO BATCH

Type of colour fault	Possible causes	Preventive action
Slow decrease in depth (over days or weeks)	Decline in dyeing capacity due to:	
	- drop in dye concentration through consumption	- Extend dyeing time - Add dye to correct deficiency
	- accumulation of deactivating contaminants in dyebath	- Prevent entrainment by thorough rinsing and by avoiding dripping from other solutions - Extend dyeing time - Increase dye concentration above the initial value - Partial or full renewal of dyebath
	- Chemical decomposition of dye through contact with non-resistant metal parts	- Check tank and fittings for possible points of contact and insulate these points
	- Slow increase in pH	- Correct pH, buffer the bath
Depth differences	Variations in dye absorptive capacity of anodic coating due to:	
	- Unconstant anodizing conditions	- Keep anodizing conditions constant (current density, time, temperature)
	- Presence of different alloys	- Adjust anodizing and dyeing parameters to the alloys present
	- Different time lag between anodizing and dyeing (acid dip, carry over into rinsing bath, retention time in water and air)	- Keep the time interval constant - Do not suspend in anodizing acid without current, rinse immediately afterwards - Do not store in water - Treat further without a break so far as possible - Level out dye absorptive capacity by brief activation in acid and a neutralizing rinse

Type of colour fault	Possible causes	Preventive action
Depth differences	- Uneven activation in the acid dip	- Keep activation time constant
	Different degrees of bleeding in the sealing bath	Equalize or minimize losses from bleeding by: - Sealing in Sealing Salt AS - Dye at higher temperature or for longer time at lower dye concentration - Use cleanest possible water for sealing - Immerse at highest practicable sealing temperature - Keep sealing pH constant, buffer bath
	Fluctuations in dyebath pH	- Buffer dyeing solution
Over-heavy dyeing at beginning of production run	Increase in the anodizing temperature	- Do not dye the first anodizing batches
Change of shade	Uneven depletion of elements in dye combinations	- Correct concentration of the element involved - Use homogeneous dyes whenever possible - Employ combinations of dyes with similar dyeing rates
	Chemical instability of the dye in solution	- Select dyes with higher stability in solution - Optimize dyeing conditions (lower temperature, more suitable pH, buffer) - Renew dye solution at frequent intervals

B. COLOUR DIFFERENCES IN THE SAME BATCH

Type of colour fault	Possible causes	Preventive action
Colour differences from part to part	Irregular current supply due to uneven contacts on the auxiliary anode frame Presence of different alloys	- Create uniformly good contact points - Clean the contact points - Dye only alloys of the same type in a given batch
Gradations of shade on marginal areas with darker edges	Irregularities in current density and heat distribution (heat accumulation at edges) in anodizing	- Improve the electrode geometry - Intensify admixture of electrolyte - Reduce current density - Intensify cooling, choose lower anodizing temperature - Reduce aluminum content of electrolyte - Select lower dyeing temperature and longer dyeing time
Lower areas of the part darker	Differences in immersion time in the dye solution between upper and lower areas due to method of immersion	- Extend total dyeing time to more than 10 min and adjust other parameters, i.e. reduce temperature and concentration - Accelerate immersion
Horizontal colour fronts in upper region of the part with striped, often wavey shaded spots	A film, usually oil containing, adheres to the work when lifted clear of the bath	- Remove film from the bath (clarify by filtration, adsorb on absorbent material, collect with a blade and skim off)

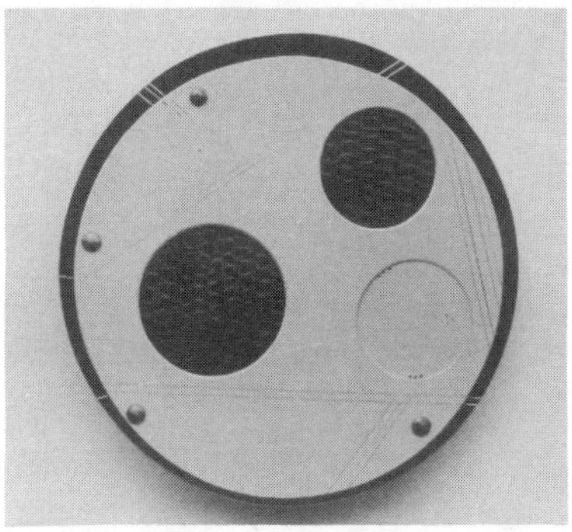

Brooch by David LaPlantz, fabricated, painted aluminum, engraved, lizard skin, actual size, 1982.

C. COLOUR DIFFERENCES IN THE SAME PART (unlevelness, coloured spots)

Type of colour fault	Possible causes	Preventive action
Clouds covering a large area beyond the single part	Anodizing temperature not sufficiently uniform (heat convection streaks)	- Intensify mixing of bath components - Reduce current density
Pale spots	Inadequate degreasing action	- More thorough degreasing - Add wetting agent to anodizing bath to form foam cushion
	Oily impurities in anodizing or dyebath	- Keep bath clear of impurities (oil-free compressed air) or remove if present - Add wetting and levelling agents as foam cushion formers
	Uneven wetting of the parts in the dyebath	- Immerse in the wet state only - Short acid treatment (activation) and rinsing prior to dyeing - Intensify mixing of bath components, or preferably agitate the parts during immersion and in the first dyeing phase - Add levelling agent to the bath
	Hydrolysed aluminum not eliminated by rinsing	- Rinse thoroughly immediately after anodizing - Follow rinsing with acid treatment (activation) or acid rinsing (pH below 2) with a subsequent neutralizing rinse
	Local over-heating in polishing (polishing clouds)	- Brief anodizing treatment, etching

Type of colour fault	Possible causes	Preventive action
Numerous pale pointed spots, especially in pale dyeings	Gas bubbles at the pore openings inhibiting dye penetration	- Immerse parts in wet state free from gas bubbles - Agitate parts on immersion until no further gas bubbles rise to the surface - Dye at room temperature
Pale spots at cavities and depressions	Occluded gas bubbles interfering with the anodizing or dyeing process	- Intensify agitation of the parts or the solution to detach bubbles as rapidly as possible
Pale spots in cavities (angles) or pale stripes running from cavities to edges	Deposits or exudations of anodizing acid which inhibit dyeing	- Plan construction and suspension of parts to facilitate draining of acid, (e.g. not completely closed selvedges, drainage holes) - Rinse thoroughly immediately after anodizing - After rinsing, neutralize, activate (acid dip), rinse
Dark spots in cavities (angles) or dark stripes running from cavities, dark stripes on horizontal edges, dark, drop-shaped points	Drip deposits & outflows from dye solutions (spots showing on drying can be partially cleared by wiping), overheavy dyeing due to local continuation of dyeing process by superficially attached dye solution	- After dyeing do not leave parts unrinsed, rinse at once - Reduce dyeing temperature and extend dyeing time - Facilitate drainage of dye solution
Dark spots	Undissolved, retained dye particles	- Do not add dry powder dye; prepare an aqueous concentrate (dissolve or disperse) and add dye through a fine gauze screen if available - Boil up dyebath with agitation - Clarify dyebath by filtration

D. VISIBLE DEFICIENCIES IN COLOUR QUALITY

Type of colour fault	Possible causes	Preventive action
Dull to chalky dyeing, often not resistant to wiping	Over-soft anodic coating, especially pronounced in marginal areas, owing to inadequate anodizing	- Reduce anodizing temperature, time current density, acid, and aluminum concentration, intensify motion of the electrolyte
	Over-heavy activation, perhaps combined with unsatisfactory anodizing	- Shorten retention time in acid bath
	Current-free suspension in anodizing electrolyte or in unrinsed state	- After switching off current, lift the parts at once and rinse
	Coating attacked owing to unduly low bath pH	- Increase the pH to at least 4
Opaque, dull to bronzing dyeing, removable by wiping	Superficial pigmentation due to premature precipitation of dye, frequently caused by excess of dissolved or hydrolysed aluminum	- Increase dyebath pH to at least 4 - Acid dip (activation) prior to dyeing to dissolve and clear aluminum

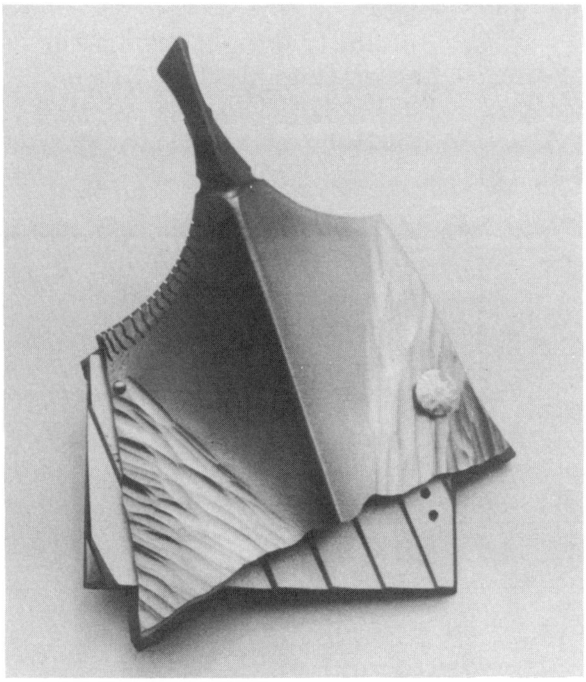

Brooch by David LaPlantz, fabricated and forged anodized aluminum, actual size, 1986.

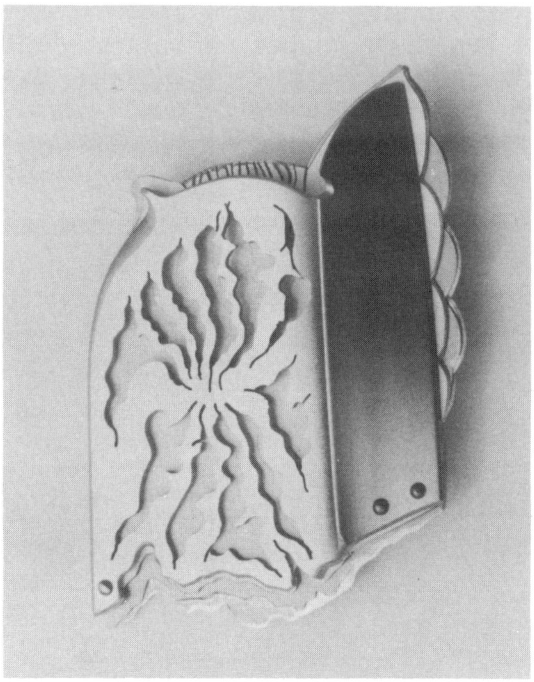

Brooch by David LaPlantz, fabricated, anodized and painted aluminum, actual size, 1987.

Sealing

Stainless steel or some other inert material must be used for your sealing container, whether you use water or nickel acetate. Do not use an aluminum container. I tried using an old aluminum pot with bad results. The nickel acetate, quite simply, etched holes through the bottom!

I use stainless steel stock pots purchased at my local restaurant supply store. Check to see if your dealer has any used restaurant equipment, or perhaps some dented stainless containers, before buying brand new stock pots. Damaged items are usually sold at a discounted price. A couple of dents won't mean too much to you, when seen in the light of the savings for your wallet.

The heat source for the sealer may also be an electric hot plate, an inexpensive electric crock pot, natural or propane gas burners. I usually use a propane gas burner.

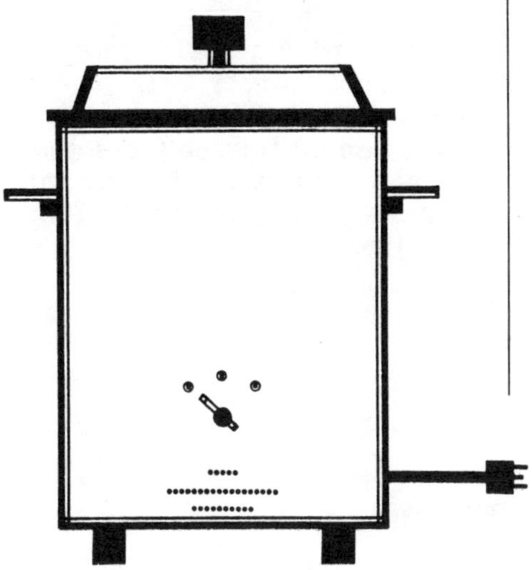

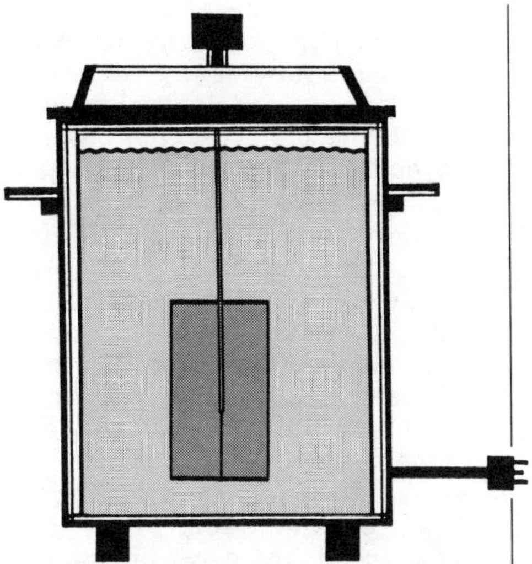

My crock pot works well for water sealing small pieces. It takes 2 1/2 to 3 hours to reach 200 degrees F. I suspend the aluminum from a bamboo skewer.

The heat source for my sealing is another item easily purchased at the restaurant supply. Single or double burner hot plates that burn propane work well. The propane tank, gauge, hose, and proper connections can be purchased at the same time.

I highly recommend purchasing a gauge for the propane tank. It is nice to know exactly how much fuel you have when you start sealing. At first, I did not have a gauge. You're right. In the middle of sealing some important pieces I ran out of propane and had to run the tank over to the propane dealer for a refill. Plan ahead.

For safety: The propane tank should be mounted outside your studio, with hose connection coming through the wall. Proper ventilation should always be observed.

My method of sealing is as follows:

Dye the aluminum. Then water rinse or spray rinse. And finally, move to the sealing bath.

If aluminum is to be clear anodized and not dyed, move directly to toe sealing bath. Otherwise, anodize and dye, then move to sealing.

I plan both anodizing and sealing to use my time and utilities (electricity, propane, etc.) to the best advantage. Therefore, I start heating my sealing solution at an approximate time to coincide with when the pieces come out of the anodizing tank, dye baths, etc.

To monitor the temperature of the sealer, I use an inexpensive candy thermometer. Keep the lid on the pot until you have reached 190-200 degrees F. To hang your pieces in the sealer, suspend the rack wire from a bamboo skewer or an aluminum wire laid across the top of the pot.

I use aluminum foil as a lid. This may sound strange, but think about it for a minute. Most lids for containers, like a stock pot, fit down onto or into the top of the pot. Once you have a thermometer, aluminum wire, or bamboo stick across the top the pot, the lid no longer fits properly to prevent heat loss. So, if you use aluminum foil, the lid can be formed to fit the pot. This way, heat stays in, the aluminum is sealed properly, the thermometer is not in the way, etc. Everything works for you.

Watch the thermometer and keep the sealing solution under a rolling boil. Make certain that the pieces do not bang against the sides of the stainless pot or each other. By maintaining a lower temperature (190-200 degrees F.) the sealing solution will be just under a boil.

Seal for a period of 20-30 minutes.

Remove the pieces from the sealer. On removal, the piece should dry almost immediately. This is due to the extreme heat of the sealer versus the temperature of the room.

Areas of the piece that do not dry should be dried with a soft, clean cloth or paper towel. A thin film may be formed on the piece from the sealer. This surface smut can be removed as you dry the piece.

I immediately oil the piece with mineral oil. This will bring up a **vibrance** in the color. Wipe off any excess oil. I am now ready to wrap the piece for shipping or finish the construction of the object.

In the final edit: Bud feels that the mineral oil probably raises the gloss of the aluminum. He feels that any residual smut "hides in the mineral oil." The oil and smut are both "water white" (clear), so the smut can't be seen. He feels, and I agree, that oiling each piece of aluminum in an anodizing shop would not be acceptable commercially. There is simply not the time to oil and wipe off the excess oil, on say, 5000 small aluminum washers. Also, keep in mind, an artist's needs and desires are miles apart from most other people's needs.

As an alternative, Bud suggests a clear acrylic lacquer might do the same thing, and might be more permanent. Since I am in the final hours of finishing this book, I will not have time to experiment. So, guess what? If you want to try this idea, you'll know the answer through your own experimentation! After all, isn't that what this workshop in all about?

If you want to polish the piece at this point, do not oil first, but go directly to polishing. Be sure to polish with care. The oxide layer is not that thick. You do not want to polish through the oxide layer. Then, clean off all traces of the buffing compound, oil, wipe off excess oil, etc.

Another alternative to the stainless steel stock pots and the crock pot is a Pyrex bowl. The Pyrex bowl works well for smaller pieces, where heating the sealer takes less time. I have fashioned a few different make-shift lids for the bowl to eliminate heat loss through evaporation. One lid is made of tempered Masonite and is sealed to be waterproof. Another is made up of discarded cereal bags made of aluminum. With these bags I have a lid structure that can be formed (sculpted) to accommodate all sizes and strange shaped racks and anodized forms.

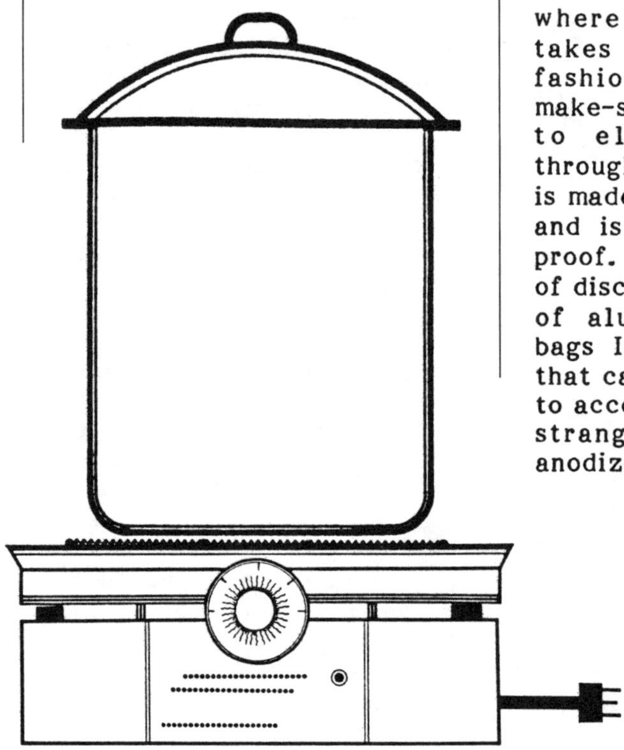

Nickel acetate/water seal or water seal alone in a stainless steel stock pot/lid on an electric hot plate.

1. Open main propane (tank) valve (A). Open gas Burner valve (B), ignite flame with spark lighter and adjust height of flame. Heat up the sealing solution to 190-200 degrees F. Monitor temperature with a candy thermometer.

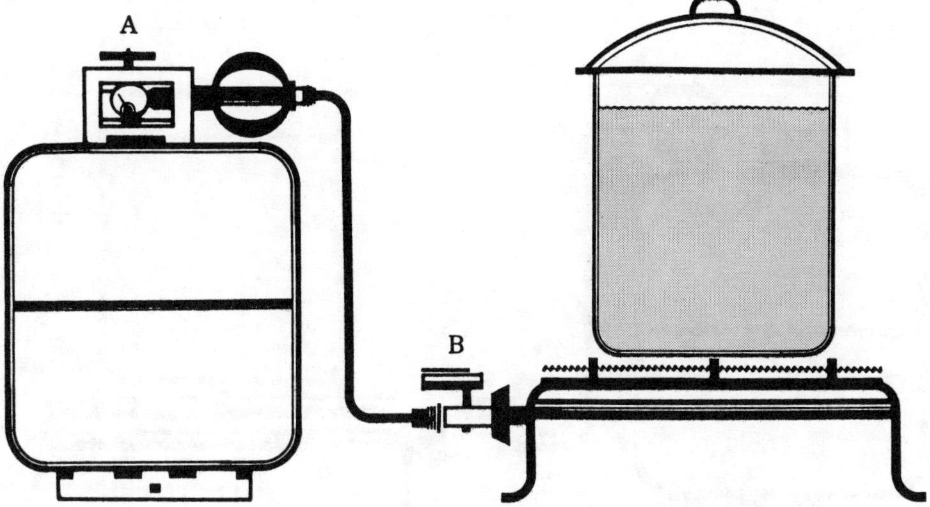

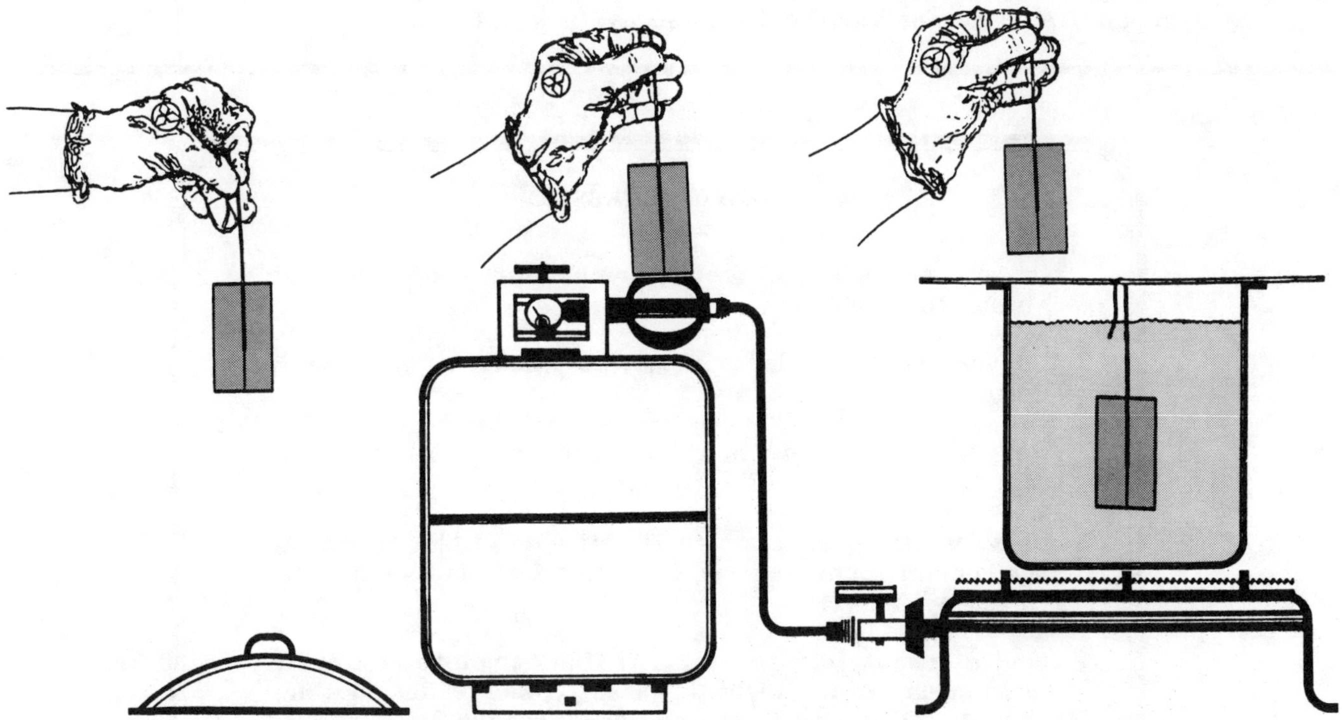

2. Remove lid. Immerse aluminum in sealer. Agitate the aluminum immediately on immersion to insure rapid and even heating. If the sealer already has some pieces in it, be careful not to touch/harm them as the new work is immersed. The drawing shows the aluminum hanging from a bamboo skewer. Replace the lid on the sealing container. The sealing time is from 20-30 minutes.

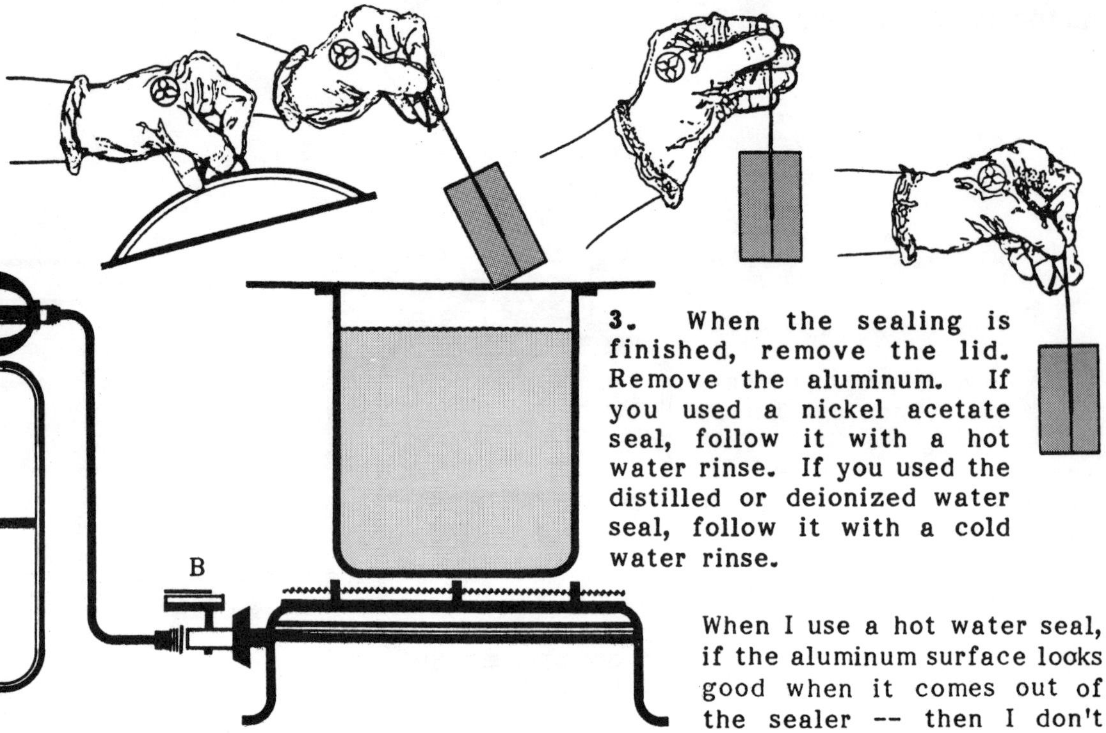

3. When the sealing is finished, remove the lid. Remove the aluminum. If you used a nickel acetate seal, follow it with a hot water rinse. If you used the distilled or deionized water seal, follow it with a cold water rinse.

When I use a hot water seal, if the aluminum surface looks good when it comes out of the sealer -- then I don't rinse. (It has only been touched by boiling water.)

---

4. When you are finished with sealing, remember to turn off the heat source (propane tank: burner valve B followed by tank valve A).

ANOTHER DIGRESSION

Do you feel **different** from other people? I do, and have always felt **different.**

Do you **laugh** out loud in movies when no one else is laughing?

Do you see a **magic** in things when most other people say, "What?" "Where?" "What?"

Do some pieces of music really **hit** you? When discussing the special sound you **hear,** do your friends look at with raised eyebrows?

I have always felt **different.** I find complete enjoyment in so much, while others, apparently, see or hear nothing at all? We all must find our way through life bringing out our own joy and happiness.

Are you different? Do you identify with my comments? I'll bet you don't feel as lonely now, do you?

DYE MANUFACTURERS/WHOLESALERS

DYE DEALERS/RETAILERS
Lines indicate which manufacturer's dyes they stock.

A. C.H. Patrick & Company, Inc.

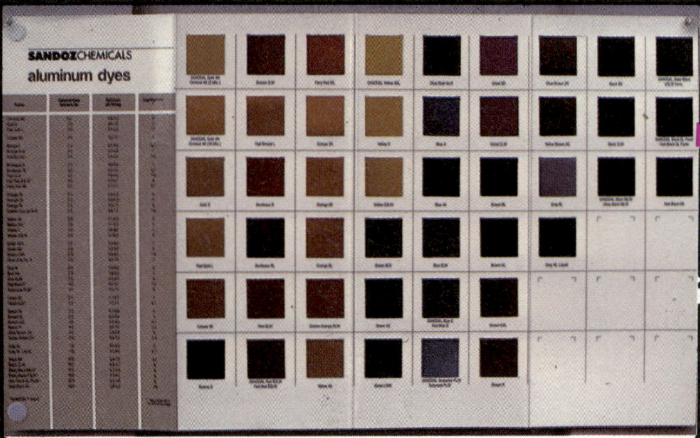

B. Sandoz Chemicals Corporation

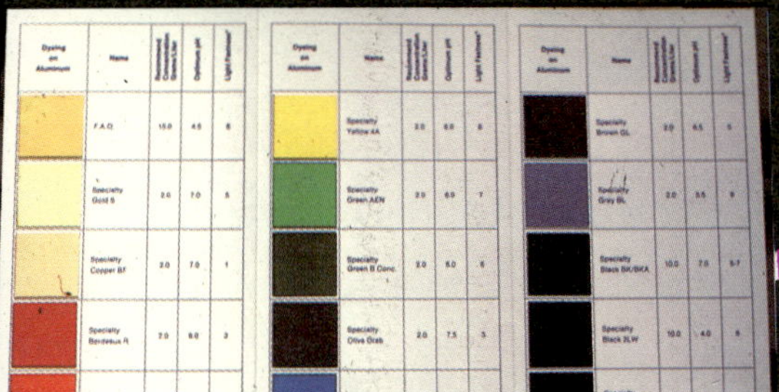

C. U.S. Specialty Color Corporation

D. EAST/WEST DyeCom
P.O. Box 12294
Roanoke, Virginia 24024
703-345-4241

MET-L-ART

P.O. Box 4611

Greenville, Delaware 19807

800-638-5278

E. CHRISTINA DEPAUL

F. CHRISTINA DEPAUL

G. THOMAS SEABURY BROWN

H. THOMAS SEABURY BROWN

I. DAVID TISDALE

J. DAVID TISDALE

K. DEBRA STONER

L. ROWENA MORALES

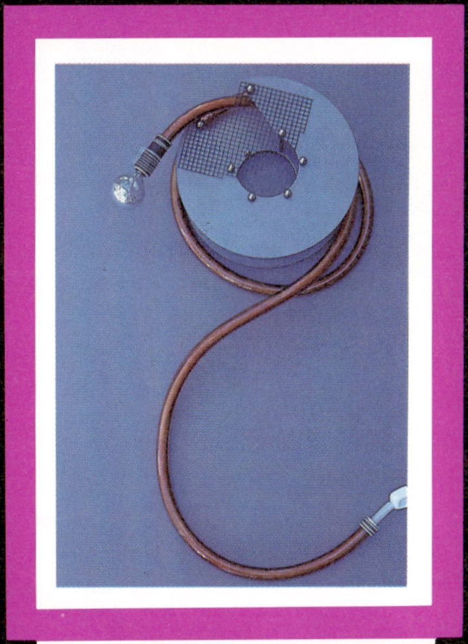

M. CAROLE S. ROBINSON

N. THOMAS SEABURY BROWN

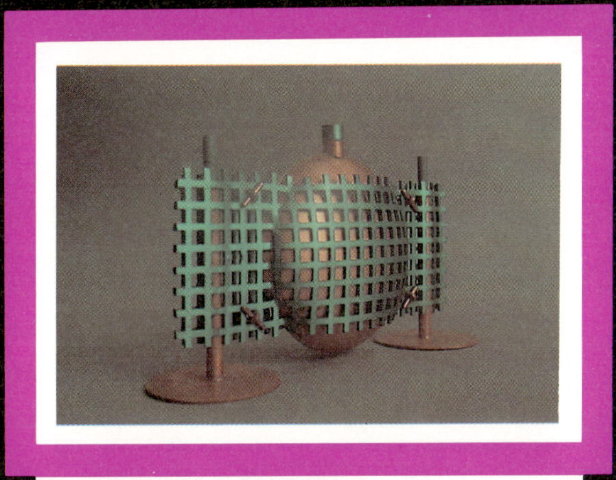

O. CATHY STAHLMAN

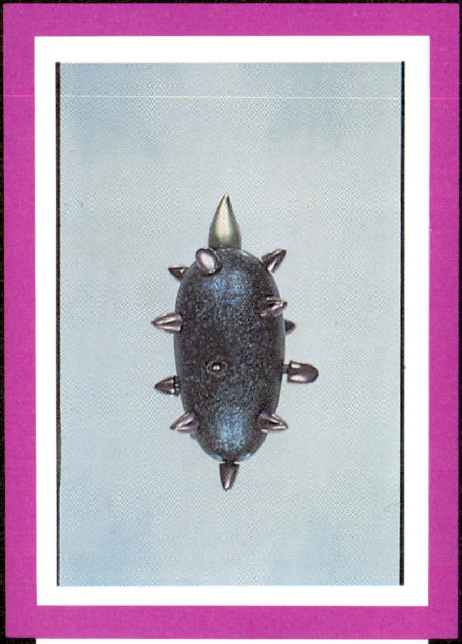

P. LAURA MARTH

Q. FUJIE BABA

R. SHARON NOVAK

S. JULIA W.P. BAKER

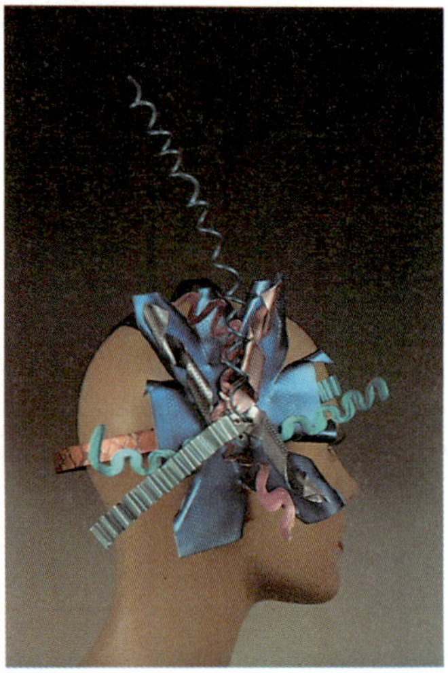

T. LENET GALLOWAY

U. BARBARA HEINRICH

V. BARBARA HEINRICH

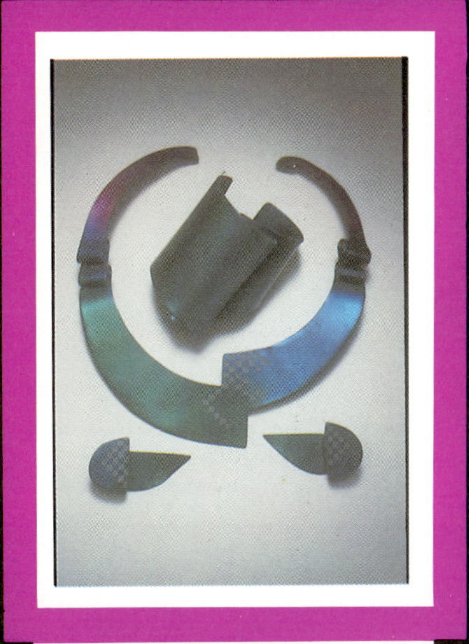

W. SYLVI HARWIN

X. SYLVI HARWIN

Y. SYLVI HARWIN

Z. ANDREA PERKINS

AA. ALEX CONNELL

BB. JANET SERATTE

CC. JAN PETERS

DD. JAN PETERS

EE. ALEX CONNELL

FF. CATHY STAHLMAN

GG. CANDACE BEARDSLEE

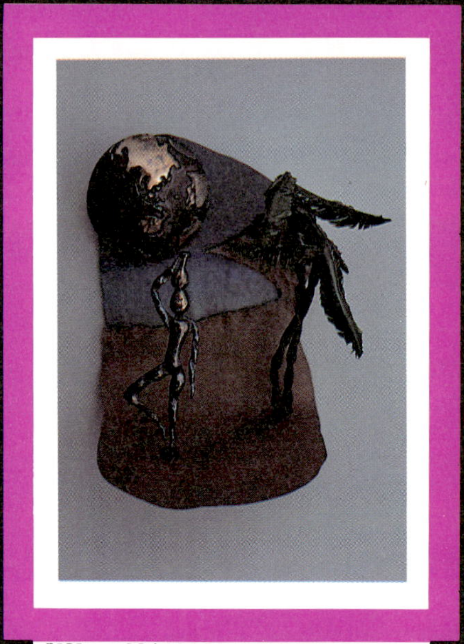

HH. NANCY BARBOUR

II. CANDACE BEARDSLEE

JJ. MICHAEL EDWARDS

KK. KATHE TIMMERMAN

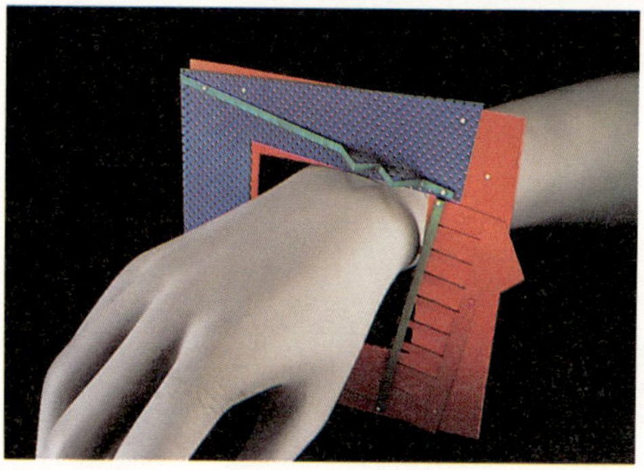

LL. LESLIE LEUPP

MM. LESLIE LEUPP

NN. LESLIE LEUPP

OO. DAVID LAPLANTZ

PP. DAVID LAPLANTZ

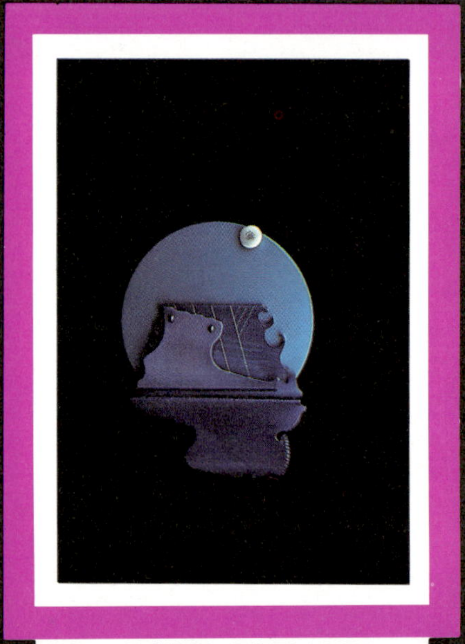

QQ. DAVID LAPLANTZ

RR. DAVID LAPLANTZ

SS. DAVID LAPLANTZ

TT. DAVID LAPLANTZ

Color Slide Show

A. C.H. Patrick & Company, Inc. Color Chip Chart

B. Sandoz Colors and Chemicals Corporation Color Chip Chart

C. U.S. Specialty Color Corporation Color Chip Chart

D. EAST/WEST DyeCom Color Chip Chart

E. CHRISTINA DEPAUL. "Table for Two," 1985. 12" X 12" X 15".

F. CHRISTINA DEPAUL. Detail- "Table for Two."

G. THOMAS SEABURY BROWN. Desk Lamp, 1986. Acrylic, delrin, florescent light source. 15" X 7" X 10".

H. THOMAS SEABURY BROWN. Salt and Pepper Shakers, 1986. Delrin, gold plated brass. 1 1/2" X 1 1/2" X 3".

I. DAVID TISDALE. Napkin Rings, 1986. 2 1/2" X 1" X 1 1/2". Photo: Rebekah Laskin.

J. DAVID TISDALE. Picnic Flatware, 1986. Left to right: 7" X 1", 8" X 1", 6 1/2" X 1 1/2". Photo: Bob Hansson.

K. DEBRA STONER."Sushi Plate and Chopsticks," 1986. 12" X 12" X 1/2".

L. ROWENA MORALES. Wall Piece, 1986-1987. Heat textured aluminum. 8" square. Photo: Rowena Morales.

M. CAROLE S. ROBINSON. "Reel Light #1.," 1986. Plastic. 12" diameter.

N. THOMAS SEABURY BROWN. Kitchen Knife, 1986. Sterling silver and stainless steel. 6" X 1/2" X 1".

O. CATHY STAHLMAN. "Volley Bottle" Perfume Bottle, 1986. Die formed aluminum, copper, silver. 4" X 2 1/2" X 5/8". Photo: John Veltri.

P. LAURA MARTH. "A Bundle of Nerve Endings" Perfume Bottle, 1987. Glass. 4" X 2". Appendages set on springs to vibrate when handling bottle.

Q. FUJIE BABA. "Nostalgia" Hat, 1987. 7" X 7".

R. SHARON NOVAK. "Conestoga Shield" Hat, 1987. Magic markers on aluminum. 24" X 11 1/2" X 2 1/2".

S. JULIA W.P. BAKER. "Fire Dancer" Hat, 1987. 3" X 7" X 8".

T. LENET GALLOWAY. "Toy Box" Hat, 1987. 5" X 7" X 7".

U. BARBARA HEINRICH. Twist-pins, 1984-85. Stainless steel springs. 1/2" diameter.

V. BARBARA HEINRICH. Twist-pins, 1984-85. Stainless steel springs. 1/2" diameter.

W. SYLVI HARWIN. "Zag" Neckpiece, Earrings, and Bracelet, 1987. Photo: Mehosh Dziadzio.

X. SYLVI HARWIN. "Window Series #2." Neckpiece, 1985. Sandblasted aluminum, photoetched silver, hinged. The triangular end piece is attached on steel springs and bobs with wearer's movement. 11" X 5" X 7/16". Photo: Mehosh Dziadzio.

Y. SYLVI HARWIN. "Flag of a Different Country" Necklace and Earrings, 1987. Sandblasted and resist dyed aluminum with ivory, silver, fire opal, cubic zirconium. necklace is kinetic. Photo: Mehosh Dziadzio.

Z. ANDREA PERKINS. Neckpiece, 1987. 8 1/2" X 4 1/2".

AA. ALEX CONNELL. Neckpiece, 1987. 8" X 4 1/2".

BB. JANET SERATTE. Brooch, 1986. Brass, porcelain. 3 1/2" square.

CC. JAN PETERS. "Alex #286." Brooch, 1986. Niobium, sterling silver. 4 7/8" X 1 5/8" X 3/16".

DD. JAN PETERS. "Folds, Angles, Curves" Brooch, 1985. Nickel silver. 1 5/8" X 2 3/4" X 1/4".

EE. ALEX CONNELL. "Fish" Pins, 1987. Larger pin-3 1/2" X 2 1/2". Smaller pins-2 1/4" to 2 1/2" X 2" to 2 1/4".

FF. CATHY STAHLMAN. "Bluegill" Fish Pin, 1985. Stainless steel wire, rubber. 4" X 2 1/8" X 1/8". Photo: Charles Lewton-Brain.

GG. CANDACE BEARDSLEE. Necktie Pin, 1986. Etched aluminum, brass piano hinges. 6 1/2" X 2" at base of tie. Photo: Eduardo Calderon.

HH. NANCY BARBOUR. Wall Sculpture. Etched copper. 9" X 4 1/2" X 3 1/2".

II. CANDACE BEARDSLEE. Bracelet, 1986. Etched aluminum, rubber center. 6" diameter. Photo: Eduardo Calderon.

JJ. MICHAEL EDWARDS. Bracelet, 1986. Ivory, colored wire. 4" diameter.

KK. KATHE TIMMERMAN. Bangles and Earrings, 1987. Sterling silver borders, gold leaf.

LL. LESLIE LEUPP. "Bracelet #WG487", 1987. 5" X 5 1/4" X 3/4". Photo: Leslie Leupp.

MM. LESLIE LEUPP. "Bracelet #NY 287" with stand, 1987. 5" X 5" X 4". Photo: Leslie Leupp.

NN. LESLIE LEUPP. "Bracelet #NY 287" without stand, 1987. 5" X 5" X 1". Photo: Leslie Leupp.

OO. DAVID LAPLANTZ. "Tombstone Series" Brooch, 1987. Painted aluminum, plastic mesh. 3" dia.

PP. DAVID LAPLANTZ. "Island Under Wraps" Brooch, 1987. 2 1/4" dia.

QQ. DAVID LAPLANTZ. "Bad Moon Rising" Brooch, 1986. 3" X 2".

RR. DAVID LAPLANTZ. "Shovel Brooch", 1986. 3" X 2".

SS. DAVID LAPLANTZ. "Architectonic" Brooch, 1987. 3" X 3".

TT. DAVID LAPLANTZ. "Tombstone Series" Brooch, 1987. Painted aluminum. 3" X 2".

ALL PHOTOS, UNLESS OTHERWISE STATED, CONTAIN FABRICATED OR CONSTRUCTED ANODIZED AND DYED ALUMINUM ELEMENTS.

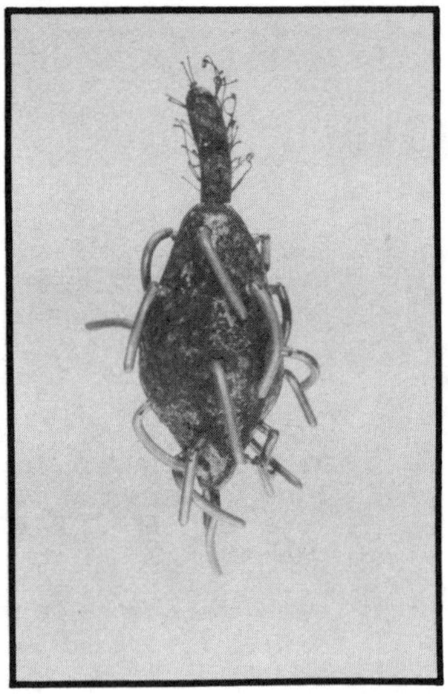

LAURA MARTH. "A Bundle of Nerve Endings" Perfume Bottle, 1987. Monofilament. 4 1/2" X 2". Appendages set on springs to vibrate when handling bottle.

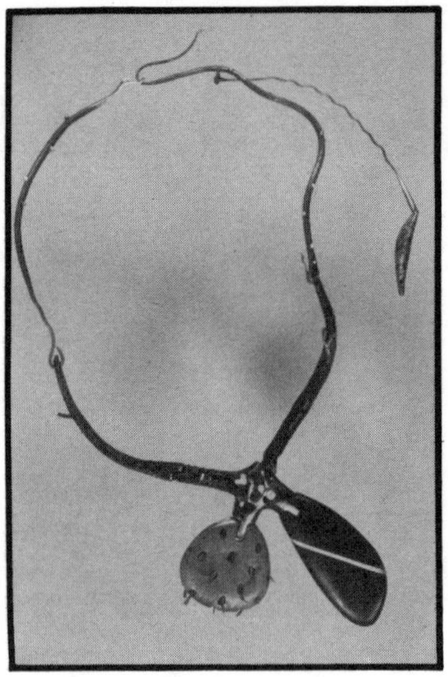

LAURA MARTH. "Indigenous Species" Neckpiece, 1987. Carved aluminum.

CANDACE BEARDSLEE. Bow Earrings, 1986. Etched aluminum. 3" X 1". Photo: Eduardo Calderon.

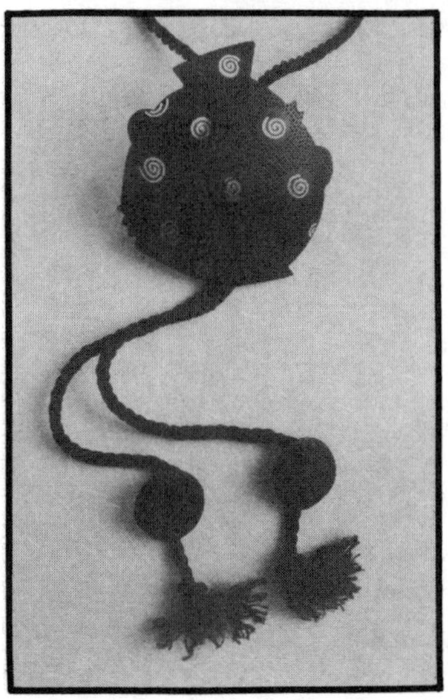

CANDACE BEARDSLEE. Necklace or Stringtie, 1986. Etched aluminum, cording. 3 1/2" diameter. Photo: Eduardo Calderon.

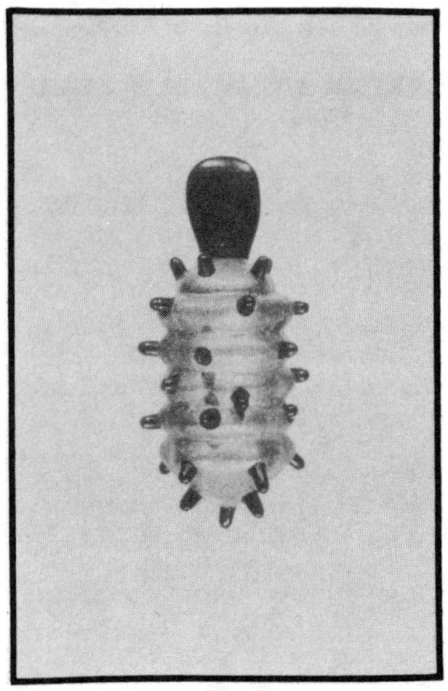

LAURA MARTH. "A Bundle of Nerve Endings" Perfume Bottle, 1987. Glass. 4 1/2" X 2".

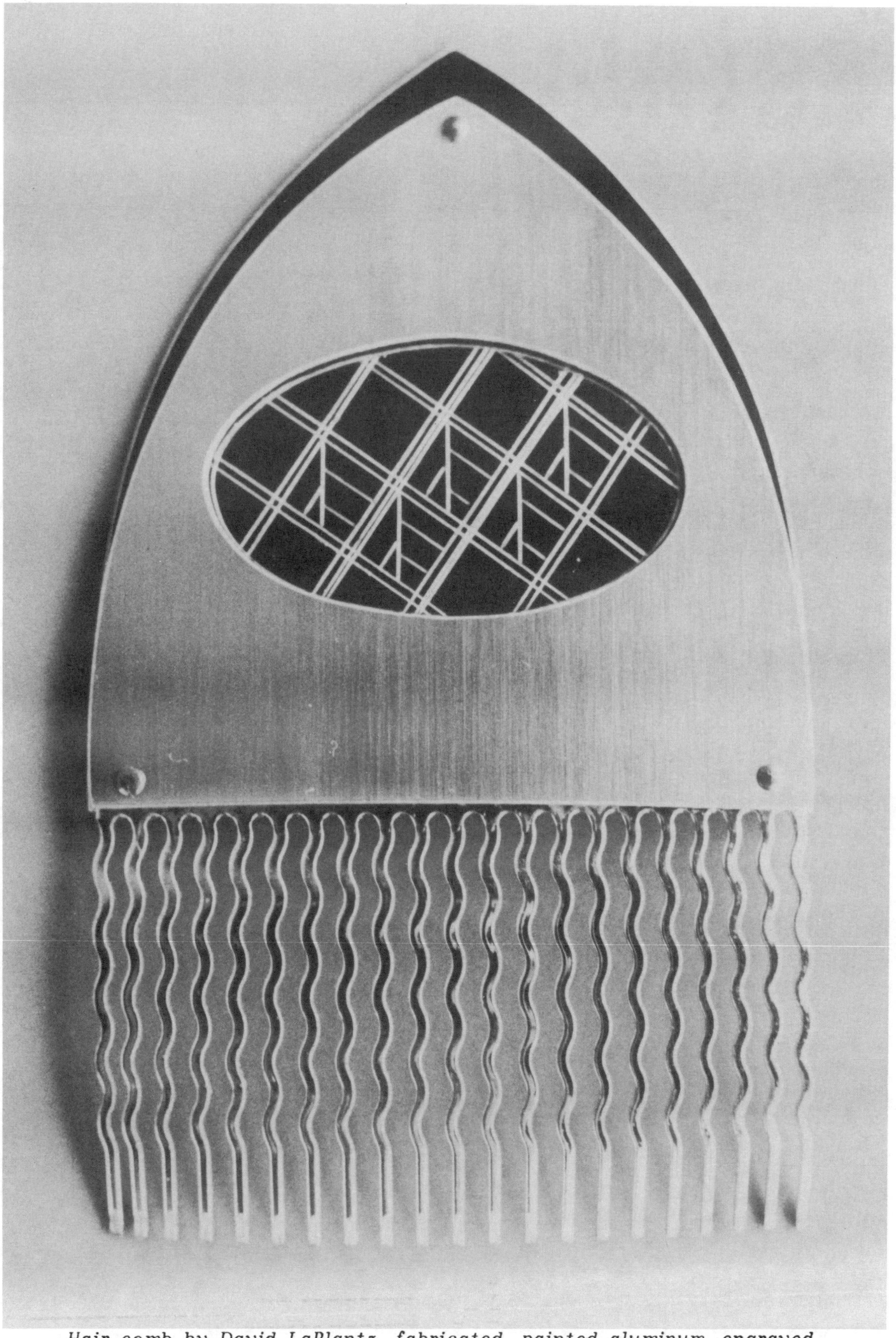

Hair comb by David LaPlantz, fabricated, painted aluminum, engraved, 3 1/2" X 2 1/2", 1982.

TESTING

Anodizing Electrolyte
For Dye Content
Thickness of Anodic Oxide Layer
For pH
Current Flow and Voltage

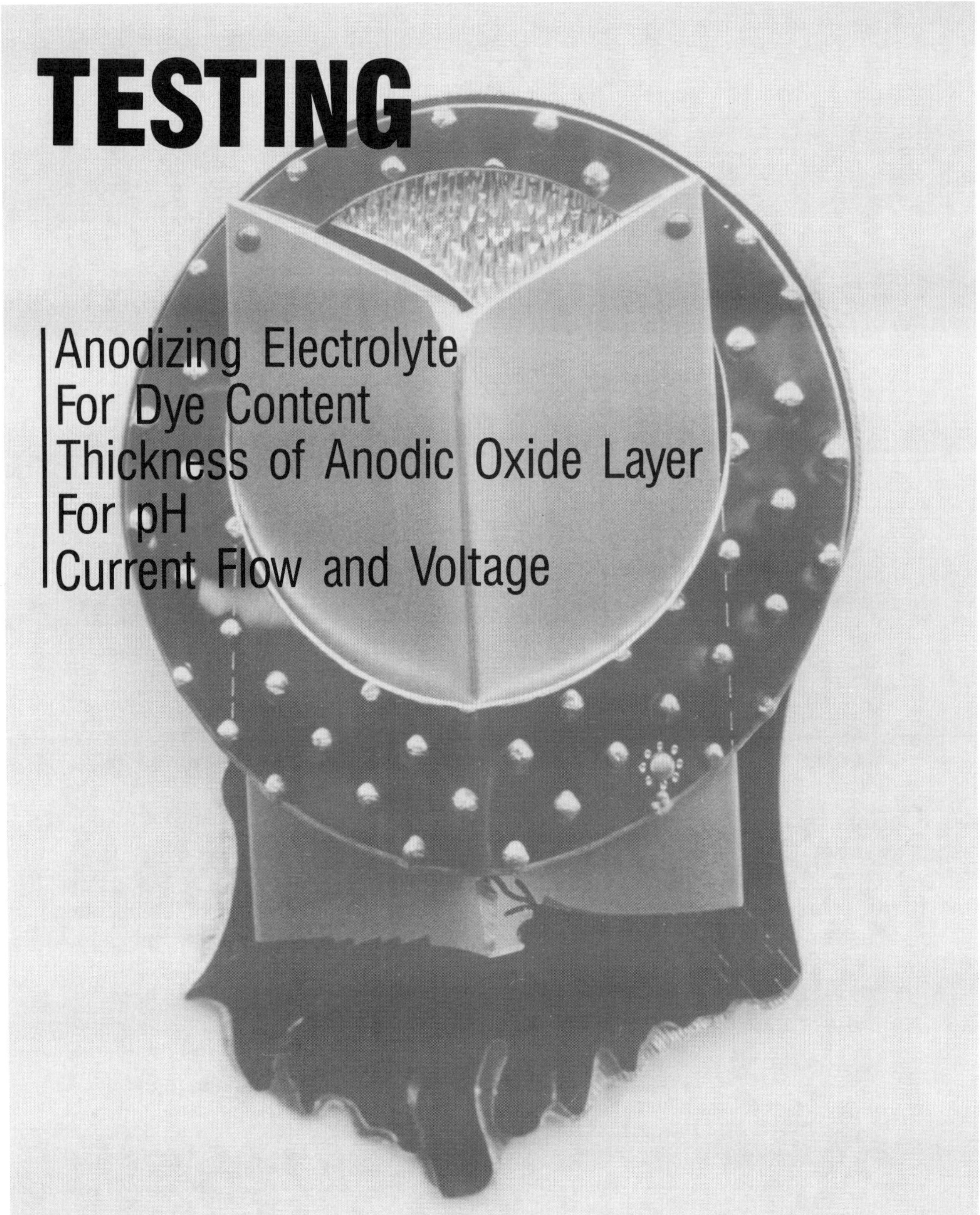

Brooch by David LaPlantz, fabricated anodized and painted aluminum, engraved, 3" X 2", 1987.

Anodizing Electrolyte

The following directions will explain how you can analyze your anodizing bath and caustic etch solution to provide maximum results. The analysis set as seen in the illustration cost about $96.00 including shipping from Chicago to California.

> Titration Explanation
> of solution for aluminum content and sulfuric acid content:
>
> I recommend, if this is a process you decide to do for yourself, that you consult your local chemistry teacher, a friend in a chemistry class for a quick hands-on demonstration of how to do this technique. Photos and/or drawings of how to do this are difficult to decipher.

It is available from:
KOCOUR COMPANY
4800 South St. Louis Avenue
Chicago, Illinois 60632
312-847-1111

KOCOUR ANALYTICAL SET, MODEL STS
PROCEDURE #279
OPERATING DIRECTIONS
ANALYSIS OF SULFURIC ACID ANODIZING SOLUTIONS

SULFURIC ACID:

1. Using a 5 ml pipet, transfer a 5 ml sample of the anodizing solution to a 125 erlenmeyer flask. Using a 25 ml graduated cylinder, add about 25 ml of water to the flask and 10 ml of Solution n-31. Now add 10 drops of PTH Indicator.

2. Fill the buret to top with Solution N-76 and bring the solution level down to the zero mark. Slowly add solution from the buret to flask, swirling the flask at the same time. Continue this until a permanent pink color is obtained. The buret reading at this point, multiplied by 2, is equal to ounces by weight per gallon of concentrated commercial sulfuric acid in the anodizing solution.

ALUMINUM CONTENT:

3. Using a 5 ml pipet, transfer a 5 ml sample of the anodizing solution to the 125 erlenmeyer flask. Using a 25 ml graduated cylinder, add about 25 ml of water to the flask. Then add 10 drops of PTH Indicator.

4. Repeat Paragraph # 2 above, and record the buret reading. From this reading subtract the buret reading obtained in Paragraph # 2 and multiply the difference by 2.76. The result is grams/liter of aluminum in the anodizing solution.

EXAMPLE:
Buret reading obtained in Paragraph # 4 is 18.5 ml.
Buret reading obtained in Paragraph # 2 is 14.0 ml.
Then: 18.5 minus 14.0 equals 4.5 X 2.76 equals 12.4 grams/liter of aluminum.

IMPORTANT NOTES:
Do not leave Solution N-76 in the buret for extended periods of time, otherwise it will freeze the stopcock. Wash buret immediately after each use.

Always rinse the buret out with distilled water after each use. Before filling, rinse it with a small amount of the solution you are going to fill it with. This prevents residual water or the previous solution from diluting the filling solution.

The buret reading obtained in Paragraph # 2, multiplied by 0.84 equals % by volume of sulfuric acid.

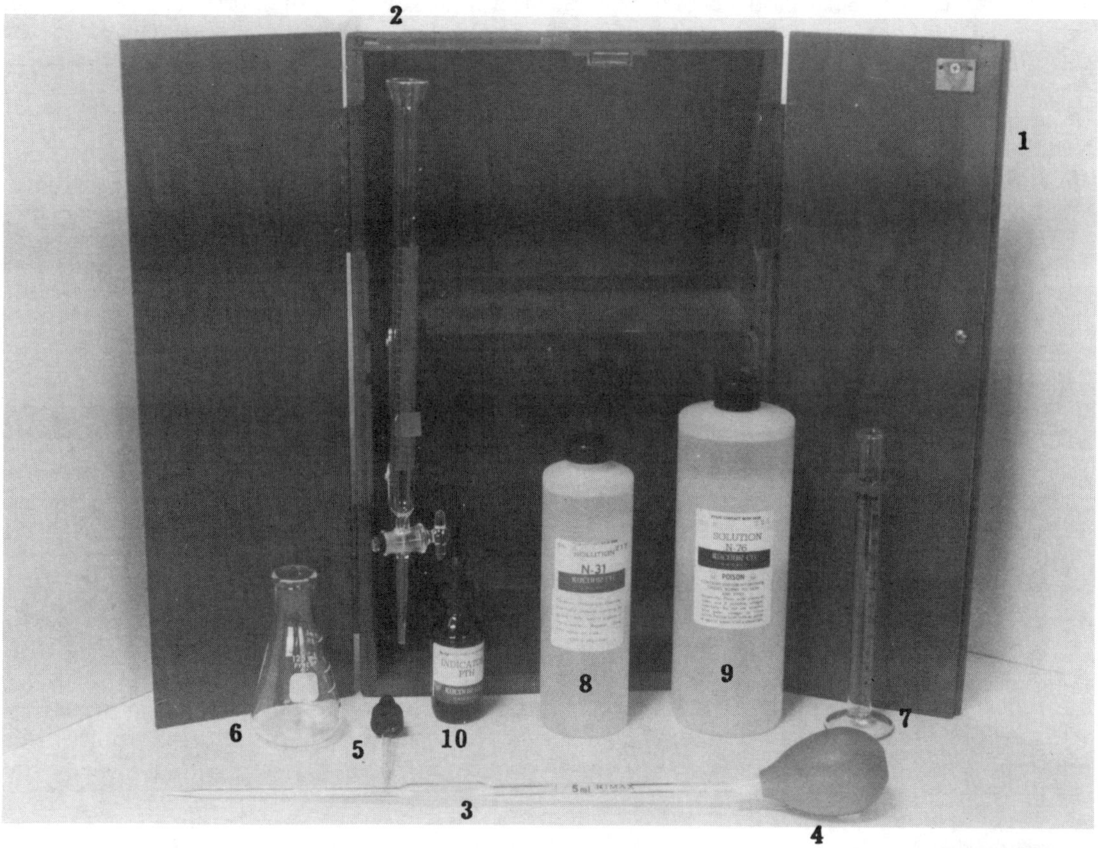

KOCOUR ANALYTICAL TEST SET MODEL STS
CONTENTS FOR PROCEDURE #279

1. Cabinet with hardware and buret holder
2. Buret, 25 ml. funnel top, replaceable plug, 15"
3. Pipet, 5 ml volumetric
4. Rubber bulb, 1 oz.
5. Dropping pipet, stopper type
6. Erlenmeyer flask, 125 ml.
7. Graduated cylinder, 25 ml.
8. Bottle, solution N-31 (500ml.)
9. Bottle, solution N-76 (1000 ml.)
10. Bottle, indicator PTH (50 ml.)

Bottle of 1N, sulfuric acid **not shown**, since it is not part of this testing set. Purchase this solution from a chemical house.

CONVERSION TABLE—5 to 25 per cent by weight of sulfuric acid. 10 to 20 per cent suggested proportions for better anodizing results.

%	grams/LITER	Avoir. OZ/GALLON
5	51.6	6.9
6	62.3	8.3
7	73.2	9.8
8	84.2	11.2
9	95.3	12.7
10	106.6	14.2
11	118.0	15.8
12	129.6	17.3
13	141.4	18.9
14	153.3	20.5
15	165.3	22.04
16	177.5	23.67
17	189.9	25.32
18	202.4	26.99
19	215.0	28.67
20	227.9	30.39
21	240.9	32.12
22	254.1	33.88
23	267.4	35.65
24	280.9	37.45
25	294.6	39.28

CAUSTIC CONTENT

These instructions are courtesy of Bud Bolmer, my fantastic, technical advisor. Thanks, Bud!

5. Using a 5 ml pipet, transfer a 5 ml sample of the caustic solution to a 125 ml erlenmeyer flask. Using a 25 ml graduated cylinder, add about 25 ml of water. Fill the clean 25 ml buret to top with 1N sulfuric acid (standard solutions of sulfuric acid are available from chemical supply houses, or do you know a friendly chemistry teacher?) and bring the solution level down to the zero mark. Slowly add solution from the buret to flask, swirling the flask at the same time. Continue this until the first, permanent cloudy precipitate forms.

6. Record the volume of sulfuric acid used as volume "A". Now add 10 drops of PTH indicator, and continue titrating with 1N sulfuric acid until the pink color disappears. Record the total volume of sulfuric acid used as volume "B".

7. Calculations:
   Free caustic in grams/liter = A X 8
   Dissolved aluminum in grams/liter = (B - A) X 5.4

For Dye Content

The following information has been kindly provided and permission given to reprint in this book by:
Sandoz Colors and Chemicals
4000 Monroe Road
Charlotte, North Carolina 28205
704-372-0210

Since the dye concentration, temperature, pH, and immersion time are specified and the dye temperature and time are physically controlled, only the dye content and pH must be periodically evaluated.

The pH should be checked daily using a pH meter and maintained within recommended limits. The dye content should be determined at least once each month or more frequently depending upon the usage of the bath.

Determination of Dye Content

1. Visual Methods

A. Dilute a 10 ml. sample of the working dye bath to 500 ml. using distilled water and a volumetric flask. (For highly concentrated dye baths such as blacks use 5 ml.) Pour 25 ml. of this solution into a 50 ml. comparison tube and compare visually with an equally diluted, freshly prepared standard dye solution. The depth of color in the 2 tubes is compared by viewing against a white background. Dilute the standard solution by additions of water until the 2 solutions appear to be at equal concentrations. Measure the final volume of the diluted standard and calculate the concentration of the working bath using the formula below.

$$CW = \frac{25\ Cs}{V}$$

CW = Concentration of Working bath (g/l)
Cs = Concentration of Standard Solution (g/l)
V = Final Volume of diluted Standard in ml.

B. Another visual method, while not as accurate as that given above, but sometimes useful for a rough estimate of strength, is to pour a sample of working solution and a standard solution side by side down a piece of filter paper held vertically. After the paper has dried it is possible to estimate the dye content by the depth of coloration.

2. Instrumental Methods

For those who have access to color measuring equipment such as a spectrophotometer, details of instrumental methods will be supplied on request by Sandoz at the above address.

"EFFECTIVE" DYE CONTENT

It should be pointed out that the preceding methods measure the amount of dye present in the bath which may be different from the amount of dye actually "effective" in dyeing the oxide film. The difference between the total dye present and the effective strength may be caused by the presence of sulfate, aluminum salts, or other contaminants that restrict the dye take-up. The effective dye strength may be determined as follows:

An aluminum panel is anodized, dried and then cut into equally sized pieces for small scale laboratory dyeings as in a test tube. Identical samples are dyed under the same conditions in the working bath and a freshly prepared standard solution. The dye absorbed in the coating is stripped using the procedure below.

Each strip is placed in a small test tube and 12 ml. of a stripping solution (the stripping solution is made up from 7.5 grams tartaric acid plus 12 grams sodium hydroxide in 1 liter of deionized water) is added and the test tube shaken for exactly 2.0 minutes. This is repeated for a total of 3 treatments for each strip. The 3 washings are combined and diluted to 50 ml. using a volumetric flask. The strength of the dye in these washings is then compared using any of the previously given methods. If the "effective" dye strength is 75% or lower than the standard or desired concentration, it is usually not economical to reinforce the bath by additions of fresh dye.

SEALING SOLUTIONS

Usually the only control necessary for the sealing tank is the pH, since the time, temperature, and nickel acetate content are usually specified. The pH should be measured daily and adjusted as required. The correct pH range is 5.4-6.0. Depending on the amount of dye leaching which occurs during the sealing process and contamination through drag-in, etc., the sealing solution should be changed every three to five weeks to insure trouble free operation.

Ordinarily it is not necessary to analyze the metal content of the sealing solution. When required, the classic gravimetric methods for this analysis are suitable. The following colorimetric method, although not as precise, is in most cases quite satisfactory and more convenient for routine use.

A Visual Colorimetric method for the Estimation of Nickel Acetate in sealing solutions.

Prepare standard solutions containing 3, 4, 5, and 6 grams/liter of nickel acetate as follows:

Weigh out 12.5 g of nickel acetate and dilute to 250 ml. of water containing 10 ml./liter of acetic acid. Take aliquots (solution divided evenly) of this solution as follows and dilute to 100 ml. yielding the following "standards":

6 ml. diluted to 100 ml. = 3 g/l nickel acetate
8 ml. diluted to 100 ml. = 4 g/l nickel acetate
10 ml. diluted to 100 ml. = 5 g/l nickel acetate
12 ml. diluted to 100 ml. = 6 g/l nickel acetate

Take a sample of sealing bath (Note 1) and 10 ml. of each of the "standards" and dilute to 1 liter. Take 10 ml. aliquots into 250 ml. beakers, add 5 drops ammonium hydroxide (28%) and 5 drops of 1% Dimethylglyoxime (Note 2). Mix and allow to stand for 5-10 minutes for red color to develop. Dilute to 200 ml. using isopropyl alcohol and compare working bath with the standards in similar containers.

Note 1 - Expected Nickel Acetate Content	Sample Size
3 - 6 grams/liter	10 ml.
1.5 - 3 grams/liter	20 ml.
0.75 - 1.5 grams/liter	30 ml.

Note 2 - Dissolve 10 g Dimethylglyoxime in 1 liter of isopropyl alcohol.

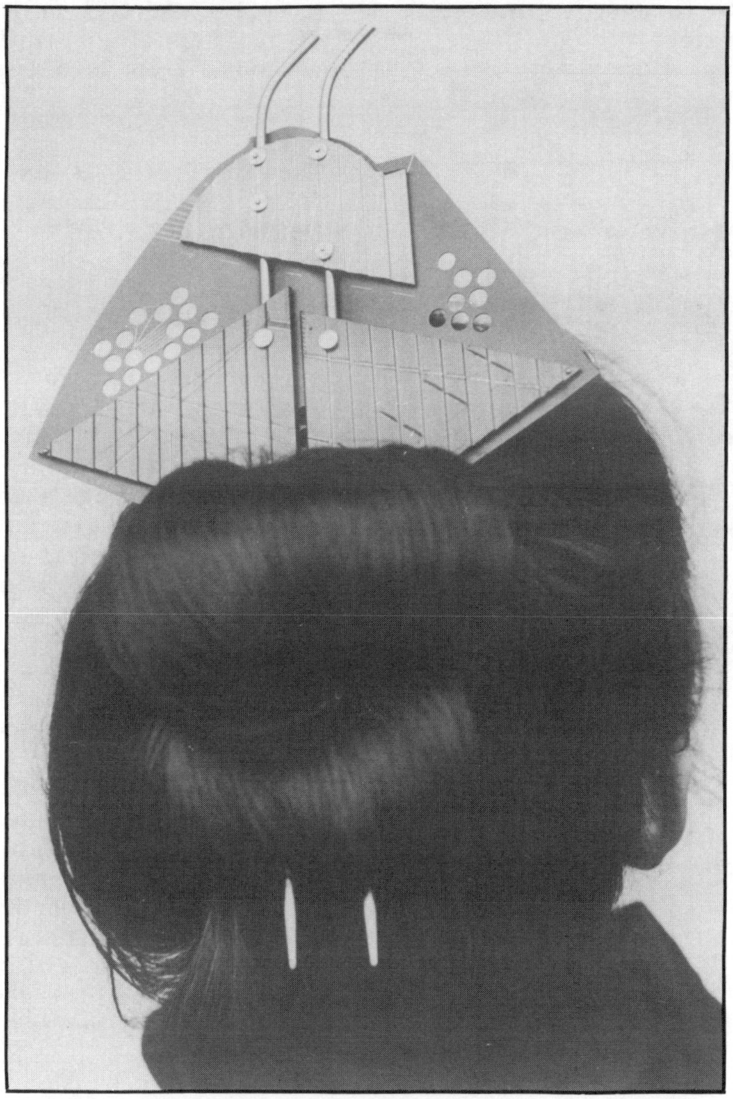

"Anodized Aluminum Hairpiece" by David LaPlantz, fabricated, 9" X 4 1/2".

Thickness of Anodic Oxide Layer

Once you start anodizing aluminum, you may be interested in knowing the thickness of the anodic coatings. I purchased a special thickness testing instrument/gage for just this purpose. It is small, light, simple to operate and quite accurate.

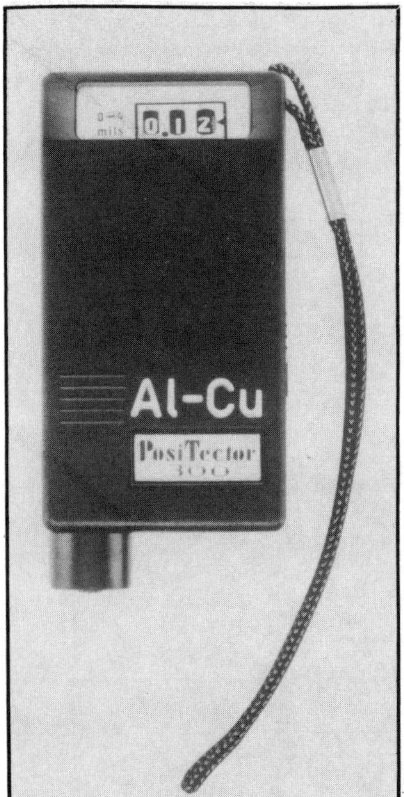

PosiTector 300 with handy carrying strap.

Thickness gages are not cheap! Prices start at about $1000.00 and go up. The instrument shown on this page is a PosiTector 300, retailing for about $1000.00 It measures the anodic oxide coating in hundredths (.00001") and (.01 mil). The digital readings indicator measures from 0 to 4 mils.

The PosiTector is available from:
Defelsko Corporation
P.O. Box 676
Ogdensburg, NY 13669
315-393-4450

The overall dimensions of the PosiTector are 2" wide X 4 1/2" long X 1" thick and it comes with a handy wrist cord. A protective leather case with a sturdy belt attachment houses and protects the gage when it's not in use. Easy to read instructions and test papers are included.

Measuring the thickness of the anodic coating on an anodized aluminum piece is quite easy. The operation can be done comfortably with one hand. First remove the protective cap from the probe. The removed cap may be used as an additional support between the gage and the surface to be measured.

PosiTector 300 with protective cap covering the probe.

PosiTector 300 measuring the oxide film on an aluminum sheet. The cap is used to steady the probe.

On flat pieces simply place the PosiTector with its probe on the aluminum surface and read the digital indicator. Be certain to hold the gage steady and perpendicular to the metal surface. On curved stock or pipe/rod, the "V" grooves in the bottom of the probe should be aligned on the curved surface. The measurement is then read on the digital indicator.

Due to the cost of these gages, I would highly recommend that you make friends with your local anodizer. Perhaps they will test your pieces periodically. This way you know you are getting the correct anodic coating thicknesses for the time, temperature, and other conditions that make up your particular anodizing technique.

Some of you may be interested in knowing, technically, how the PosiTector thickness gage works. Personally, I just measure the thickness of my anodic coatings and go along my way merrily.

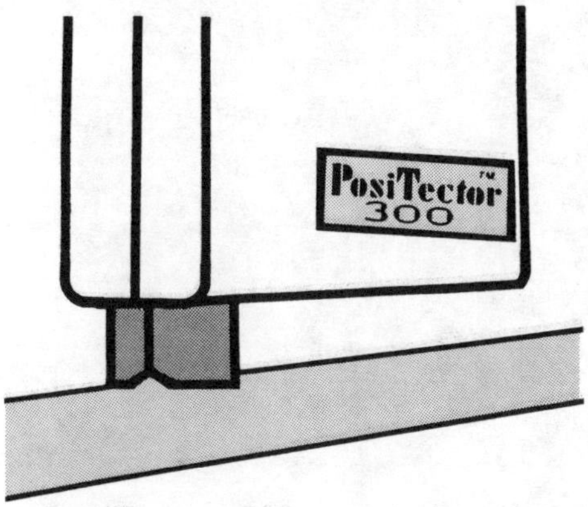

PosiTector 300 measuring the oxide film on a rod or tubing.

The following technical explanation of how the PosiTector works has been kindly supplied by the Defelsko Corporation.

In contrast to existing coating thickness gages the Positector uses a compensation process. This means that in addition to the main measuring probe a second probe has been installed in the interior of the gage in a mirror-like position. This secondary probe also remains in thermal contact with the primary probe. By means of a servo motor via a micrometerspindle the reference plate is moved to a given distance opposite the secondary probe.

When the distance between the reference plate and the compensating probe equals the thickness of the layer to be measured, the voltage signal at both probes are also equal. Through an electronic control the micrometerspindle comes to a stop. The mechanical position of the micrometerspindle is a direct measurement of the thickness of the layer. The measurement result is displayed digitally. The measurement signal that is displayed is not a processed value. The linearity throughout the whole measuring range from 0-1800 um occurs automatically and corresponds to the accuracy of the micrometerspindle. All error causing influences that have been mentioned above influence both probes and are compensated. Thus, they do not influence the measurement result. Therefore, the various equalizing and adjustment knobs and the inconvenient routines of calibration become unnecessary. Because the measuring probe is integrated in the apparatus whose total size is that of a cigarette pack, the PosiTector motto is: Touch and read.

This ml to um conversion chart is located on the back of the PosiTector. This chart shows the relationship of mils to um.

mil to um Conversion Chart
1 mil = 25.4 um = .0254 mm

mils/um

.1= 2.5
.2= 5.1
.3= 7.6
.4=10.2
.5=12.7
.6=15.2
.7=17.8
.8=20.3
.9=22.9
1.0=25.4
1.1=27.9
1.2=30.5
1.3=33.0
1.4=35.5
1.5=38.1
1.6=40.6
1.7=43.2
1.8=45.7
1.9=48.3
2.0=50.8

2.2=56
2.4=60
2.6=66
2.8=71
3.0=76
3.2=81
3.4=86
3.6=91
3.8=97
4.0=102
4.2=107
4.4=112
4.6=117
4.8=122
5.0=127
5.2=132
5.4=137
5.6=142
5.8=147
6.0=152

For pH

pH PAPER

pH paper is a quick, inexpensive, fairly accurate way to measure pH. The following is one company's pH testing procedure.

1. This is an example of the pH paper container I use in my studio. This particular company sells a two part plastic container, with sample pH color charts covering pH readings from 0 - 13.

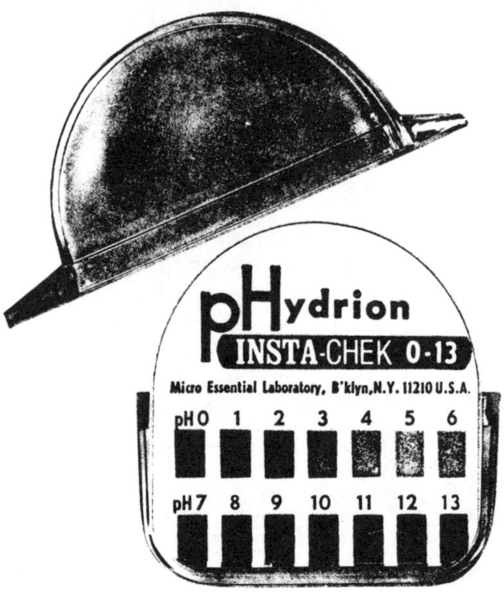

2. To test for pH, start by removing the plastic lid from the base.

3. Inside you will find two golden colored, foil wrapped coils of pH paper.

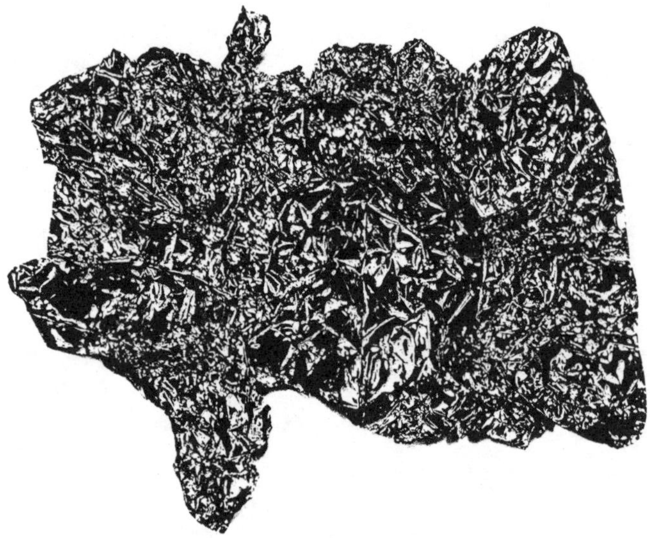

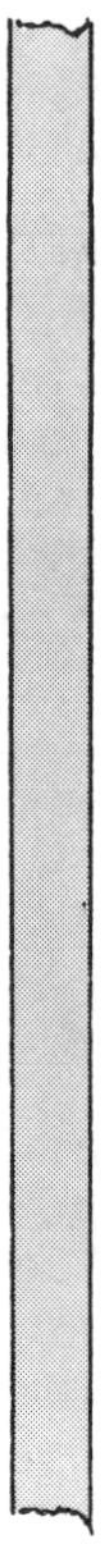

4. Open the foil, unroll the coil, and cut off about four inches of pH paper.

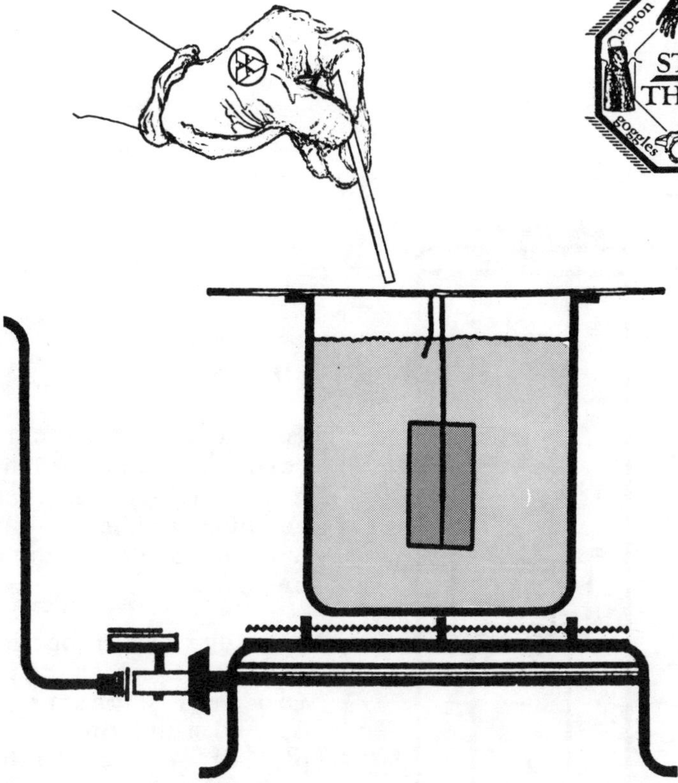

In this illustration, I am testing the pH of the sealing bath. The solution is made up from distilled/deionized water or sealing salts.

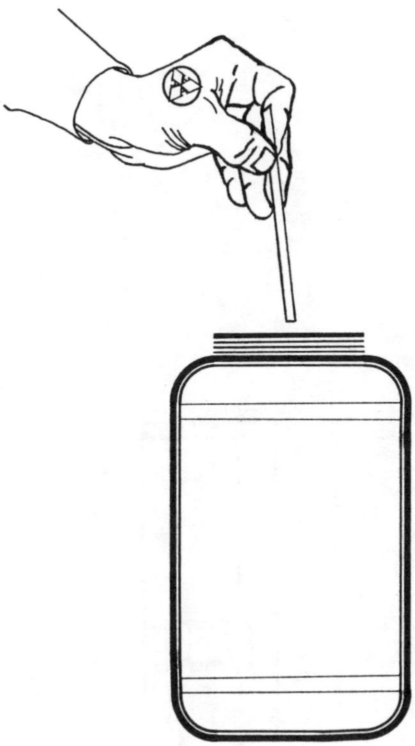

5. Lower the pH paper into the solution you wish to test. As the pH paper comes in contact with the solution, you should begin to see a reaction, that is, the paper will change color.

In this illustration, I am testing the pH of distilled/deionized water prior to mixing the dye powder. If you attempt to test the pH of the dye and distilled/deionized water, the dye will color the paper and obscure the reading. A pH meter can test a colorful dye bath.

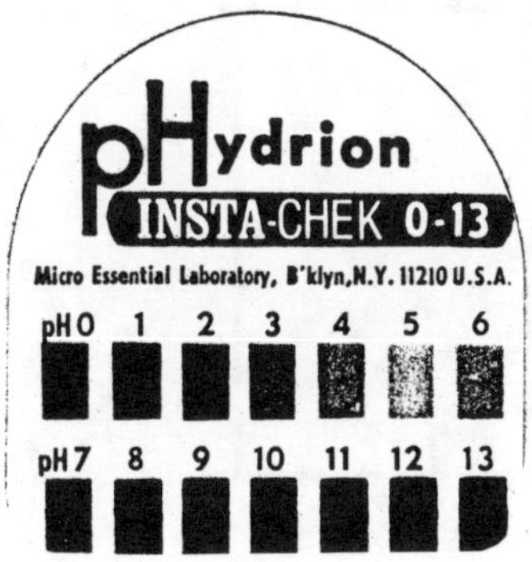

6. Remove the pH paper and compare it with the chart. Match your color with the chart's colors as closely as possible. This is your pH reading. If you have a doubt, cut off another four inch length and repeat the entire process.

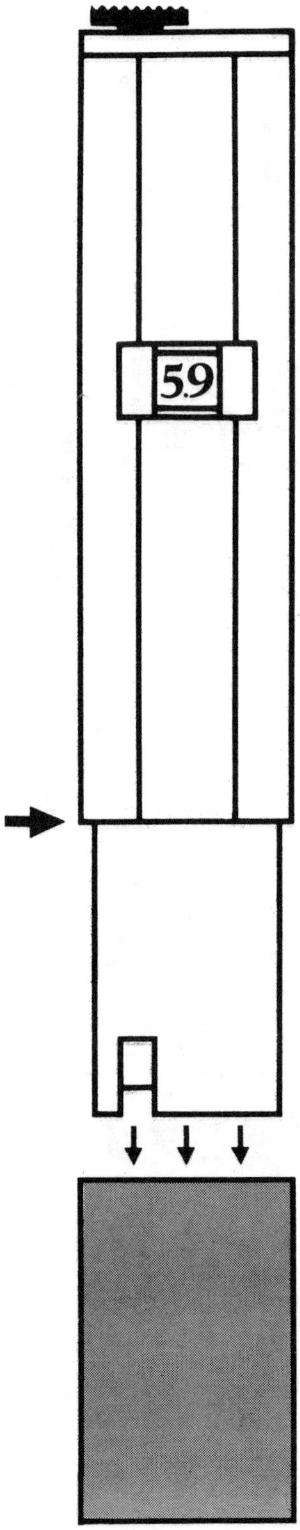

pH METER

To get a more accurate pH reading, you may wish to buy a pH meter. Accurate pH readings are/may be necessary for your dyes and sealing bath.

Each dye has an optimum pH. The three major dye manufacturer's charts in the Appendix list the appropriate pH for each of their dyes. If you keep your dyes at the appropriate pH, and if you use the suggested proportions, your color will match those of the manufacturer.

In sealing, the correct pH is vital. Optimum pH for a water seal is 5.7 and for nickel acetate (Sealing Salt AS) 5.3-5.5. Check the pH daily of the sealing bath and adjust it accordingly. Too low a pH will cause leaching out of the dye.

The drawing to the left is an actual size illustration. The following shows how simple the pH meter is to use.

The switch on the top left turns on the pH meter by sliding to the left. The window with the number 5.9 is where you will take the pH reading. The cap has been removed so the electrode can be immersed. The arrow, on the left side, is the uppermost level the meter should be immersed into a solution.

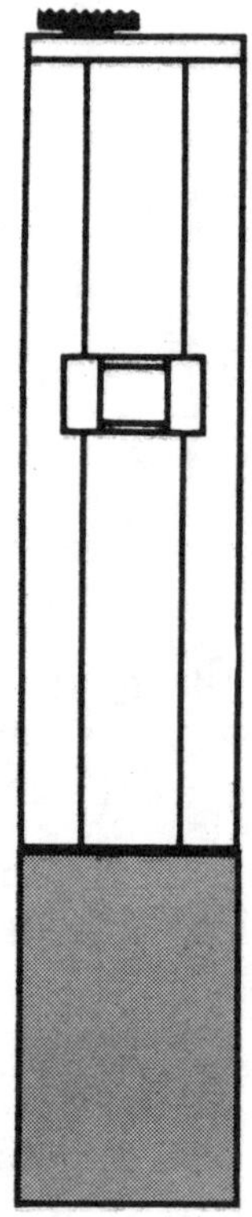

This drawing illustrates the pH meter with the protective cap in place. The switch is turned "off" and the meter reading window has no numerical display.

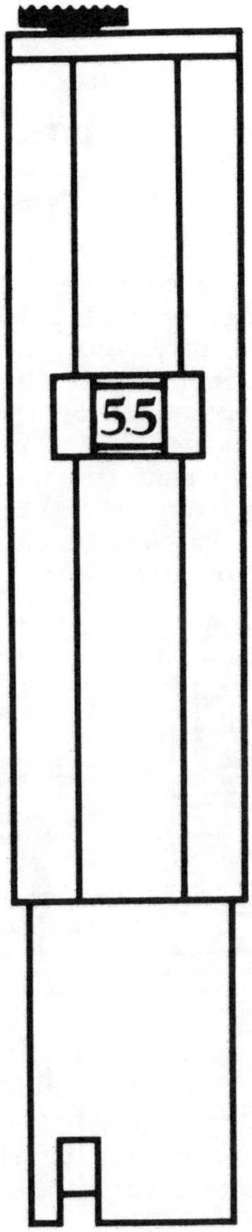

This particular pH meter is a Hanna Model 0624-00, pHep-pH Electronic Paper. The range of this meter is 0.0-14.0 pH, with an accuracy of -0.2 pH. It is available from Presto-Tek and costs about $40.00. See Supply Sources chapter.

To recalibrate the pH meter, a buffer solution with a specific pH value is required. Buffers are available from Presto-Tek and rated for either a pH of 4, 7, or 10. The buffer rated at a pH of 7 is appropriate for checking this meter. It is wise to periodically recalibrate your pH meter.

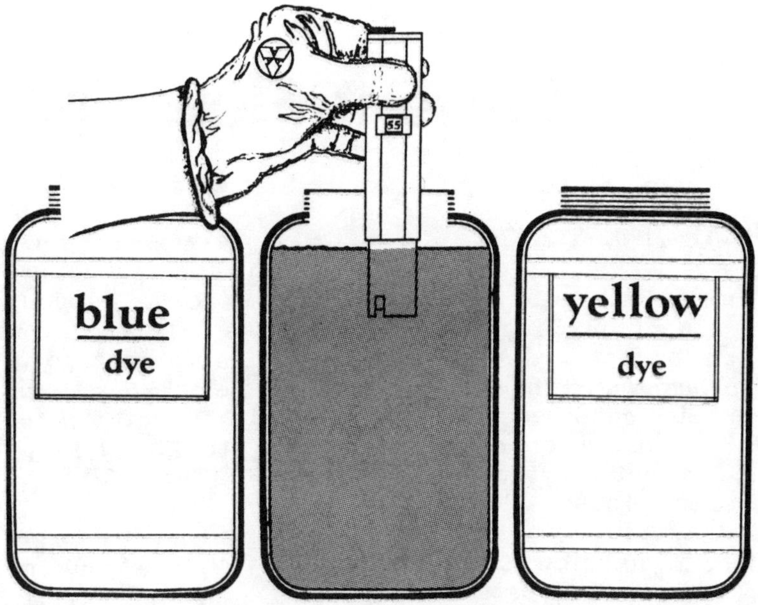

An illustration of the pH meter checking cold dye.

Acetic acid lowers the pH and sodium hydroxide will raise it.

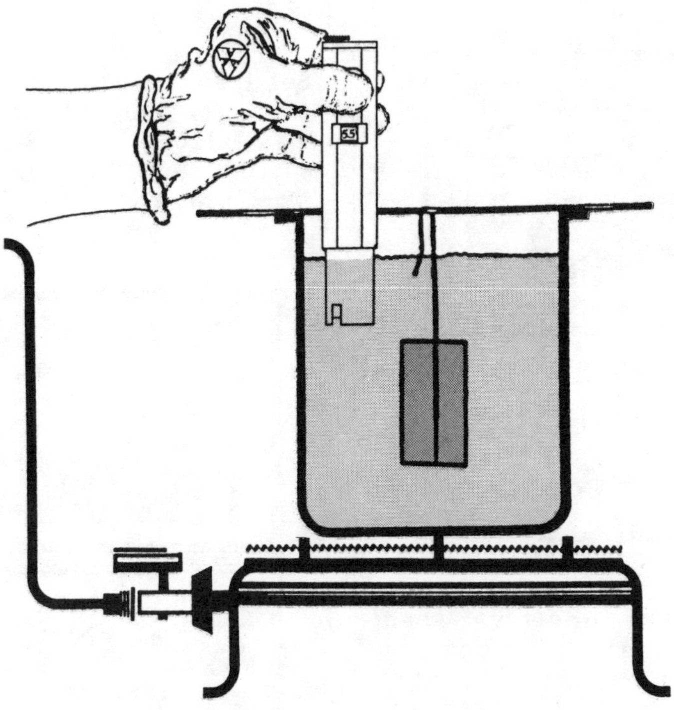

An illustration of the pH meter checking the hot sealing bath.

Current Flow and Voltage

MULTIMETER/MULTITESTER

A multimeter/multitester tests if an aluminum piece is anodized or not. It also tests voltage while the aluminum is being anodized. This meter is capable of doing other tests, but they do not relate to anodizing.

The illustration below is of a small, inexpensive, battery operated multimeter/multitester available on the market today. Prices range from about $12.00 to $16.00. Watch for sales at hardware stores and other places like Radio Shack. Larger and more expensive models are also available. This model, however, should be enough to get you started. Trade up only when you need the next better model.

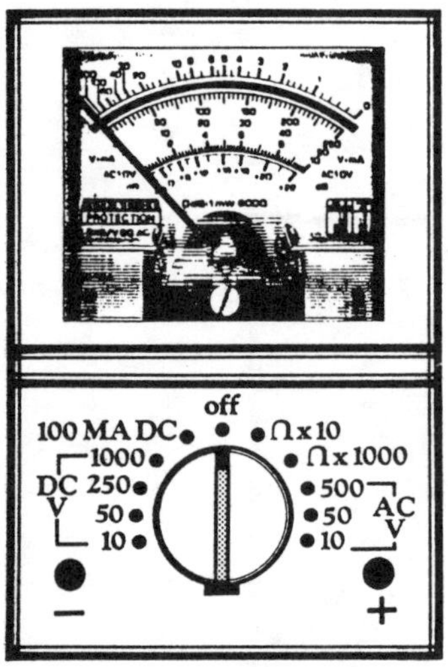

In the illustrations on this page, the positive (red wire/probe) and negative (black wire/probe) probes have been eliminated. You will, however, see them when specifics tests are being discussed later.

The multimeter "on/off" switch is located below the meter window. As you can see in the drawing, the "off" position is straight up or at the 12 o'clock position. If the switch is rotated to the left or to the right, the meter is turned "on."

It is very easy to forget to turn "off" the multimeter when you are done testing.

TEST FOR CONTINUITY

This is a test to see if the aluminum is anodized or not. Remember, an anodized piece of aluminum will not pass a current. In other words, anodized aluminum is an insulator, as mentioned earlier. An un-anodized piece of aluminum will pass a current.

Set the multimeter switch on the **●Ωx1000** setting. You may also use the **●Ωx10** setting, but the **●Ωx1000** is more sensitive.

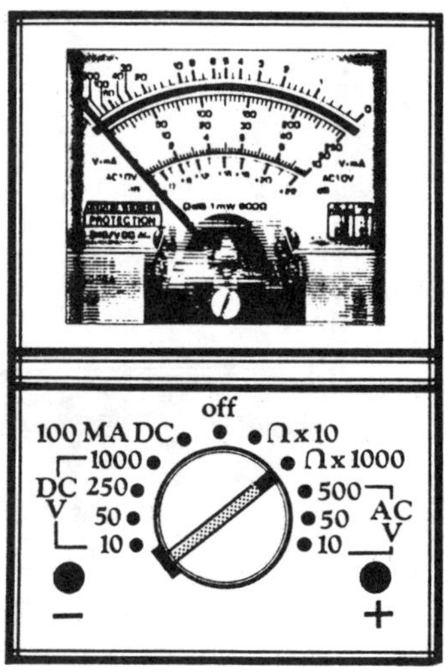

Place the positive (red) and the negative (black) probes on two different places on the aluminum. If the piece of aluminum is not anodized, or not anodized very well, the indicator arm will move to the extreme right on the scale. Be sure to check 2-3 places on the piece.

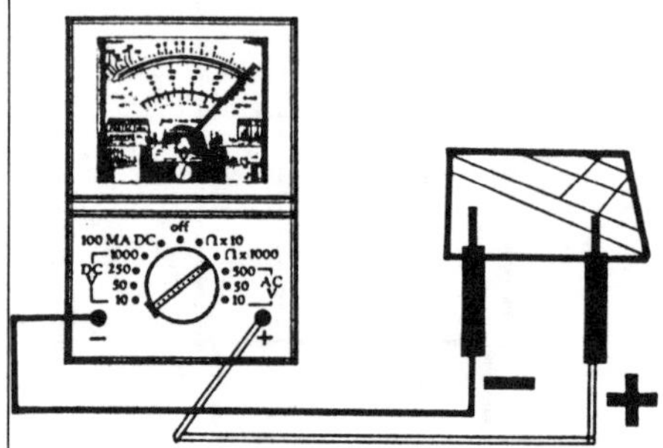

If the piece is anodized, no reaction will take place. The indicator arm will remain at the zero setting on the left side of the scale.

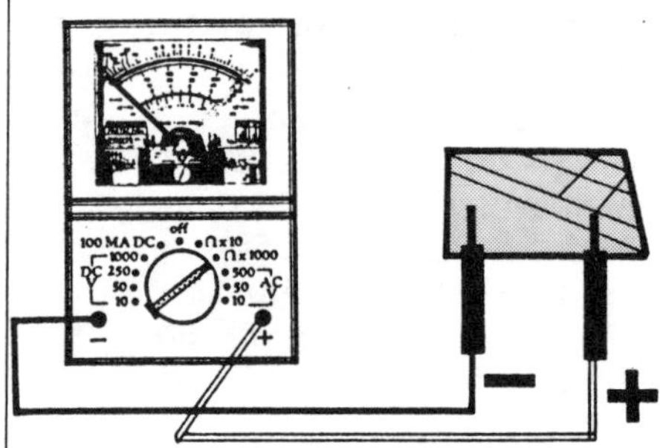

Turn the switch to the "off" position when finished.

TEST FOR VOLTAGE

This is a test to determine the voltage across the electrodes of the electrolytic cell. In other words, testing the voltage across the bus bar and the lead cathode(s) -or- is there enough voltage for proper anodizing?

Set the multimeter switch to the 50 DC voltage position or 9 o'clock. This is because you are measuring voltage above 10 volts and below 250 volts. Observe the middle scale (0-50 volts).

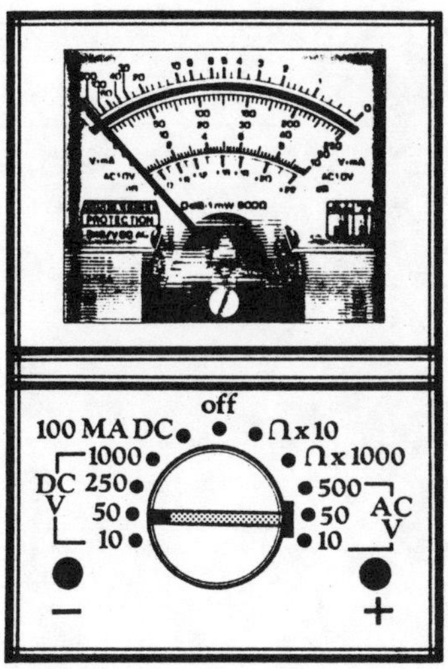

Place the positive (red) probe on the bus bar or aluminum rack while you touch the negative (black) probe to the wire that connects the two lead cathodes. If you only have one lead cathode touch the lead cathode only.

The meter reading should be between 12-16+ volts, with the probability of 14 volts being the standard voltage reading, if you are using a battery charger. More expensive power supplies will create higher voltages (18-20 volts).

Take a second reading by keeping the positive probe on the bus bar/rack while you move the negative probe to the other lead cathode. This is to insure that both cathodes are receiving the same voltage. Sometimes the connection wires will fail, for one reason or another, to carry the proper current. Perhaps one end of the wire has become disconnected or lead sulphate has built up between the wire and the lead. This build-up will reduce or stop current flow. When this happens no anodizing will take place. It is wise to know that all of your equipment is working properly each time you anodize. Time wasted is time lost forever.

Turn the switch to the "off" position when finished.

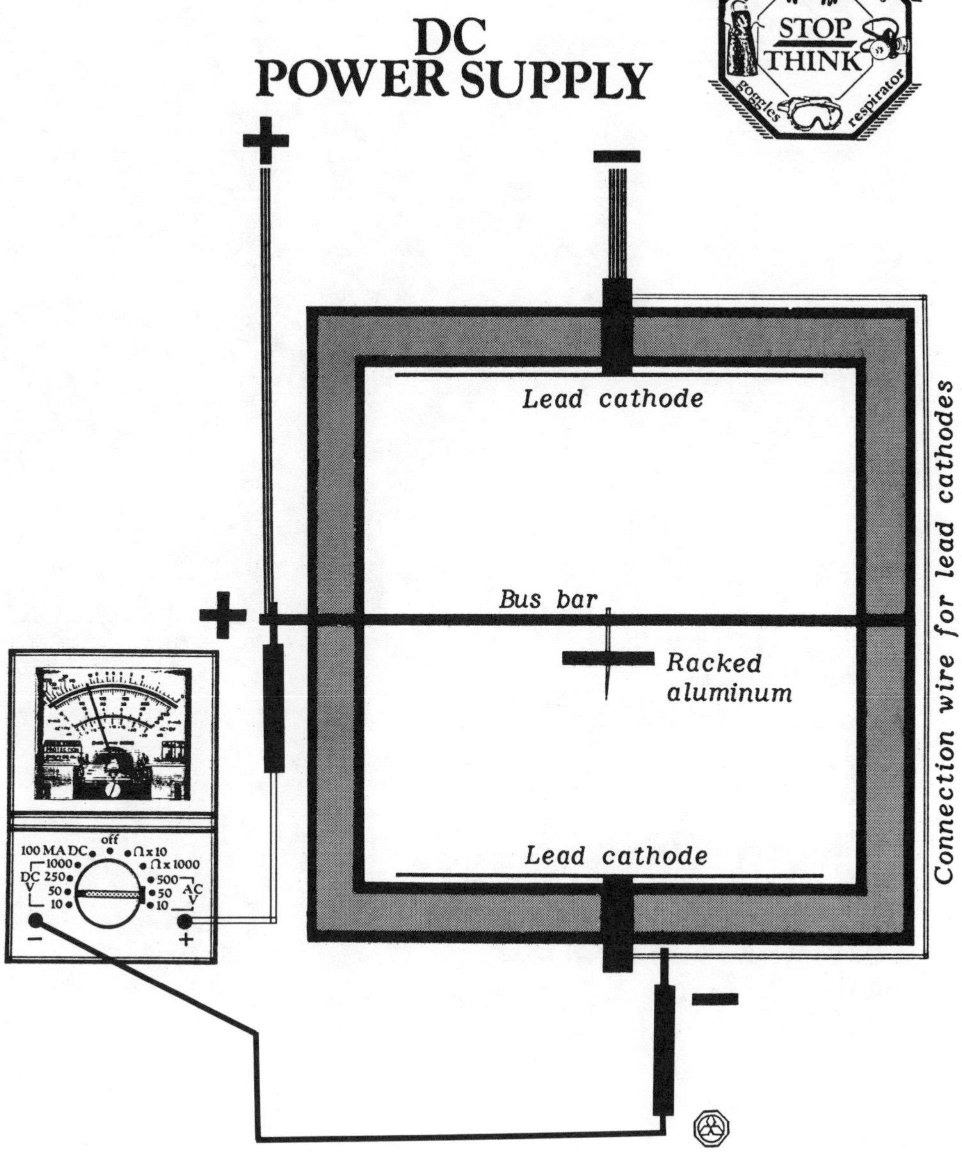

VISITING ARTISTS & SCHOLARS

Christina DePaul
Chuck Evans
John Verhoeven
Mike Edwards
Earl Thomas Yeager, Jr.

Brother (from Family Portrait series) by Christina DePaul, anodized aluminum, acrylic, wood, 12" X 10" X 6", 1987.

VISITING ARTIST: CHRISTINA DePAUL

Multi-Color Anodizing & Resist Technology

It is my pleasure to introduce our Visiting Artist, Christina DePaul, Assistant Professor of Art, University of Akron, Akron, Ohio. Christina will discuss and demonstrate "Multi-Color Anodizing and Resist Technology."

Multi-color anodizing is the process in which more than one color is applied to the surface of the aluminum. This technique is based on the resist and stripping process (also referred to as the bleaching process). Through various resist techniques, different colors and designs can be combined on the same piece of aluminum. By selectively applying a resist, desired patterns and colors can be masked and a variety of results obtained.

Modern Love by Christina DePaul, anodized aluminum, steel, and stone, 13 1/4" x 3 3/4", 1987.

RESISTS

Several different types of resists can be used. Certain resists work better for the various applications. The following resists can be used in aluminum anodizing: etching resists such as asphaltum, silk screen inks, paint markers, grease pencils, 3M electroplater's tape #470, lead tapes and lacquers.

This is an example of multi colored anodizing.

Anything that can hold up to a 60/40 solution of nitric acid, which is the dye stripper, will function as a resist.

To keep from getting confused when masking several colors, use a resist which is available in several different colors, such as nail polish or use a clear lacquer. Peel away resists are also available that are used in electroplating. These are very easy to remove and not as messy as the other resists mentioned because they do not require a solvent to remove. When using resists that do require a solvent, stay away from solvents that contain oil (mineral spirits, paint thinner, etc.). The oil will get into the pores of the anodic coating and cause streaking and spotting to occur. Lacquer thinner and MEK (Methyl Ethyl Ketone) are satisfactory for all resists mentioned. Use proper ventilation and wear gloves when using these solvents.

The following photo series is a chronological sequence used to produce multi-color effects.

1. Once the anodizing cycle is completed, the aluminum is rinsed thoroughly in cold water. The surface of the anodic coating is very fragile at this time.
DO NOT TOUCH THE SURFACE OF THE METAL.

2. Dye with the first color desired. Rinse well in cold water and dry. Dry with compressed air or drip dry.

3. The areas with the first color desired are resisted. See following page for resist possibilities.

4. Strip the unprotected areas with the Nitric dip (60%/40%) solution.
RINSE WELL IN COLD WATER.

5. Re-immerse in the second color desired. Rinse well in cold water and dry. Repeat stages 2 through 5 for as many colors as desired.

6. Using a soft cloth, remove the resist with a suitable solvent for the resist used. Be sure to wear protective gloves, etc. Beware of solvent fumes.

7. Seal the anodic coating.

Thanks for watching my demonstration. I hope it has opened some new avenues to the endless possibilities in multi-colored anodizing!

THE FOLLOWING PHOTOS SHOW SEVERAL DIFFERENT TECHNIQUES FOR APPLYING RESISTS

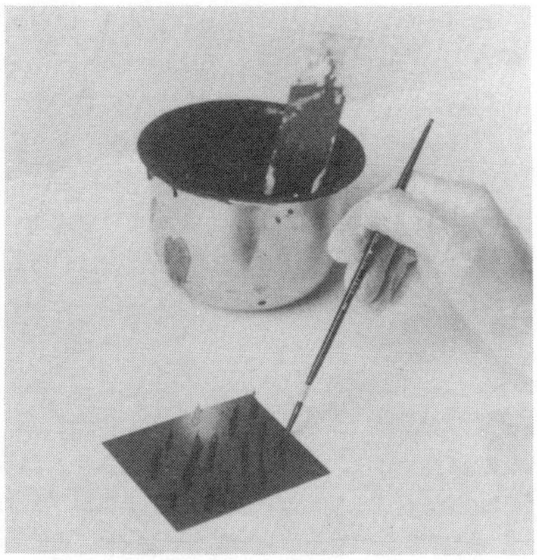

A. Resist applied with a paint brush.

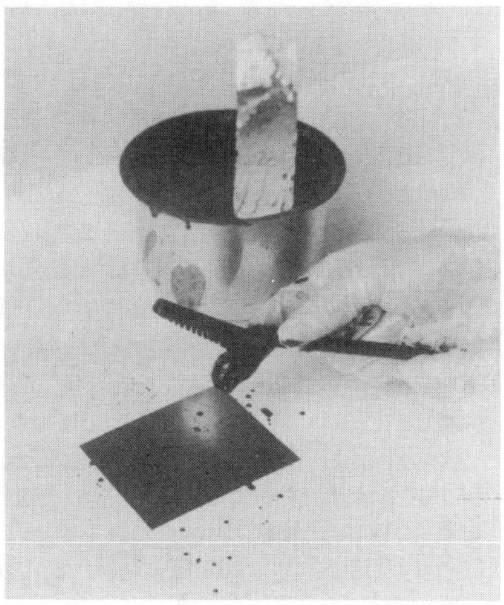

B. Resist splattered with a tooth brush.

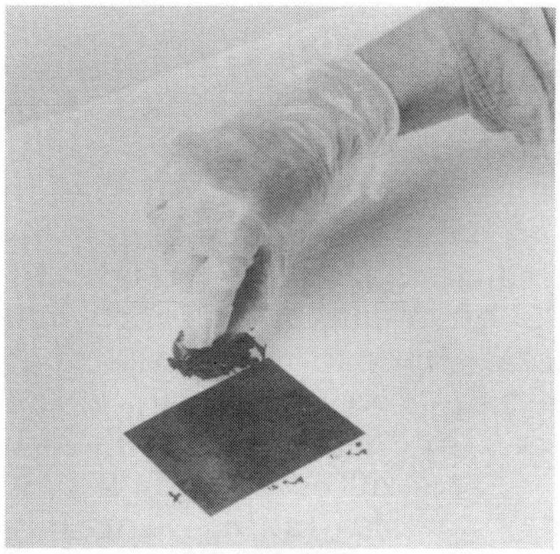

C. Resist applied with a sponge.

D. The resist dripped onto the surface and then manipulated.

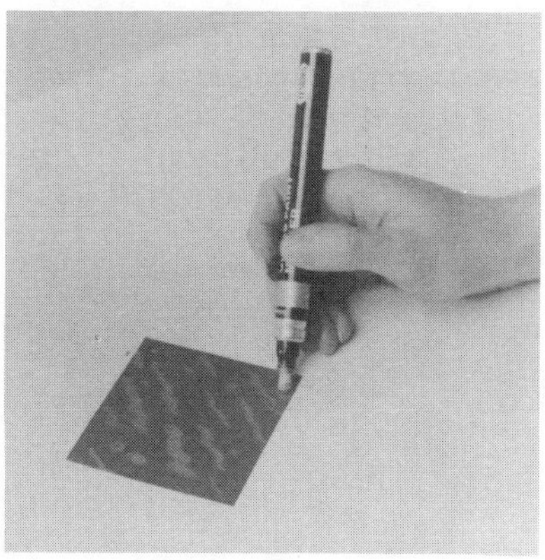

E. Paint markers used as a resist.

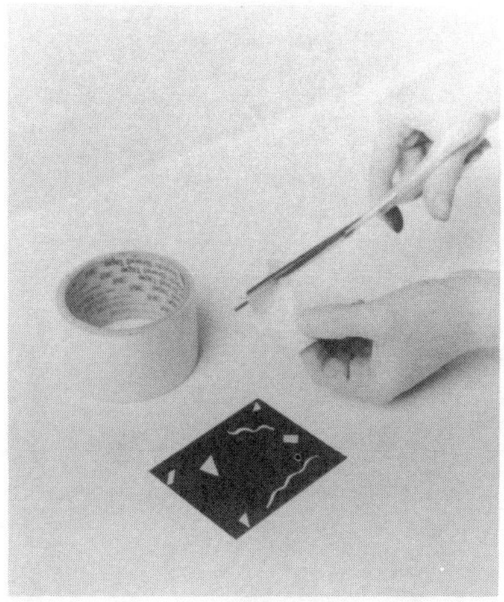

F. Tape used as a resist.

G. The resist is silk screened onto the surface of the aluminum.

Web Torch Texture on Aluminum

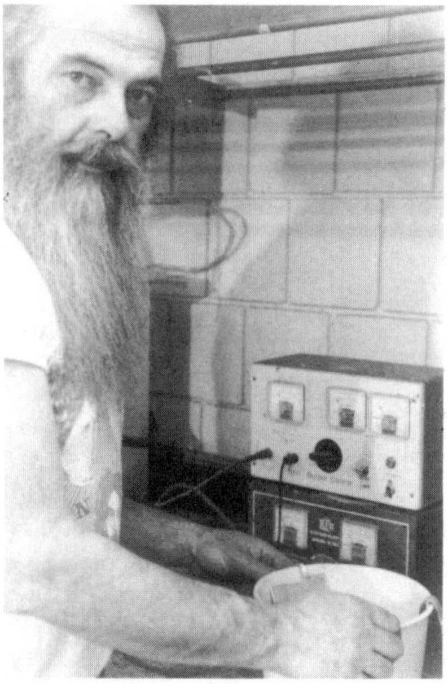

It is a real treat to introduce Chuck Evans, our Visiting Artist, from the Department of Art and Design, Iowa State University, Ames, Iowa. Chuck will discuss "Webbed Torch Texture on Aluminum."

My work with anodized aluminum began in late 1981, as part of the research I was gathering for a book I would later publish. During this period I developed a very simple and extremely unscientific "facility" for anodizing and coloring aluminum. From an artistic standpoint, I produced only a handful of finished jewelry pieces. My interest was centered primarily on setting up an uncomplicated studio sized system and to begin experimenting with the metal in terms of surface effects and color applications. I was also drawn to the weight of the aluminum, availability and relative cost. Up until that time, few people were looking into alternative surface effects and I felt most were taking a fairly straight forward approach to the metal's surface and the application of dye colors. I enjoyed "playing around" with the material and the number of trial and error and sample pieces grew, as I began trying to push the processes to every limit I could imagine. I basically threw away all the books and "proper" procedures relative to bath strength, filtration, cleaning, rinsing, current rate, temperature control, time, cathode size, etc. In the back of my mind I was saying, "Well, you can always mix another bath and go by the book." During this *scientist goes berserk* phase, I was curious why aluminum would not take on torch or heat textures similar to that of "reticulated" silver and nickel silver alloys. After several attempts with the torch it was quickly obvious that no such textures would appear on untreated aluminum. Any effects achieved on raw metal would only appear at the molten or puddle state and were not at all what I was seeking. With that, I went on to other things.

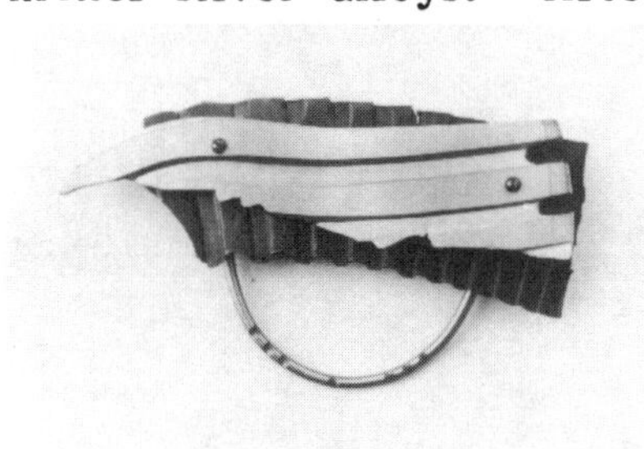

Pin by Chuck Evans, embossed and anodized aluminum, 3" X 2 3/4". Collection of Kathy Palmer.

In late 1984, the bug bit again and the old question popped up once more, "Why won't the damned stuff reticulate?" Like the good scientist I am not, and uncharacteristically, I decided to go back to the basics. I began talking to myself, "Why does silver reticulate?" From earlier work and published data, I knew the presence of a layer of cupric oxide was critical to the occurrence of the reticulation like surface on silver alloys. In order to dye or color aluminum, the surface of the piece is first made microscopically porous or oxidized electrochemically in a bath which develops an oxide layer over the base aluminum. The piece will then absorb dye colors into the oxide layer. After asking the last question, I quickly applied the torch to a piece of anodized scrap and a very obvious texture appeared. It wasn't the classic reticulation texture, but one of web or net-like veins running in all directions like a bullet hole through a windshield. The oxide layer was the answer, and similar to silver, the critical component. Since that day, I have used this technique many times, with both torch and kiln on all types of aluminum, mostly scrap of unknown alloy composition and thicknesses ranging from foil to heavy plate, with the same web textures developing.

The actual application is quite simple. Prepare the aluminum as you would to color (dye). I generally place the piece in the anodizing bath for 10-20 minutes and rinse. For jewelry scale work, the Prest-O-Lite acetylene torch is used to heat the metal. Lacking proper measuring devices, I cannot say exactly at what degree of temperature the webs begin to appear, but it is sufficiently below the melting point so that the melting or puddling is not a problem. This is much less of a problem with aluminum than with silver and one would have to try hard to blow a hole through the metal as is often the case with silver. A general, overall heating is not necessary. Simply start at one end, hold the torch tip over an area until the webs begin to appear and slowly advance the flame. Webs will grow behind or around the flame as it moves on. After the texture occurs, the piece can be colored by conventional means or quenched in a dye bath as the berserk scientist often does!

Thanks for listening. Watch out for the webs you torch!

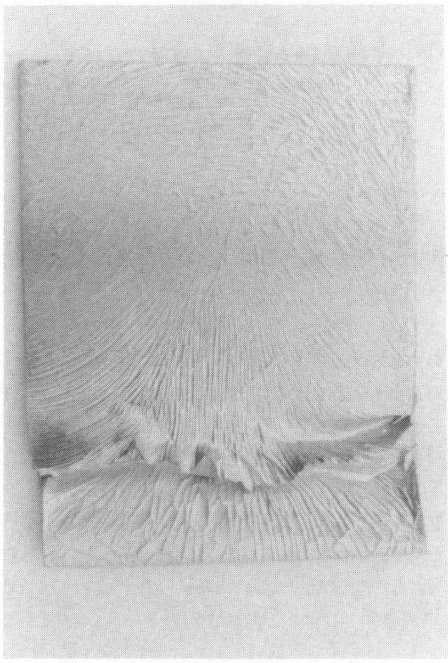

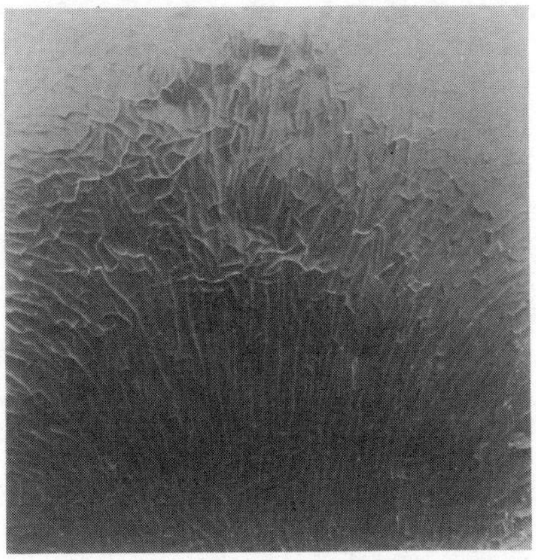

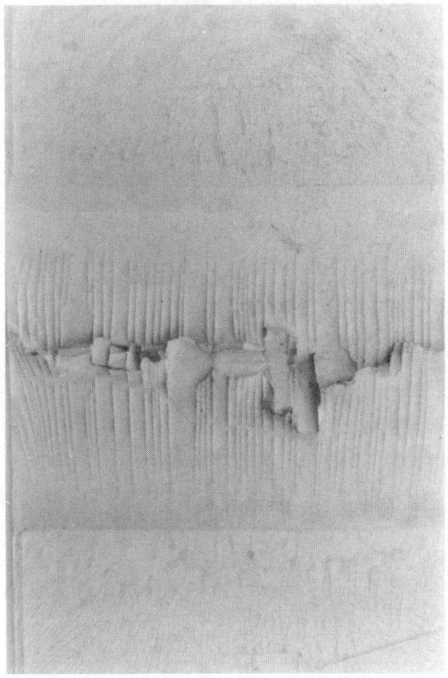

Aluminum samples by Chuck Evans display the potential of the torch created web surface textures.

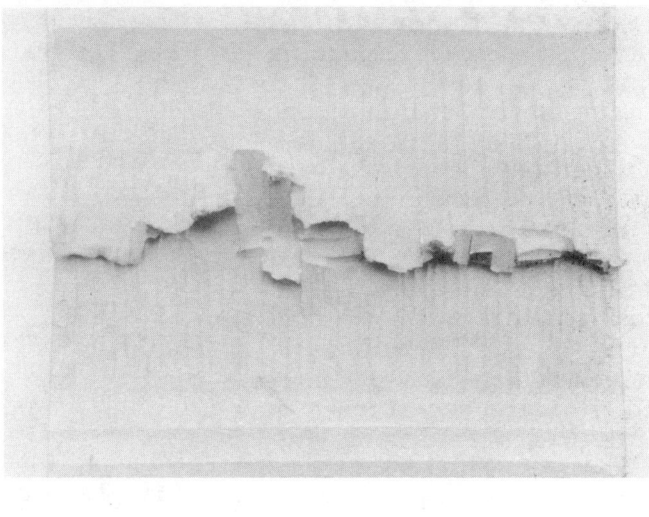

The Cause of Wrinkling on Aluminum

The following synopsis of information has been written by Professor John Verhoeven, Senior Metallurgist, Ames Laboratory, Iowa State University, Ames, Iowa for Chuck Evans. This explains the cause of wrinkling or web effect on anodized aluminum.

Professor Verhoeven's conclusions are based on samples Chuck provided as seen on the opposite page in Figures #1 and #2. The samples are of 3003 aluminum alloy, .102" in thickness and anodized for approximately 15 minutes at 12-15 volts, D.C. in a solution of 15% sulfuric acid to water. The samples were dyed dark blue. Thin cross sections were cut off, secured to an adhesive mounting compound, polished and photographed through a 250 power microscopic lens to reveal the nature of the aluminum material.

The thermal expansion of aluminum oxide is smaller than that of aluminum. On heating, the aluminum will expand more rapidly than the relatively thick oxide layer developed by the anodizing treatment and this causes the surface oxide layer to be pulled into tension and therefore break along the lines perpendicular to the direction of maximum expansion. At the lines, or webbed ridges, where the oxide layer fractures, fresh aluminum is exposed and will immediately oxidize, but its freshly oxidized layer will be much thinner than the surrounding electrochemically anodized oxide layer and will also be colorless.

On cooling, the aluminum inner structure will shrink more than the oxide layer and this will cause the oxide layer to compress. Because the oxide layer is much thinner along the cracked lines, the layer will be much weaker also. In areas away from the cracks, the thick oxide layer will exert a stronger restraint on the inner aluminum during cooling than right at the cracked regions where the oxide is quite thin.

Therefore, at the cracked region the restraining force of the oxide is weaker and the inner aluminum structure bulges outward, possibly producing further cracking in thin layers along the previously cracked locations.

In Figures #1 and #2 the tips of the protrusions have a jagged appearance similar to that of a volcano crater and this is typical of all samples examined and provides evidence for the above explanation.

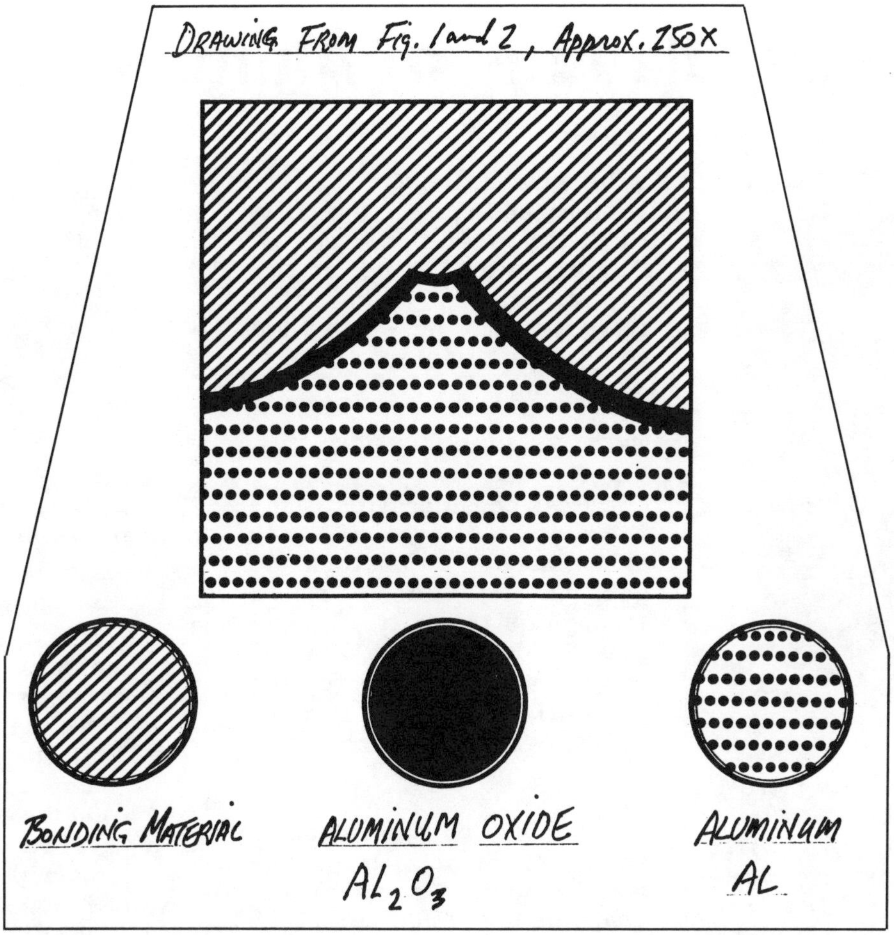

Figure #1. Transverse cut through a "wrinkle" on aluminum sample anodized and dyed blue, 250X. Photo: John Verhoeven.

Figure #2. Same as Figure #1. Photo taken at a different location. Photo: John Verhoeven.

VISITING ARTIST: MIKE EDWARDS

Hot Manipulations in Aluminum

It is my pleasure to introduce our Visiting Artist, Mike Edwards, who will discuss and demonstrate some "Hot Manipulations in Aluminum."

Hello, my name is Mike.

Shortly after I started working with aluminum, I began to look for interesting and less time consuming methods to texture the metal. On one trip to the salvage yard, I found a piece of partially melted aluminum sheet.

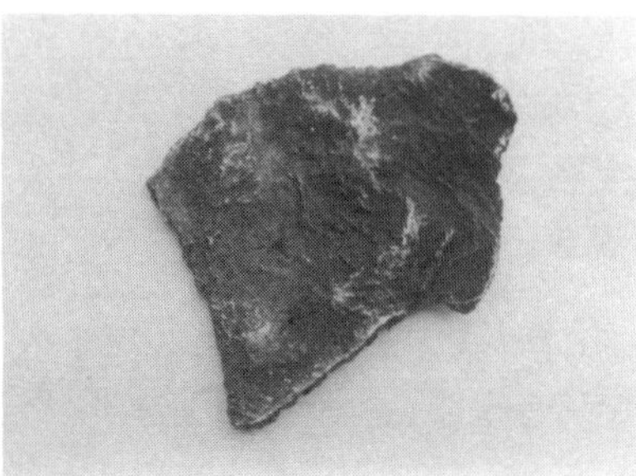

I became intrigued by the texture and skin-like quality of this piece of metal. During the course of my investigations I found that aluminum may be worked hot to produce different textures and changes in form.

TEXTURE DEVELOPMENT

The first consideration is the thickness and type of stock to be used. Thin stock (16 gauge or less) tends to burn through quite easily. Thicker gauge aluminum allows a larger safety margin for the prevention of burnt through holes, although such holes may be exploited effectively.

The second consideration is the surface finish of the metal, anodized and sealed or unanodized. Anodized stock produces a ridged, facet-like texture, while unanodized stock produces a more molten, bubbly surface texture.

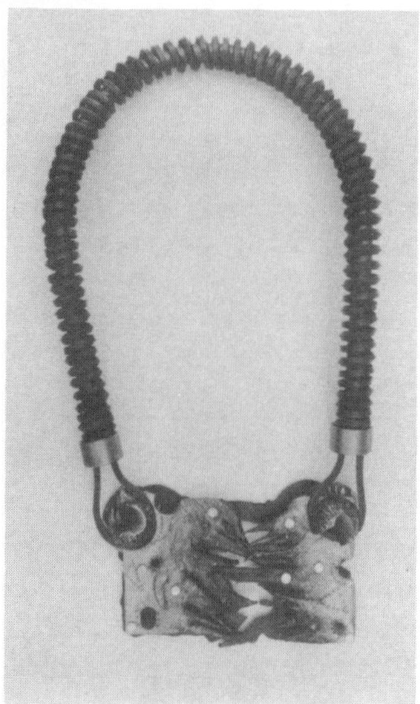

Necklace by Mike Edwards, anodized aluminum and brass, 4" X 15", 1987.

Detail of Necklace by Mike Edwards, notice the spiky projections.

1. A minimum of tools are required to utilize these techniques.
   - an easily manipulated heating surface--a firebrick will do, but a charcoal block is better
   - a heat source (in this case a Prest-O-Lite acetylene torch) preferably hand held, but a kiln may be used
   - a steel poker. It is wise to have an extra poker in reserve in case the first one becomes too hot!

Items in photo below as seen left to right:
- charcoal block on fire brick
- 2 pieces of aluminum stock
- steel poker--1/4" dia. rod
- acetylene torch

Two pieces of similarly prepared stock can be seen in this photo. Each piece

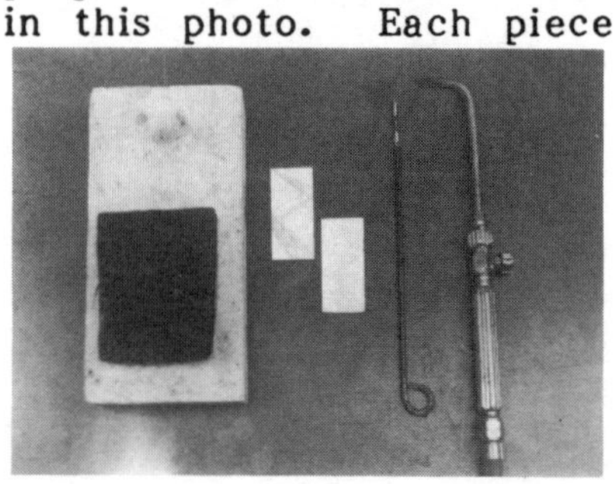

has the same alloy composition and the same pattern scored into the metal surface with a steel etching point (double pointed etching needle--common to jewelers and printmakers). The scored patterns remain even after the surface texture has been developed with the heat from the torch flame. The principle difference is that the piece of stock on the left is anodized, while the piece on the right has been stripped of its oxide coating in a caustic soda (lye) etch bath.

2. The anodized metal blank is placed on a heat resistant surface. With a soft flame, heat is applied at one end of the blank. Aluminum transfers heat well, so start with localized heat and continue moving the flame as a texture develops.

3. After the heating begins, watch for a surface texture to develop on the metal surface. The aluminum below the anodized oxide coating will expand and break the oxide coating above, creating a faceted, ridged texture.

4. As the texture forms, quickly move on to the untextured areas. If the metal below the surface becomes totally molten, the surface tends to flatten out.

5. After the texture is developed, allow the metal to air-cool or quench it in water.

6. The same steps are repeated with the unanodized piece of aluminum. Notice the bubbly surface texture.

ACCENTUATING TEXTURE BY AGITATION

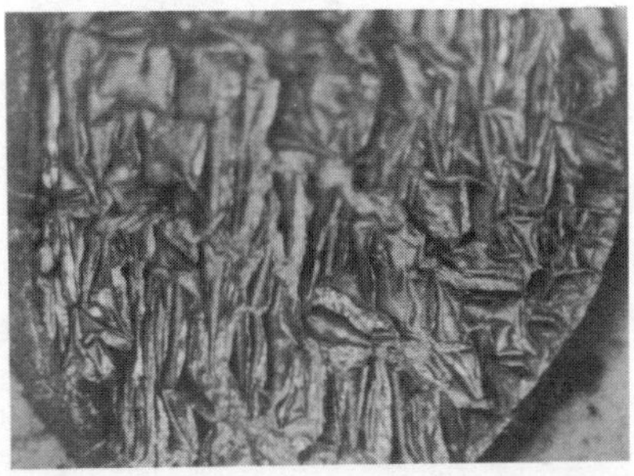

By manipulating the aluminum blank, exaggerated textures are possible.

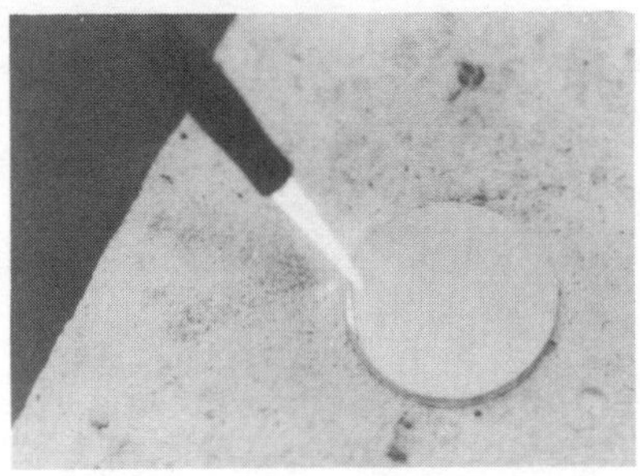

1. Begin heating the metal as described in the previous demonstration, allowing the metal to texture. Use an oxidizing flame.

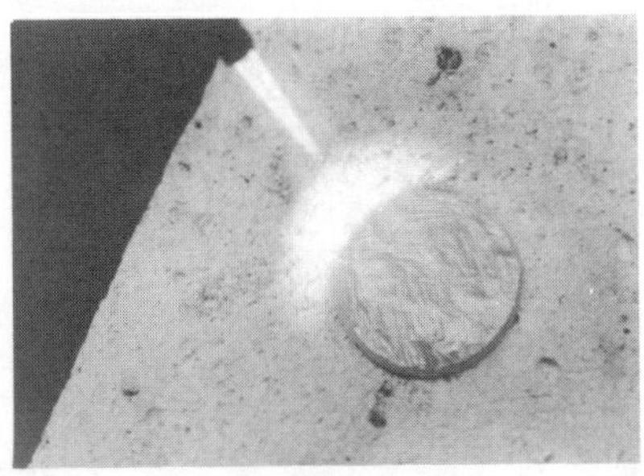

2. Continue heating after the texture has developed. Heat until the aluminum becomes fully fluid beneath the oxide coating.

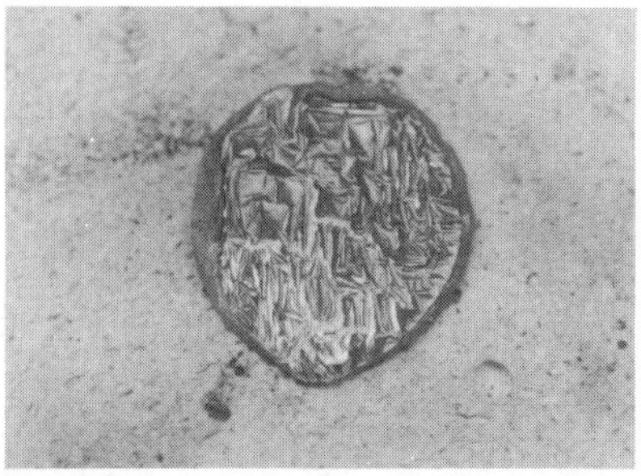

3. Discontinue heating, then immediately agitate the molten aluminum beneath its oxide coating by carefully tipping and rocking the heating surface (firebrick). Continue agitation until the metal solidifies.

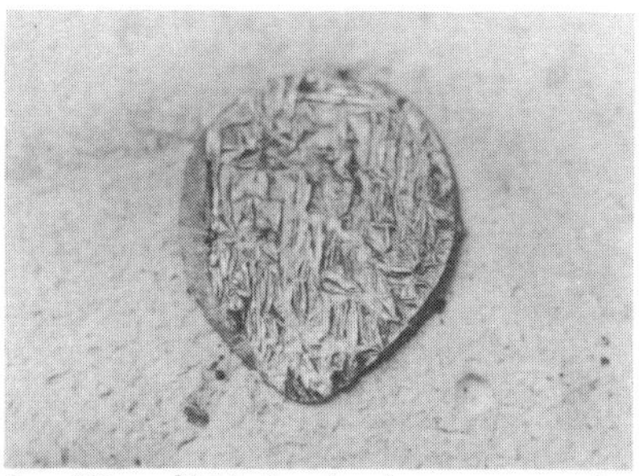

4. The finished piece may be air-cooled or quenched. If you do not like the texture, simply reheat and agitate for another texture pattern.

DRAWING MOLTEN ALUMINUM

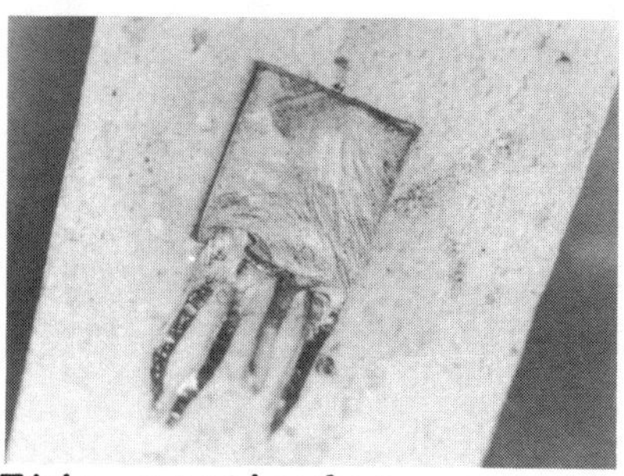

This particular process utilizes the oxide coating of aluminum as a drawplate to draw out spiky, wire-like projections.

1. First, place a small object, such as a pencil, under the edge of your heat resistant surface (firebrick). This will cause the surface to have a slight incline. Place the piece of aluminum on the higher side, with the edge you wish to draw from, on the downhill side of the incline.

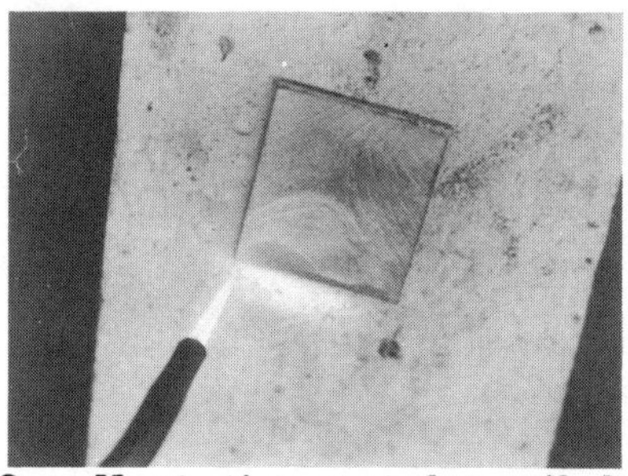

2. Heat the metal until it begins to melt below its oxide film. As localized melting occurs, a small bulge of molten metal will form on the downhill side of the blank. Be careful not to heat the metal to a fully fluid stage; localized heating and melting is more desirable.

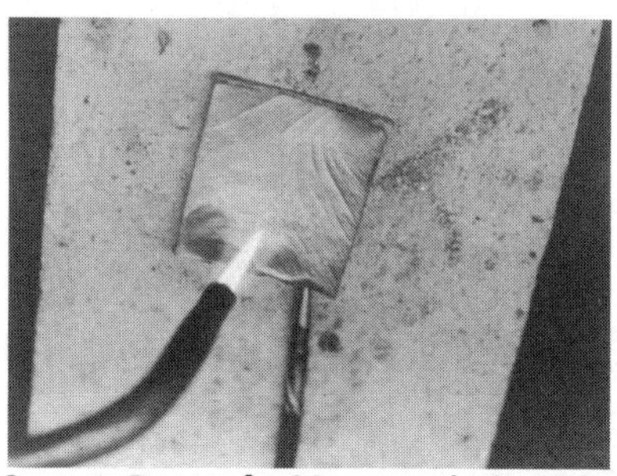

3. Carefully prick the molten pool of aluminum on the lower side of the blank with a cold steel poker. The size of the hole in the oxide skin will determine the diameter of the spike at its base.

4. Carefully pull the poker downhill. As this is done a spiky projection will be drawn from the reservoir of molten aluminum. Draw the projection out to the desired length.

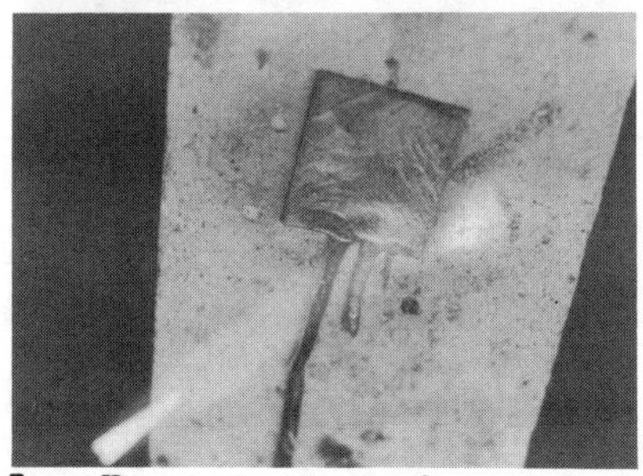

5. Repeat steps 3 and 4 as desired. If the pool of molten aluminum becomes depleted, reheat the top of the blank, allowing more material to refill the empty pool.

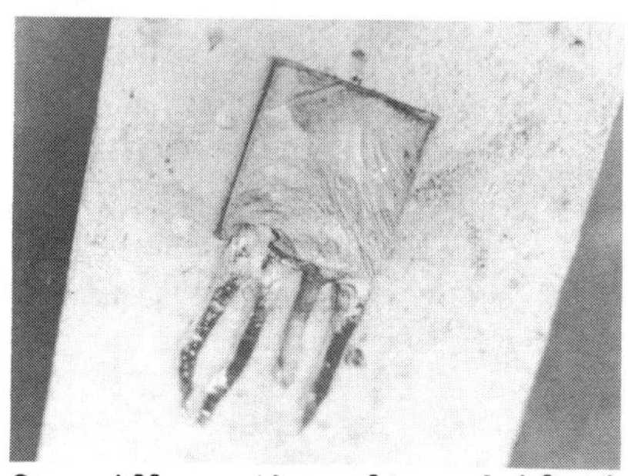

6. Allow the altered blank to cool and solidify. After it has reached a solid state, it may be quenched.

SLUMPING ALUMINUM

Aluminum may also be slumped into different contours or shapes. This is accomplished by heating the metal shape until it reaches its plastic range, and allowing the metal to collapse into the negative space around it. If the metal reaches a fully molten state, the positive areas tend to thin and become weak, perhaps making the blank unusable.

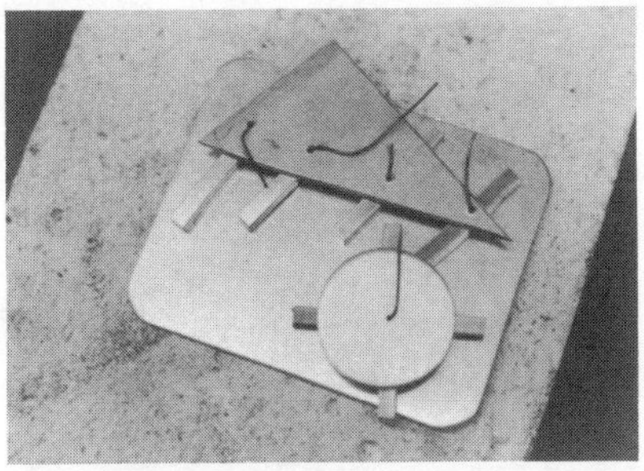

1. A set of aluminum shapes are prepared by cutting and drilling registry holes through them. Matching holes are drilled in the support sheet below. Lengths of square aluminum wire (any non-combustible object could be used) have been cut and drilled to raise the sheet shapes above the base plate. Iron binding wire is inserted into the drilled holes to maintain registry throughout the operation. The drilled holes may be used for rivet holes in the final assembly.

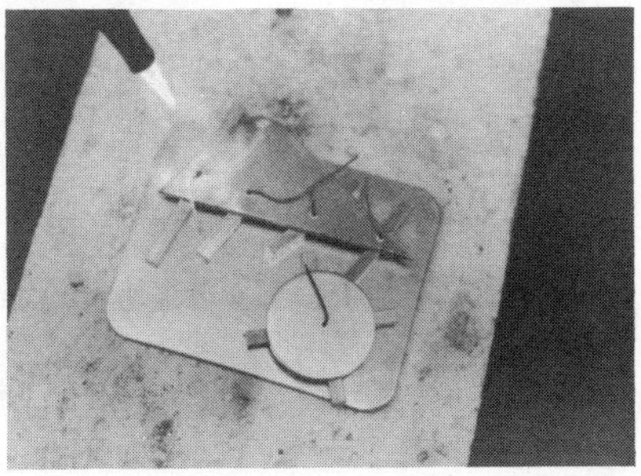

2. Carefully begin heating one end of the top element of the assembly. As the top element begins to texture, it will also begin to slump into the negative, unsupported areas beneath. **DO NOT OVERHEAT!** If the metal becomes fully molten, the positive, supported areas will thin out, possibly making the top shape unusable.

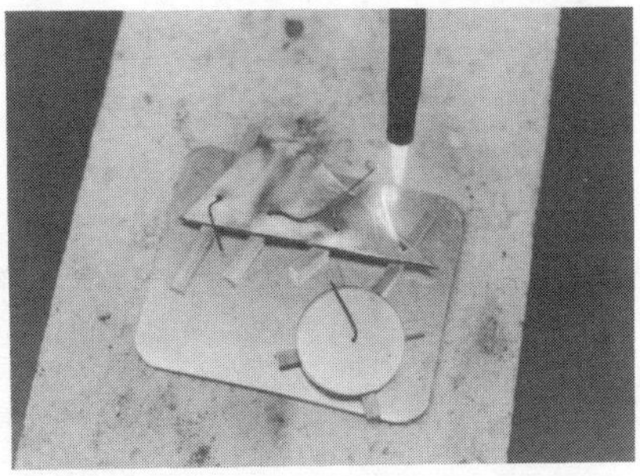

3. Slowly and carefully move on to the other untextured and unslumped areas until the surface is fully contoured and textured.

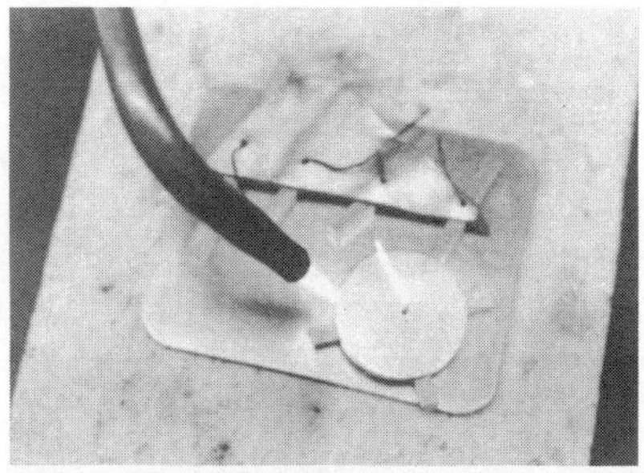

4. Repeat steps 2 and 3 on the other untextured elements. If texture is desired on the base plate, allow the assembly to cool, then turn it over and repeat steps 2 and 3 again.

STONE SETTINGS

Heat may also be used to set stones in aluminum. Use only heat-resistant stones for this technique. Cubic zirconium, synthetic corundum, synthetic tourmaline, and y.a.g. are satisfactory for this process. Man-made synthetic stones have fewer impurities and flaws than their natural counterparts. Consequently, synthetics are less likely to crack or craze. Do not use this technique with turquoise, opals or other stones with a high water content. It is important that the stones be chemically resistant since they may be subject to further processing in the anodizing, dyeing and sealing baths.

1. Choose a piece of aluminum sheet at least 1 1/2 times thicker than the distance between the girdle and table of the largest stone. Drill the holes slightly smaller than the diameter of the stones. Enlarge the holes with a tapered bur until each hole is as large as its stone. With the appropriate setting bur, cut a shoulder on which each stone will sit. The table of the stone should be co-planer with the metal surface.

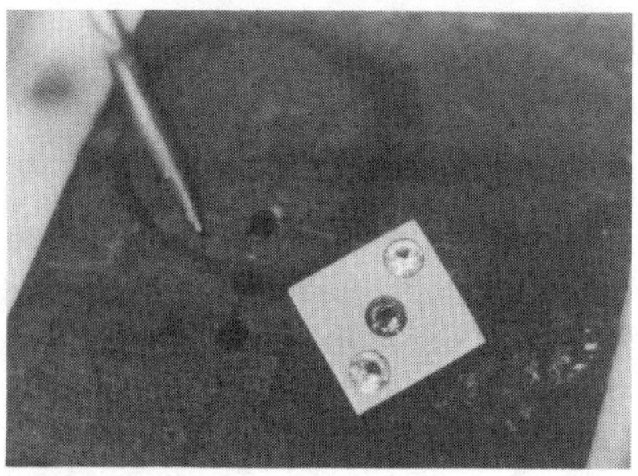

2. Place the metal on the charcoal block. Using the metal as a template, poke holes into the charcoal block below. These holes will accommodate the stones' pavilions should they protrude beneath the metal's surface. The holes allow the stones to lie flat in their seats.

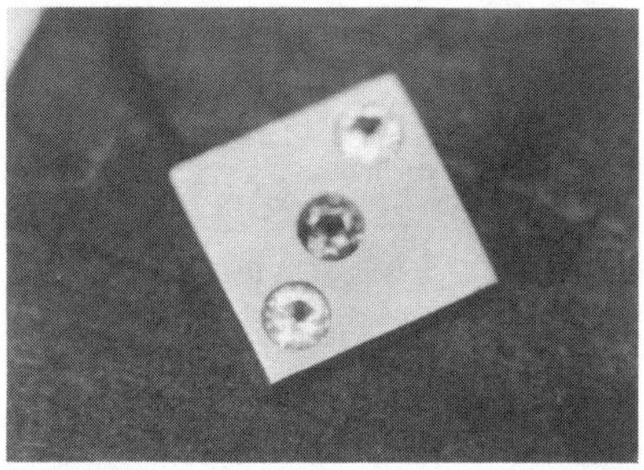

3. Place the stones in their seats and position the metal over the previously made holes.

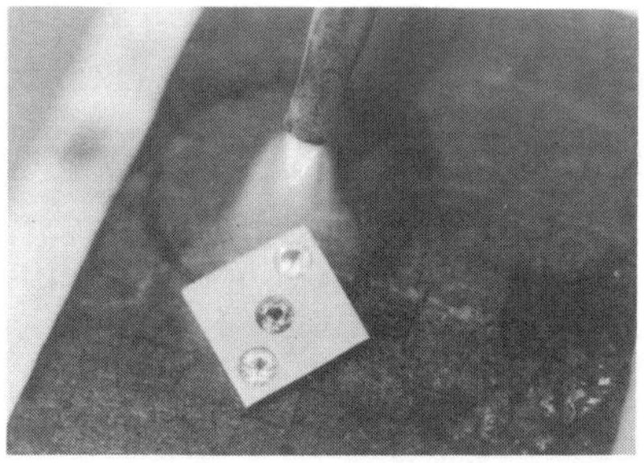

4. Begin heating the metal slowly with a cool, soft flame. Slow heating will prevent the stones from undergoing more thermal stress than is necessary.

5. Slowly increase the intensity of the torch flame until the texture appears and the hard straight edges of the metal become round. When the edges become slightly rounded, a small bit of slumping occurs around the girdle of the stone, permanently securing it in place.

6. The finished piece is allowed to **AIR-COOL.** Quenching will result in broken or cracked stones.

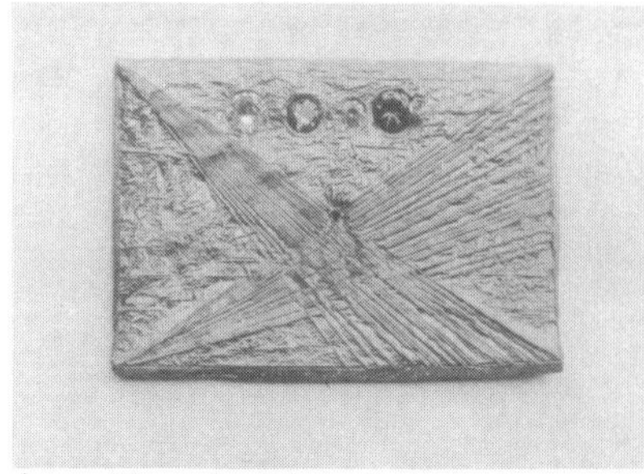

Anodized aluminum brooch, synthetic stones with a scored, manipulated texture. 1 1/2" X 2".

Architectural Proposition #3, anodized aluminum, bronze, copper, brass, CZ and amethyst, 3 1/2" X 2 1/2".

Thank you for attending my demonstration. May your textures be good ones and your stone settings bright and clear!

VISITING SCHOLAR:
EARL THOMAS YEAGER, JR

Anodizing and Coloring Aluminum

EARL THOMAS YEAGER, JR.
Industrial Arts Instructor
Troy Junior High School
Troy, Pa.

The following article was first published in February, 1966 in Industrial Arts/Vocational Education Magazine (IA/VE).

Kind permission has been granted to reprint this fine article in this book by:
The Cummings Publishing Company
26011 Evergreen Road
Suite 204
Southfield, Michigan 48076
313-358-4900

In case you are interested, the magazine "Industrial Education" has a yearly subscription rate of $17.00.

How the industrial process of anodizing and coloring aluminum may be adapted to the secondary school level at a minimum of expenses and additional facilities.

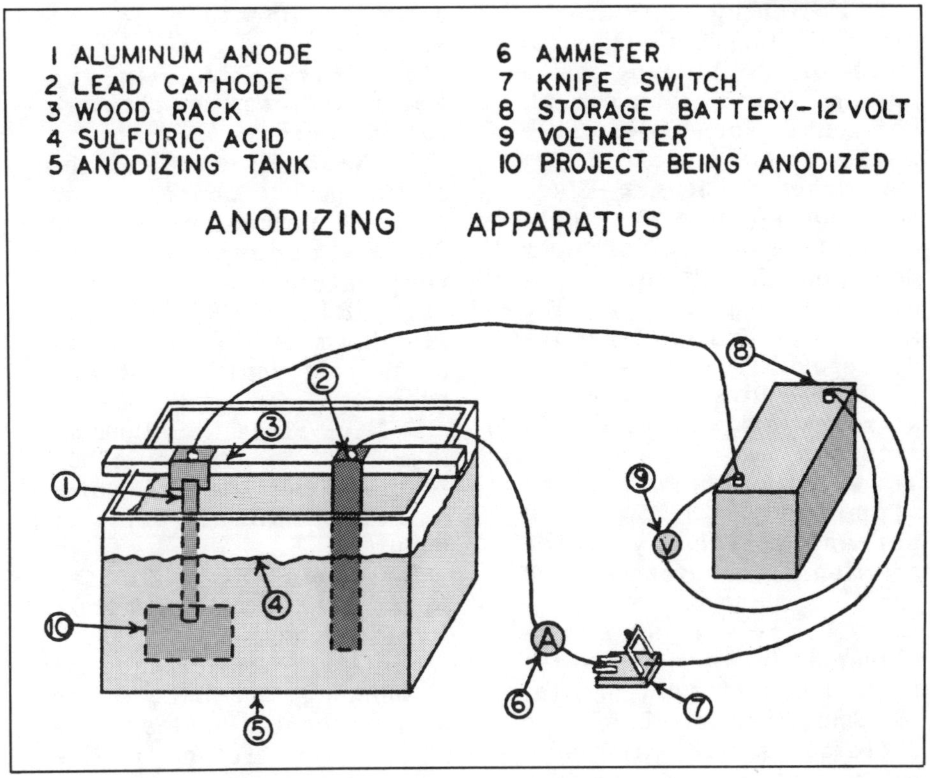

There are many aluminum projects made in secondary school laboratories and an aluminum anodizing and coloring apparatus will provide a different attractive, and practical method of finishing these projects. It is especially applicable in the art-metal and metal-spinning areas of an industrial arts metal laboratory.

The anodizing apparatus may be constructed easily using the following materials: 12-volt storage battery, ammeter (0-5 a.), voltmeter (0-25 v.), knife switch, No. 12 house wire, acid-resistant tank, aluminum anode, lead cathode, wooden rack for supporting the anode and cathode, and sulfuric acid. The materials are assembled very much like an electroplating apparatus.

The materials needed are: nickel acetate, caustic soda, detergent, nitric acid, phosphoric acid, aluminum dye, thermometer, plastic tongs, quart jars, water supply, and aluminum to be anodized.

Wide-mouth glass jars may be used as storage containers for these chemicals. They may then be arranged in sequential order as shown in the procedure used for anodizing.

When aluminum is made the anode in a liquid oxidizing solution, large quantities of free oxygen migrate to the anode where they combine with the aluminum to form a thin uniform coating of aluminum oxide. As a result, the aluminum will absorb dye, and upon sealing the color becomes very durable.

In a school laboratory, such industrial features as an air agitation system in the anodizing tank may be eliminated. On this small scale, temperatures are not critical and the operations may be performed at room temperatures (70 to 78 deg. F.). The following 20 steps are in sequential order and must be followed closely for effective results. When transferring the aluminum to be anodized from one step to the next, use plastic tongs and avoid touching the metal with your fingers.

Anodizing Sequence

1. **Polishing**--Polish the piece of aluminum with steel wool or by buffing on a buffing wheel. Remove all scratches or undesirable marks from the surface.
2. **Nonetch Cleaner**--Place the aluminum in a 25 percent alkaline soap solution or detergent for three minutes at room temperature. The purpose of this is to remove dirt or soil from the surface.
3. **Rinse**--Rinse in tap water at room temperature for 20 seconds.
4. **Caustic Etch**--Place the aluminum in a 5 to 10 percent sodium hydroxide solution at room temperature for five minutes.
5. **Rinse**--Rinse the aluminum in warm tap water at 95 deg. F. for one minute.
6. **Smut Removal**--Place in a 30 to 50 percent nitric-acid solution at room temperature for one minute and 45 seconds.
7. **Rinse**--Rinse in tap water at room temperature for 45 seconds.
8. **Rinse**--Rinse in tap water at room temperature for 45 seconds.
9. **Chemical Brightener**--Place the aluminum in a solution that has a concentration of 80 percent phosphoric acid, 3 percent nitric acid, and 17 percent water. Subject the aluminum to these conditions for two minutes at room temperature.
10. **Rinse**--Rinse in warm tap water (95 deg. F.) for 45 seconds.
11. **Rinse**--Rinse the aluminum in tap water at room temperature for 45 seconds.
12. **Anodize**--Place the aluminum in a concentration of 15 percent sulfuric acid for 15 minutes. The concentration should be room temperature with 11 volts d.c. and .95 to 1.00 amperes.
13. **Rinse**--Rinse the aluminum in tap water at room temperature for 45 seconds.
14. **Rinse**--Rinse in tap water at room temperature for 45 seconds.
15. **Neutralizer**--Place the aluminum in a concentration of 30-50 percent nitric acid for 45 seconds at room temperature.
16. **Rinse**--Rinse the aluminum in tap water at room temperature for one minute.
17. **Dye**--Place the aluminum in a dye concentration of .1-20 grams per liter of water at room temperature for 15 minutes.
18. **Rinse**--Rinse in tap water at room temperature for 30 seconds.
19. **Color Seal**--Place the aluminum in a concentration of 1 percent nickel acetate at room temperature for five minutes.
20. **Final Seal**--Place the aluminum in tap water (distilled water is preferred) at 120 deg. F. for 15 minutes.

The listed steps show the sequential order in which the chemicals and equipment should be placed. When all the chemicals have been secured and arranged in this order, the anodizing and coloring aluminum becomes a very simple process.

There are four various anodized finishes which may be performed using the 20 steps previously stated. Each finish requires a different sequence through the 20 steps. These four finishes are known as:
(1) normal finish
(2) frosted, etched, or matte finish
(3) specular diffuse finish
(4) specular reflective finish.
They may all be effected successfully on either smooth or embossed aluminum if the

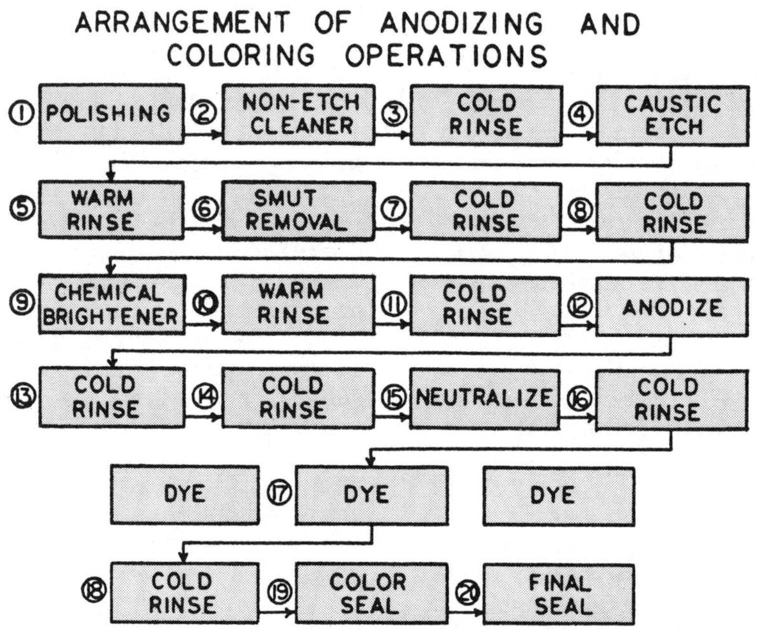

Note that three blocks are shown for the dyeing operation indicating that various colors may be used. However, anodized aluminum is placed only in color desired.

proper sequence for the desired finish is correctly followed.

Attach Object to Anode

The aluminum object to be anodized must be fastened securely to the anode. It is advisable to leave a small tab on the object into which a hole is drilled. The end of the anode may be tapered and forced into the hole in the object. Upon being anodized, the tab may be removed from the object. Aluminum or titanium bolts, nuts, and straps may be used if available. Caution must be taken to suspend the object in the solution so that it does not tough the sides or the bottom of the tank.

Darkness of color is directly proportional to the length of anodizing time. A 10-minute anodizing period will result in very little dye being absorbed. A 20-minute period will give a medium darkness, and a 30-minute period will result in a very dark color. Regardless of the time that the anodized object is left in the dye beyond 15 minutes, the darkness of the color will not increase. The anodizing period must be increased if a darker color is desired.

After anodizing, certain areas of the aluminum object may remain undyed by deliberate plan, if so desired. Masking tape, paste wax, wax crayon, grease, and paint are excellent block-out materials. These substances may be applied to the anodized object in any desired pattern before placing it in the dye. The area that is covered with one of these substances will not absorb the dye and when removed from the dye, the substance may be cleaned off, exposing the undyed pattern.

SEQUENCE OF OPERATION FOR VARIOUS FINISHES

NORMAL FINISH

1-2-3-6-7-8-12-13-14-20.	NATURAL
1-2-3-6-7-8-12-13-14-17-18-19-20.	COLORED

FROSTED, ETCHED, OR MATTE FINISH.

1-2-3-4-5-6-7-8-12-13-14-20.	NATURAL
1-2-3-4-5-6-7-8-12-13-14-17-18-19-20.	COLORED

SPECULAR DIFFUSE FINISH

1-2-3-4-5-6-7-8-9-10-6-7-8-12-13-14-20.	NATURAL
1-2-3-4-5-6-7-8-9-10-6-7-8-12-13-14-17-18-19-20.	COLORED

SPECULAR REFLECTIVE FINISH

1-2-3-6-7-8-9-10-6-7-12-13-14-20.	NATURAL
1-2-3-6-7-8-9-10-6-7-12-13-14-17-18-19-20.	COLORED

Commercial Dyes Available

Commercially prepared dyes are available for the sources listed in the bibliography near the end of this article; however, fabric dye, leather dye (water-soluble neat dye) or liquid shoe polish may be used as substitutes. They will not produce the uniform color that may be accomplished with commercial dyes, but they are good substitutes if commercial dyes are not available.

Anodizing is of great commercial importance today. It is used in a large tonnage of refrigerator and appliance trim, automotive trim, and architectural installations. In industry, the oxide films are actually measured according to how they are going to be used: .1 to .2 mil oxide films are applied for interior work; .2 to .4 mil films are applied to exterior parts such as auto trim; .8 to 1.0 mil films are used for permanent architectural products such as store fronts.

Durability

A project that has been anodized and dyed has a very durable surface that will not be affected by such things as salt water, the sun or any of the elements of weather. The dyed surface will resist the flame of a propane torch until very shortly before the melting temperature of the aluminum is reached. At this time the surface begins to crack or peel.

The following safety precautions must be observed when working with the anodizing apparatus:

1. Safety glasses and protective clothing (rubber gloves and aprons) should be worn when working with sulfuric acid.
2. Extreme care must be taken when making up the sulfuric-acid solution. The acid must be added to the water in small increments and complete agitation of the bath made before the next increment of acid is made.
3. The anodizing tank electrodes should be insulated or arranged in such a fashion that the students cannot unintentionally touch both anode and cathode simultaneously.
4. Caustic etching and acid brightening solutions are hazardous to eyes, skin, clothing, and the respiratory system. They should be used only in hooded or well-ventilated areas with suitable eye, skin, and clothing protection.

The procedure outlined is based on information supplied by industrial concerns.

Avenues for Study

Anodizing and coloring aluminum in school laboratory opens many different avenues for further study. There are many phases involved in anodizing and coloring aluminum that are worthy of further study; however, some of the more important phases are the following:

1. Try to anodize with other electrolytes such as nitric, oxalic, or chromic acid in place of the sulfuric acid.
2. Try to place two or more colors of dye on one piece of anodized aluminum.
3. Devise a method whereby you can accurately measure the thickness of the anodized coating and make a comparison of the coating thickness of each of the four methods of anodizing used in this article.
4. Anodize a piece of aluminum by each of the four methods used and compare the hardness of each of the coatings.
5. Anodize several pieces of aluminum each in a different electrolyte temperature and observe the effect of dye adsorption.

There are many additional phases of anodizing and coloring aluminum that would be very educational in being studied further.

The following list of industrial companies is very helpful in setting up an anodizing apparatus. Many of these companies will furnish, upon request, various samples of dye and information concerning the anodizing procedure.

(Author's note: since this article was printed over 20 years ago many of the addresses may be out of date and dye samples may not be available. Try anyway-all they can say is no.)

Bibliography

Aluminum Company of America, P.O. Box 840, Lancaster, Pa.

American Cyanamid Company-Pigments Division, P.O. Box 567, OA, Chicago, Ill. 60680.

Carbic-Hoechst Corporation, Mountainside, N.J.

Ciba Company, Inc., 7535 Lincoln Avenue, Skokie, Ill.

E. I. Dupont De Nemours and Company, Inc., Organic Chemicals Dept., Dyes and Chemical Division, Wilmington 98, Del.

Koppers Company, Inc., Chemicals and Dyestuffs Division, Pittsburgh, Pa. 15219.

Reynolds Metals Company, Reynolds Metal Building, Richmond, Va. 23218.

Sandoz Inc., 61-63 Van Dam Street, New York, N.Y. 10013.

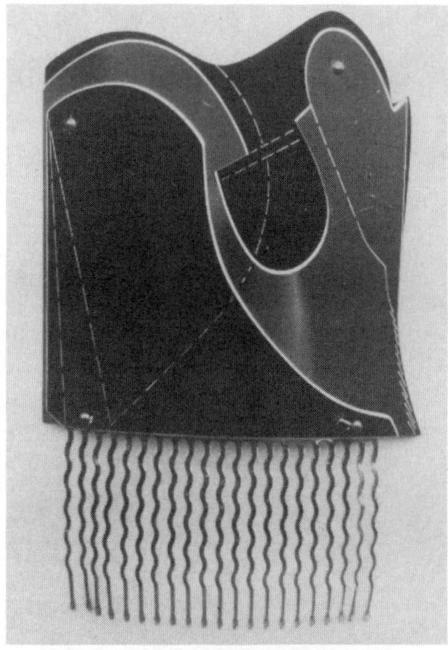

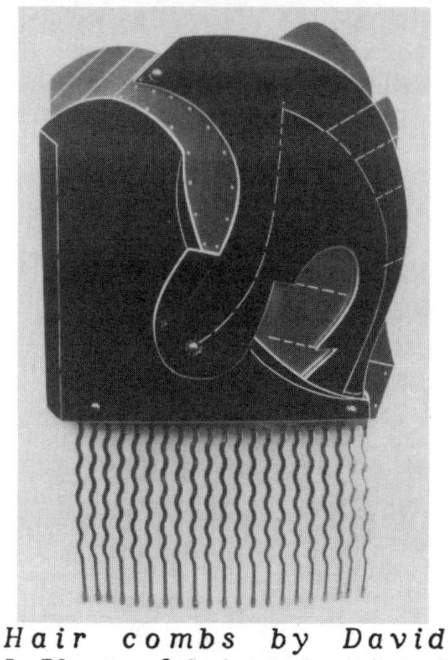

Hair combs by David LaPlantz, fabricated, painted aluminum, engraved, plated steel comb, 4" X 2 1/2", 1984.

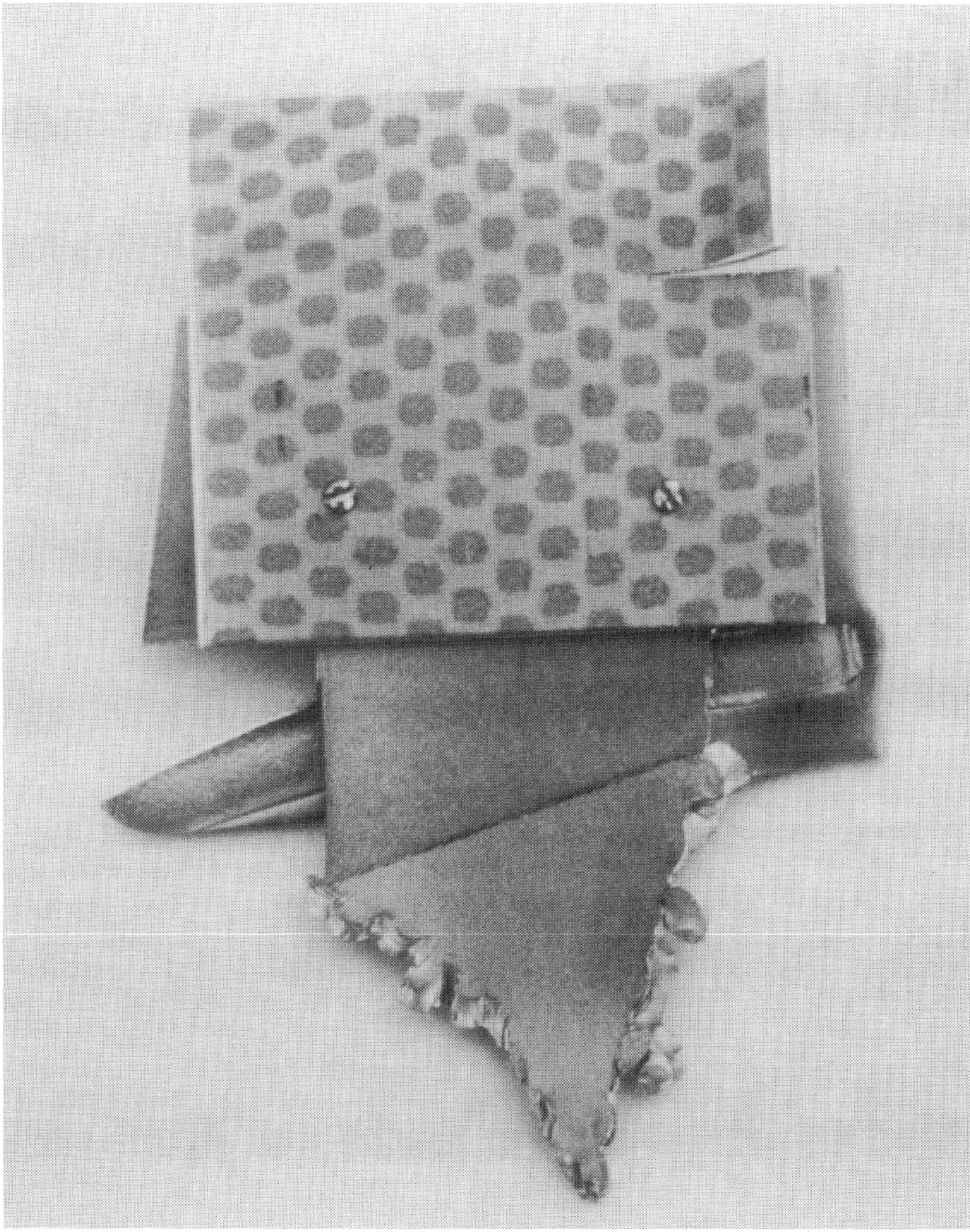

Pin by Chuck Evans, embossed, sandblasted and anodized aluminum, 3 1/2" X 2".

FIELD TRIPS

Milner's Anodizing

Do you enjoy going on field trips?

Let's hope your answer is "YES!"

Every good workshop should contain at least one field trip. This one has two!!

Therefore, to expand your knowledge on industrial anodizing, let's first travel to Milner's Anodizing in Santa Rosa, California. Kirk Senseney will be our tour guide at Milner's. Then we will travel to the Anomatic Corporation in Westerville, Ohio.

I know what you're thinking, and you're right.

You are reading this book miles from California and Ohio. How do you take part in these field trips?

Easy, simply turn the page, and as if by magic, you will be transported. Enjoy! Did you remember to bring your bag lunch?

A special thanks to Terry Burson, owner and operator of Milner's Anodizing for the opportunity to photograph his anodizing business, and to Kirk Sensensey for all of his help during the photography sessions. Their combined courtesy, openness and kindness made this portion of my book a "breeze."

This opportunity to visit industrial anodizing businesses may pique your interest to visit one in your local area.

Do not be disappointed if the anodizer is not too interested in giving you a guided tour. After all, they are in business to make money which usually does not leave time for the tour bus crowd.

Consider taking in some aluminum parts to be anodized. You may even have the opportunity to watch your pieces being anodized, dyed, and sealed. You might even get to ask a few questions. After all, you are now paying the operator to work for you. Money talks at times or at least deserves a few answers. Good luck!

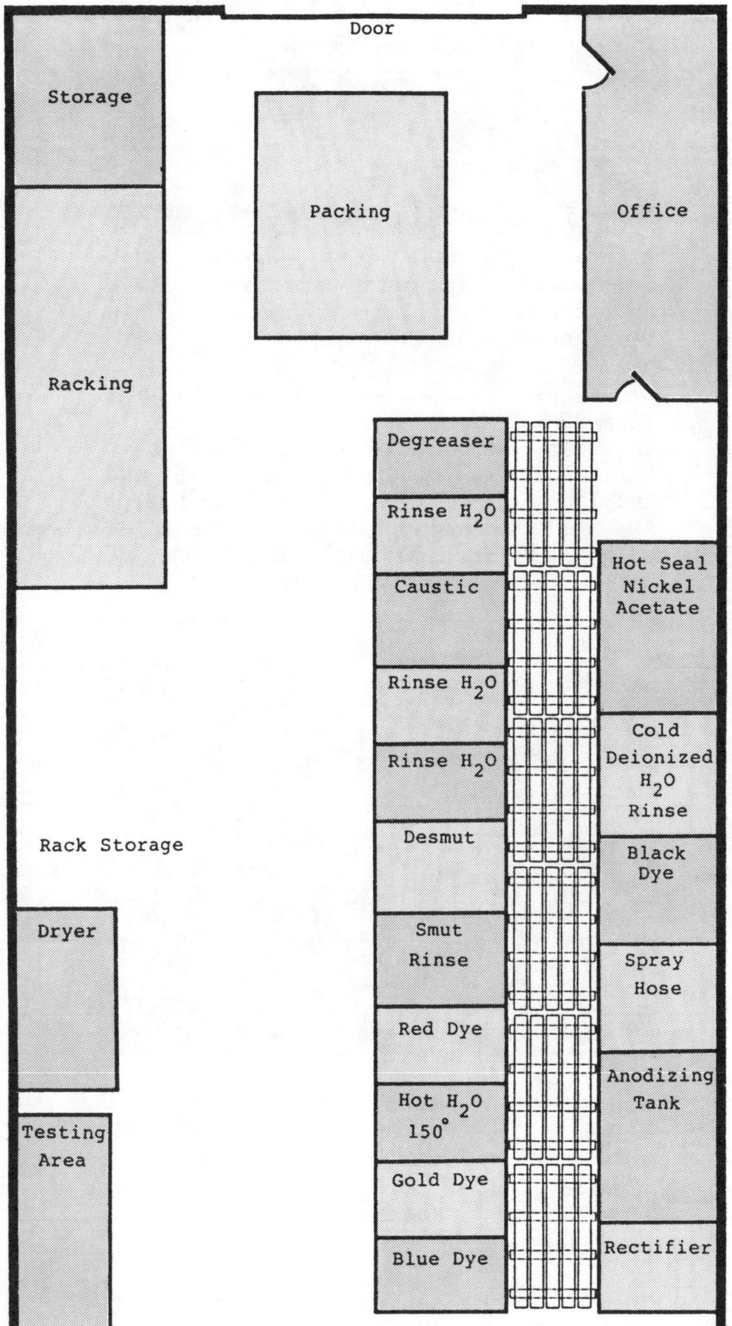

An overhead view of Milner's Anodizing.

Illustration credit: Jane Sullivan

As we arrive at Milner's Anodizing, Kirk Senseney is removing a rack of black anodized aluminum parts from a water rinse tank en route to the dryer.

He suggests that we look around until he is free. He cautions us, however, not to touch anything, and to be careful where we lean or put down our cameras, purses and the like. The spills on the floor boards and tanks throughout the shop are filled with chemicals that may cause harm. "Be alert and be careful," he suggests.

The sub-floor between the tanks is constructed of wood slats (nailed to 2 by 4's) with open spaces so the water/chemicals can drop down onto the drain in the concrete floor. This reduces slipping hazards from excess water and prevents electrical shocks.

The anodizing operator is **always** wearing rubber boots, just in case, to insulate against accidental electrical shocks.

As we walk around the shop, an orderly layout of tanks containing various acids, water rinses and other solutions catches our attention.

This is an overall view of the anodizing tank, vent system directly behind tank, and the D.C. power supply to the right.

This view shows the caustic tank with the lid closed. Note the angle of the lid which provides air draft to the vent pipe. Caustic acid creates hydrogen gas which is not nice to our lungs. The caustic solution is also dangerous if it comes in contact with the skin or the eyes!

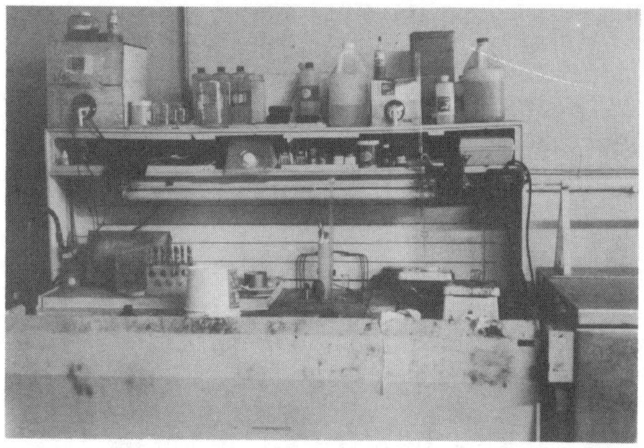

Milner's testing area for chemicals and dyes is located in the corner of the shop away from chemical contamination. The free sulfuric and aluminum content in the anodizing tank, along with the dye concentrations, are tested here.

The anodizer's racking area is neatly laid out for maximum efficiency. Organization of racking components pays big dividends. Note the overhead rod with "U" shaped hooks which the racks will hang from to speed up racking time. Various sizes of notch tipped discs, mounted on rods set at an angle, await racking assignments.

To better explain racking, Kirk shows us a few examples of aluminum pieces already anodized/dyed/sealed or ready for anodizing.

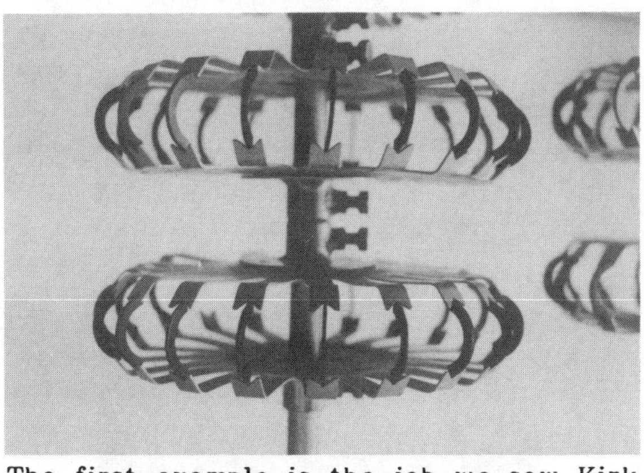

The first example is the job we saw Kirk working on as we arrived at Milner's.

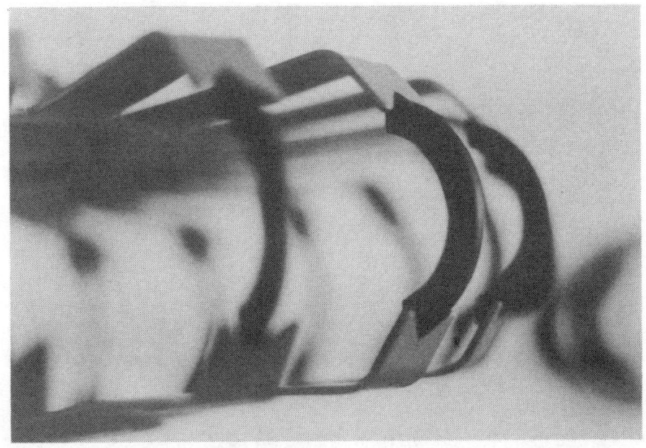

These are small aluminum parts that have already been anodized, dyed black, and sealed. They are ready to be unracked. Look closely. Can you see that each part is less than a half circle? The parts were hard to rack. It's difficult to insure precision holding, without falling out, during one step or another in the complete anodizing, dyeing, and sealing process. We are told, that more parts than the customer requires are anodized, to compensate for any which end up in the bottom of some tank! Planning ahead pays dividends later.

These are two large electrical chassis plates (15" long) racked and ready for anodizing. Each chassis is held by 2 notch tipped disc fingers on each end. This provides maximum current flow and the proper support for the aluminum throughout the anodizing process.

This view displays the dual connection on the top and bottom of the rack. Notice how the aluminum sets into the top and bottom notch.

Notice the firm connection between the aluminum piece and the notched tip disc.

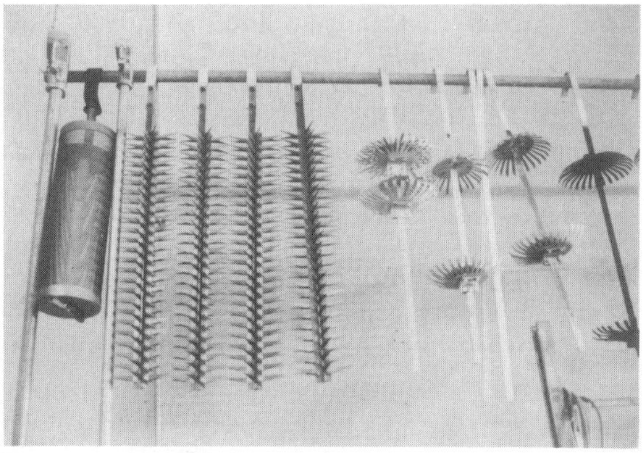

Here are additional racks ready for use. The anodizing barrel rack on the far left is made of titanium and is used to hold small parts such as rivets. The titanium racks in the middle have tapered points, while the racks to right are the notch tipped disc type and are made of aluminum. Titanium racks will outlast the aluminum racks because they do not require stripping after each anodizing. They are stronger, have more spring, but are also more expensive.

Some racks are made from titanium sheet with split fingers while others use heavy (1/8" diameter) titanium wire.

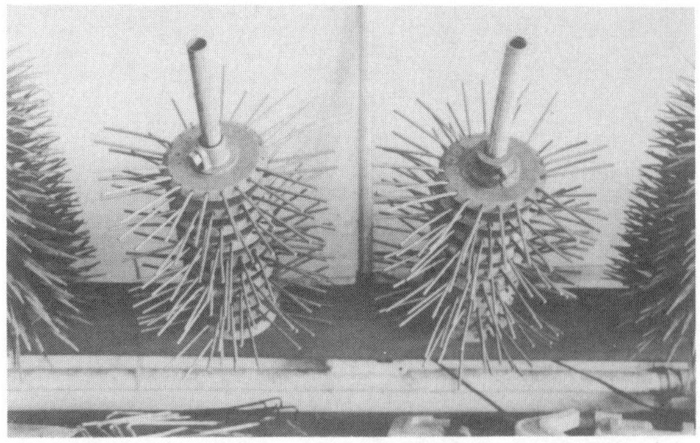

Notice that below the angled mountings for the notch tipped discs is a small, handy shelf for miscellaneous racking parts and tools.

One of the jobs for this day was to clear anodize these electrical component parts.

After attaching the upper and lower titanium disc racks to the spline, Kirk measures the correct distance between the discs with the aluminum piece. The individual notched fingers must be in alignment to provide proper current flow. Proper alignment also produces a sturdy connection to support the aluminum piece.

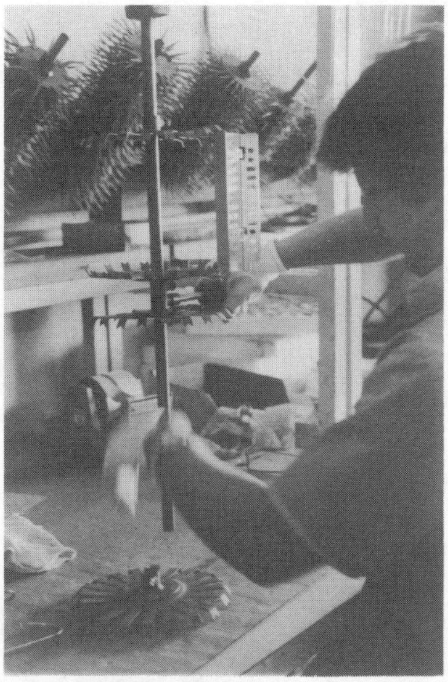

The second tier of disc racks is being attached. A screwdriver is used to tighten the notched tip disc rack to the titanium spline. Note that Kirk is wearing cotton gloves during the entire racking operation.

White cotton gloves reduce skin oil and down time caused by having to degrease aluminum parts prior to anodizing.

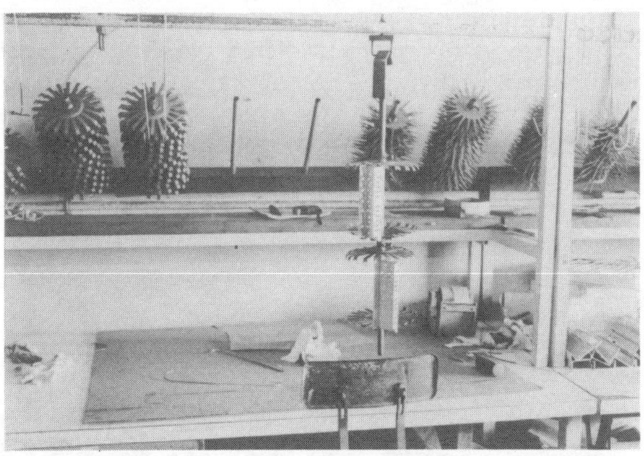

Additional components are suspended between the notched tip fingers until the upper and lower racks are completely filled. The more parts that can be processed at one time, the more efficient the operation. Time is money.

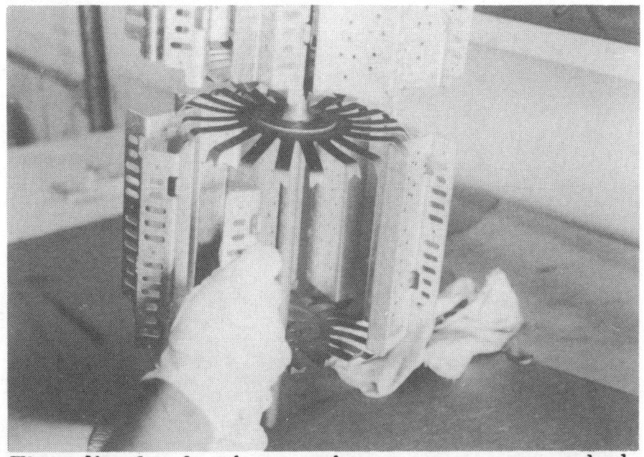

The final aluminum pieces are suspended between the titanium disc racks. Notice connection of upper and lower notch tip fingers.

The finished racked parts are ready for the next step in the anodizing cycle.

For time efficiency, two double tier racks are processed through the anodizing cycle simultaneously. This photo shows the two racks being lowered into the caustic etch tank.

A bubbling activity indicates that the proper reaction (i.e. etching) is taking place between the aluminum and the caustic soda (sodium hydroxide/lye). Hydrogen gas is given off by this reaction. Proper ventilation is required. Beware of breathing the fumes!

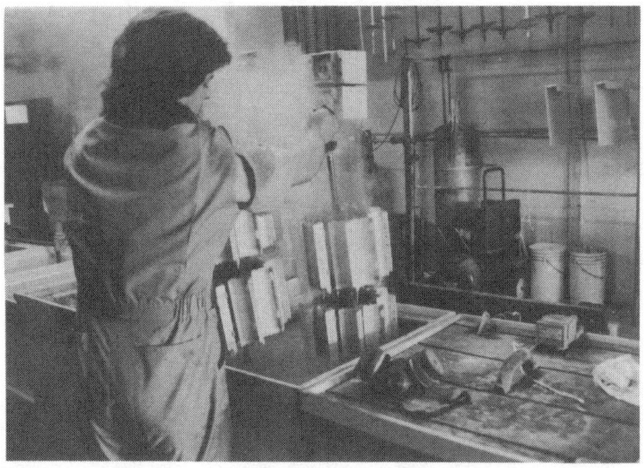

Both racks are removed from the caustic tank and moved through a water rinse tank.

To reduce potential drag-out of the caustic soda, the racks are put through yet another water rinse. Each rack unit goes into a separate water rinse tank.

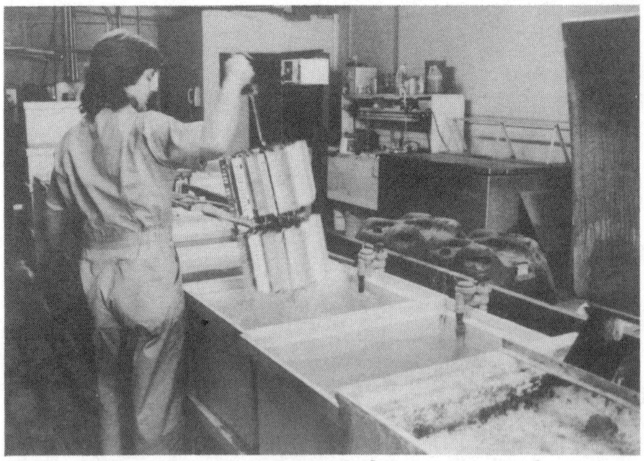

The racks are removed from second water rinse tanks.

The racks now spend a few minutes in a proprietary desmut tank. This removes the light gray to dark black colored smut on the aluminum surfaces. Smut is caused by the various metals that make up the aluminum alloy. The racks will remain in the desmuter until the aluminum pieces are white once again.

Then there is another water rinse for each rack.

A high pressure rinsing with a hose and spray nozzle, insures against chemical drag-out. A spray booth is built into a deep sink with drain.

The unit that the racks hang from is an interesting concept. The rod that is attached to both sides of the booth rotates around while the attachment in the middle of the bar spins both clockwise and counter-clockwise. This unique feature allows the pressure of the water from the spray nozzle to spin the rack and provide a most efficient rinsing operation.

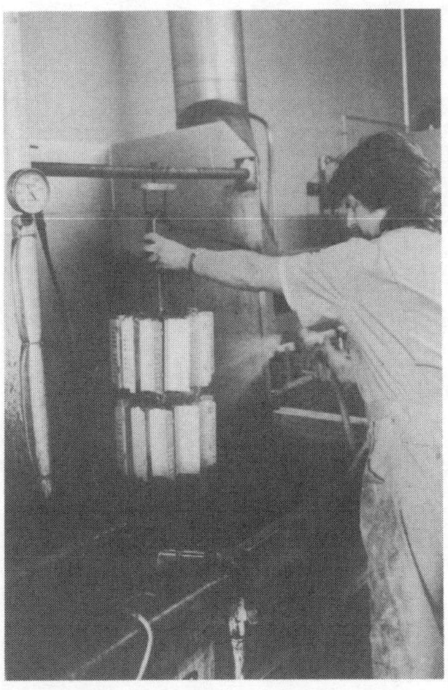

Kirk begins to spray the rack with the high pressure water hand nozzle.

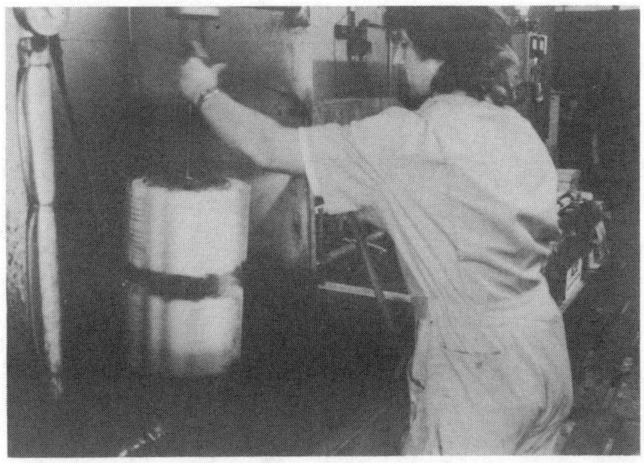

As the pressure increases, the rotation of the rack allows for a more complete water rinsing.

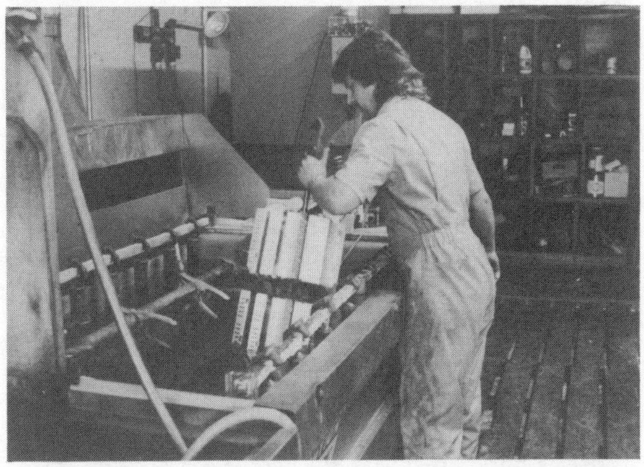

The rinsed rack is now ready for the anodizing tank.

The rack is lowered into the anodizing tank and the hook is attached to the copper bus bar.

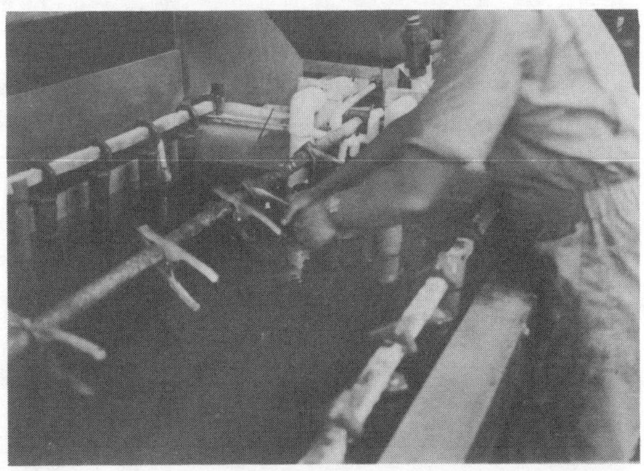

This bus bar is about 1 1/2" in diameter. The current is not turned on until all of the racks are attached or hanging from the bus bar.

Powerful spring clamps are used to provide additional contact support between the rack hook and the bus bar.

The power is turned on and the anodizing begins.
The D.C. voltage is brought up slowly (ramped). Notice the temperature of 62 degrees F. The optimum temperature for anodizing is about 70 degrees F. If the electrolyte temperature goes above 70 degrees, Kirk turns on the chiller to cool the tank down. 72 degrees F. and above is too high a temperature for optimum anodizing results.

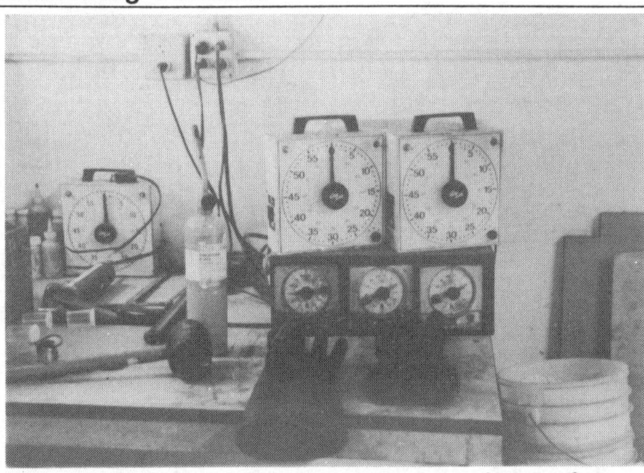

The aluminum pieces are anodized for a period of 1 hour. Time clocks are set to monitor all operations in the shop that require specific time periods for optimum results. A buzzer informs Kirk when the particular operation is completed. Some shops employ both buzzers and blinking lights to alert the workers that a particular operation is completed.

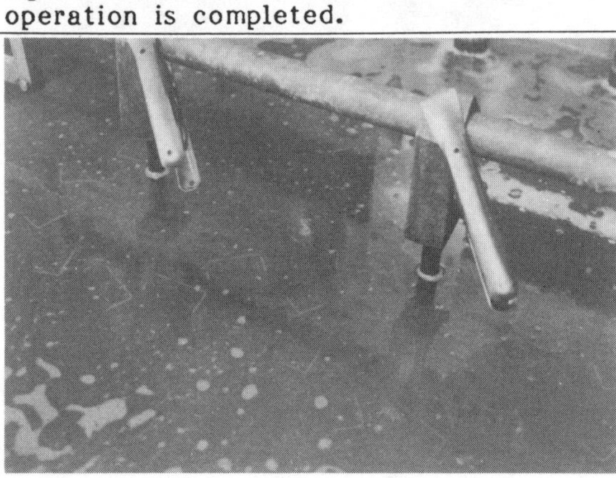

As the anodizing begins, hydrogen bubbles rise up from the cathodes. This gassing indicates the anodizing is taking place.

The anodizing is now complete and the racks are removed from the anodizing tank.

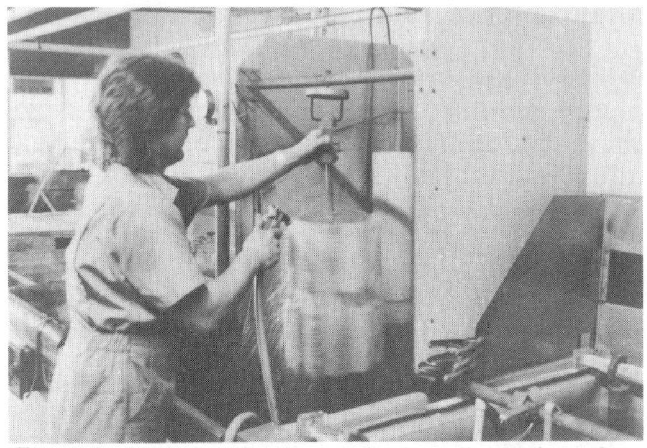

The aluminum pieces are rinsed again with the high pressure hose in the spray booth. For optimum results in anodizing, complete and thorough rinsings are absolutely necessary to eliminate drag-out. Note the racked items are spinning. This insures a more complete rinsing.

The racked items, if clear anodized, are suspended in a hot nickel acetate sealer for 10 minutes. If the aluminum pieces are to be dyed, they will go into the dye tank for a set period of time depending on the dye and color required. Rinsing and nickel acetate sealing would then follow.

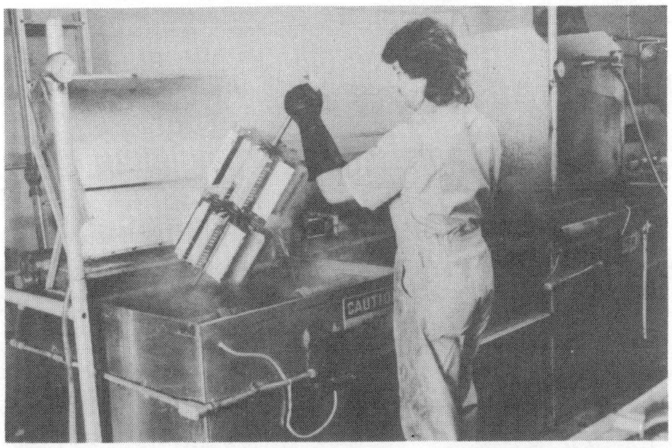

The racks are removed from the nickel acetate sealer. Kirk is wearing rubber gloves because the 10 minute sealing renders the metal too hot to handle.

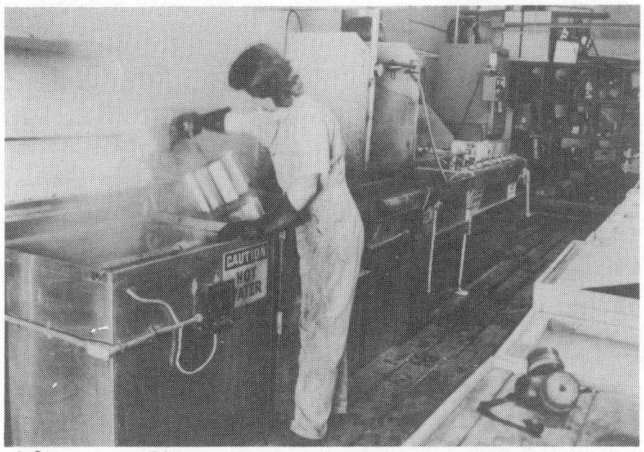

After sealing, the racks are placed in a room temperature, deionized water rinse tank.

A view of the deionized water rinse tank with racks.

Next step: hot water rinse.

To facilitate a complete drying of the anodized parts, Milner's uses a hot box, or dryer, essentially a large metal container with hot air. After lowering the racks into the box, Kirk closes the lid and the drying begins.

After a few minutes in the dryer, the racks are removed. The individual aluminum items will be unracked, wrapped in brown paper, boxed and made ready for the customer.

As Kirk removes the racks from the dryer, our field trip to Milner's Anodizing comes to a close. We thank Kirk, and Terry Burson for their hospitality.

During all of the visits made to various anodizers in the Oakland, California area, I generally found a casual attitude toward health hazards and safety practices. I do not agree with a causal or sloppy attitude toward safety precautions in the anodizing shop. Milner's was such a joy to visit because of their obvious concern for safety.

Anomatic Corporation

The Anomatic Corporation of Westerville, Ohio has developed a most unique and dramatic departure from conventional anodizing methods.

So far you have seen single and/or multiple racked items set to be anodized. The people at Anomatic utilize a highly automated processing system that provides low cost, high volume production of highly reflective anodized aluminum parts. A continuous titanium belt carries piece part racks, rather than having the racked items suspended from the traditional bus bar. The only manual operations required are the racking and unracking of the parts from the conveyor belt on each end of the line.

The Anomatic, as the machine is called, is a patented development of William B. Rusch, Sr. It is a smooth continuous system that produces high quality, air/gas pocket-free, anodizing, with no waste.

Aluminum pieces from less than 1 inch to parts having a diameter up to 4 inches and lengths to 10 inches can be anodized on the Anomatic machine.

Luggage key

The Anomatic Corporation produces such items as: large cartridge cases for automatic weapons used aboard military aircraft, lipstick tubes and other cosmetic cases, metal buttons and commemorative medallions, decorative hardware items, and various automotive, appliance and other OEM parts requiring uniform quality coatings.

Reader's Digest Lucky Sweepstakes coin

Commemorative object

The titanium strips that comprise the conveyor belt are joined to form an endless loop. The belt has spring clips riveted in place along each side to grip the interchangeable racks. The belt, after racking, travels around upper and lower rollers through a succession of cleaning, rinsing, bright dipping, rinsing, anodizing, rinsing, dyeing, rinsing, and sealing tanks. The unracking and drying take place at the end of the 50 foot machine.

The day I took a tour, Mardi-Gras coins and parts for the cosmetic industry were being anodized, dyed, and sealed. This factory tour was one of the most exciting I have ever taken!

Mardi Gras coin

Star coin

Mardi Gras coin

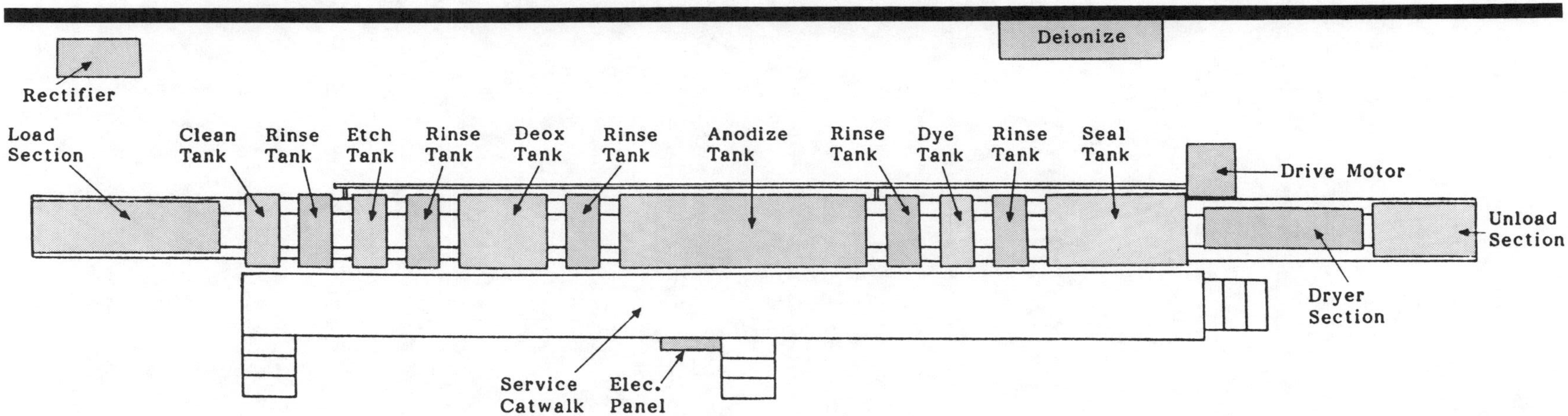

Overhead and side view of The Anomatic, the patented, smooth, continuous anodizing machine, developed by William B. Rusch, Sr. The dashed lines indicate the conveyor belt carrying the aluminum pieces through all the tanks. The loading and unloading tables, at the beginning and end, are where the aluminum pieces are attached and removed from the conveyor belt.

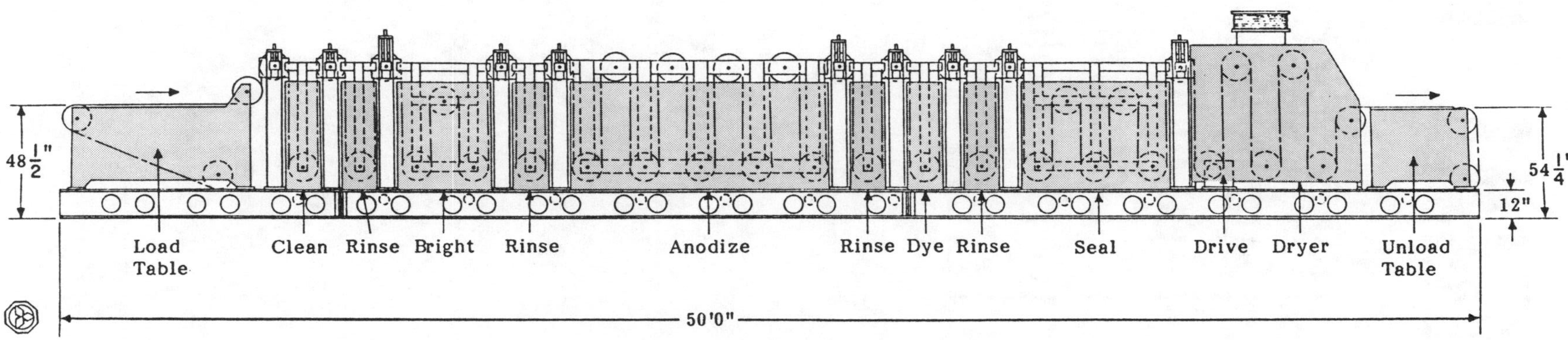

FINAL PAGES

Bibliography

Fabricated, painted aluminum, 2" dia., 1988.

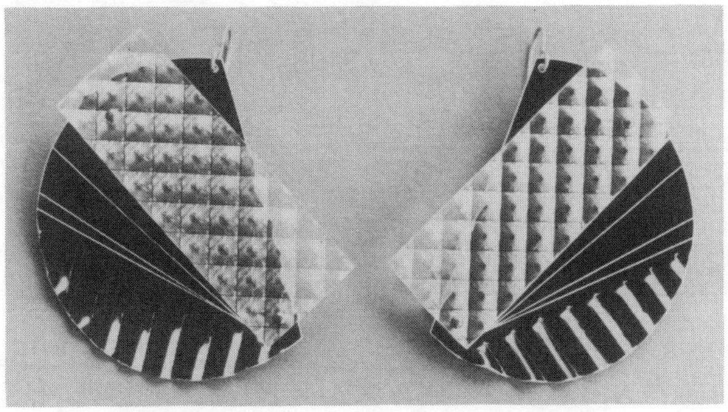

Earrings by David LaPlantz. Fabricated, painted aluminum, plexiglas, 2" dia., 1982-83.

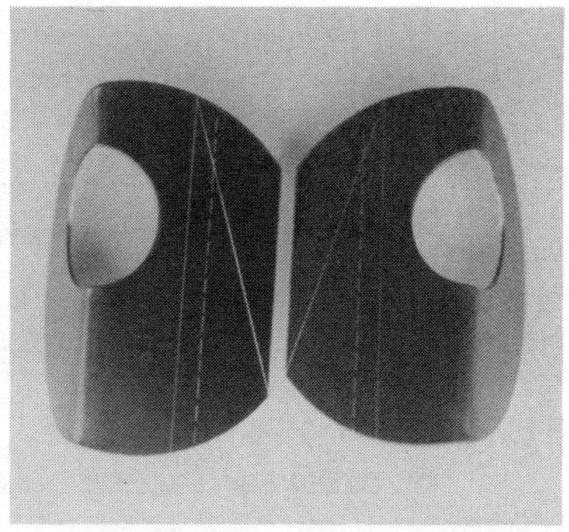

Fabricated, painted aluminum, 2" dia., 1988.

This bibliography is divided into four major areas:

1. **PRACTICAL/TECHNICAL**
2. **SAFETY**
3. **HEALTH**
4. **VISUAL/AESTHETIC**

This bibliography includes some titles I feel will help you to clarify and expand your knowledge of health/safety issues, anodizing aluminum, some related techniques, and visual/aesthetic considerations. It is not meant to be the last **word** on the subject or to include **all** of the titles ever published. After all, half of the fun of learning is to discover something no one told you about. Of course other titles exist, wouldn't it be fun to think you discovered them?

1. PRACTICAL/TECHNICAL

Aluminum Technology, Book 5: Finishing Aluminum. Aluminum Development Council of Australia Limited, Sydney, Australia, 1980.

Arrowsmith, D. J. "Seeing is Believing on Surface Treated Aluminum," Transactions of the Institute of Metal Finishing. 1986, Volume 64, pp. 6- 15. Published in England.

Auld, J. Leslie. Your Jewelry. Sylvan Press, London, 1951.

Boikess, Robert S., Edelson, Edward. Chemical Principles. Harper and Row, Publishers, New York, 1978.

Brace, A. W. and Sheasby, P. G. The Technology of Anodizing Aluminum. Robert Draper Ltd., Teddington, England, 1968.

Brand, Erick D. Band Instrument Repairing Manual. Erick D. Brand, Elkhart, Indiana, 1978.

Chamberlain, Walter. Etching and Engraving-The Thames and Hudson Manual of. Thames and Hudson Ltd., London, 1972.

Dull, Charles E, Metcalfe, H. Clark, Brooks, William O. Modern Physics. Henry Holt and Company, U.S.A., 1955.

Edwards, Rod. The Technique of Jewelry. Charles Scribner's Sons, New York, 1977.

Evans, Chuck. Jewelry-Contemporary Design and Technique. Davis Publications, Inc., Worcester, Massachusetts, 1983.

Newman, Lee Scott and Newman, Jay Hartley. Electroplating and Electroforming for Artists and Craftsmen. Crown Publishers, Inc., New York, 1979.
Rodgers, Thomas M. Handbook of Practical Electroplating. The Macmillan Company, New York, 1959.
Ross, Robert B. Handbook of Metal Treatments and Testing. Halsted Press, John Wiley and Sons, New York, 1977.
Thompson, G. E., Furneaux, R. C., and Wood, G. C. "Electron Microscopy of Anodic Films Formed on Aluminum in Sulfuric Acid," Transactions of the Institute of Metal Finishing. 1977, Volume 55. pp. 117-128. Published in England.
Tisdale, David. "Anodizing Aluminum." Metalsmith. Spring, 1985, Volume Five, Number Two. pp. 27-29. Published in the United States.
Untracht, Oppi. Jewelry Concepts and Technology. Doubleday and Company, Inc. New York, 1982.
Untracht, Oppi. Metal Techniques for Craftsmen. Doubleday, New York, 1968.
Van Horn, Kent R.,Ed. Aluminum, Volumes I, II, III. American Society for Metals, Metals Park, Ohio, 1967.
Warburton, L. Electro-Plating for the Amateur. Argus Books Limited, Watford, Hertfordshire, England, 1976.
Wernick, S. Pinner, R. and Sheasby, P. G. The Surface Treatment and Finishing of Aluminum and Its Alloys. Volume 1 and 2. Fifth Edition. Robert Draper Ltd. Teddington, England, 1987.
Wilson, Henry. Silverwork and Jewelry. Pitman Publishing Corp., London, 1902.

2. SAFETY

Chemical Safety Data Sheets. Manufacturing Chemists Association, Washington, D.C.
Ellis, J. G. and Riches, N. J. Safety and Laboratory Practice. Macmillan Press, London, 1978.
Fischer Manual of Laboratory Safety. Fisher Scientific Company, Pittsburgh, Pennsylvania, 1972.
Freeman, N. T. and Whitehead, J. Safety in the Chemical Laboratory. Academic Press, Subsidiary of Harcourt Brace Jovanovich, Publishers, San Francisco, 1982.
Green M. E. and Turk, A. Safety in Working with Chemicals. Macmillan Publishing Company, New York, 1978.
Hawley, G. A. Condensed Chemical Dictionary. Van Nostrand Reinhold Company, New York, 1977.
Metal Finishing Guidebook Directory. 56th Guidebook-Directory Issue 1988, Volume 86, No. 1A. Metals and Plastics Publications, Inc., Hackensack, New Jersey.
Pipitone, David A., ed. Safe Storage of Laboratory Chemicals. John Wiley and Sons, New York, 1984.
Sax, Newton Irving. Dangerous Properties of Industrial Materials. Van Nostrand Reinhold Company, New York, 1979.

3. HEALTH

Clark, Nancy and Cutter, Thomas and McGrane, Jean-Ann. Ventilation, A Practical Guide for Artists, Craftspeople, and Others in the Arts. Nick Lyons Books/Center for Occupational Hazards, New York, 1984.
International Labor Organization. Encyclopedia of Occupational Health and Safety. McGraw-Hill, New York, New York, 1972.
McCann, Michael, Ph.D. Artist Beware. Watson-Guptill Publications, New York, 1979.
McCann, Michael, Ph.D. Health Hazards Manual for Artists. Nick Lyons Books, New York, 1985.
National Institute of Occupational Safety and Health. The Industrial Environment--Its Evaluation and Control. U.S. Government Printing Office, Washington, D.C., 1973.
Sax, Newton Irving. Dangerous Properties of Industrial Materials. Van Nostrand Reinhold Company, New York, 1979.
Stellman, Jeanne, and Daum, Susan. Work is Dangerous to Your Health. Vintage Paperbacks, New York, New York, 1973.

Finishes for Aluminum, Process and Characteristics. Reynolds Metals Company.
Fisch, Arline. Textile Techniques in Metal. Van Nostrand Reinhold Company,New York, 1975.
Fishlock, David. Metal Colouring. Robert Draper Ltd., Teddington, England, 1962.
Glasstone, Samuel. The Fundamentals of Electrochemistry and Electrodeposition. American Electroplaters' Society. Franklin Publishing Company, Inc., New Jersey, 1960.
Goldstein, J. I. and Yakowitz, H., Eds. Practical Scanning Electron Microscopy. Plenum Press, New York, 1975.
Graham, A. Kenneth. Electroplating Engineering Handbook. Reinhold Publishing Corporation, New York, 1955.
Gray, Harry B. and Haight, Gilbert P., Jr. Basic Principles of Chemistry. W.A. Benjamin, Inc., New York, 1967.
Hatch, John E., Ed. Aluminum Properties and Physical Metallurgy. American Society for Metals, Metals Park, Ohio, 1984.
Hegmann, W. and Brace, A. W. Aluminum Workshop Practice. Technicopy Limited, Stonehouse, England, 1978.
Henley, V. F. The Anodic Oxidation of Aluminum and Its Alloys. Pergamon Press, Oxford, England, 1982.
Hiscox, Gardner D., Editor. Henley's Twentieth Century Formulas, Recipes, and Processes. The Norman W. Henley Publishing Company, New York, New York, 1930.
Hobbs, Douglas B. Working with Aluminum. The Bruce Publishing Company, Milwaukee, Wisconsin, 1947.
Hopkins, Albert A., Editor. The Scientific American Cyclopedia of Recipts, Notes and Queries. Munn and Company, New York, New York, 1906.
Hubner and Schiltknecht. The Practical Anodizing of Aluminum. MacDonald and Evans, London, 1960.
Hubner, and Speiser, Die Praxas Der Anodischen Oxidation Aluminiums. This is available ONLY in German from Aluminum Verlag, GMBH, Dusseldorf, West Germany.
Jenny, Alexander, Winifred Lewis-translator. The Anodic Oxidation of Aluminum and Its Alloys. Charles Griffin and Company, Ltd., London, 1940.
Jensen, Marvin. "Color, Color, Color-More on Aluminum & Aluminum Anodizing." Metalsmith, Fall, 1986, Volume Six, Number Four. pp. 32-35. Published in the United States.
Kennedy, Sylvia S.J. "David Tisdale." Ornament, March, 1984, Volume Seven, Number Three. pp. 20-23, 44-45. Published in the United States.
Knutson, Linda. Synthetic Dyes for Natural Fibers. Madrona Publishers, Seattle, 1982.
Koppes, Wayne F. Metal Finishes Manual-For Architectural Metals and Metal Products. Third Edition (Revised). National Association of Architectural Metal Manufacturers, Oak Park, Illinois, 1975.
Kuhn, A. T. Industrial Electrochemical Processes. Elsevier Publishing Company, New York, 1971.
Lindsay, Robert Bruce. Basic Concepts of Physics. Van Nostrand Reinhold Company, New York, 1971.
Lowenheim, Frederick A. Electroplating. McGraw-Hill Book Company, New York, 1978.
Lowenheim, Frederick A. Modern Electroplating. John Wiley and Sons, Inc., New York, 1963.
Maryon, Herbert. Metalwork and Enameling. Dover Publications, Inc., New York, 1955.
McCreight, Tim. The Complete Metalsmith. Davis Publications, Inc. Worcester, Massachusetts, 1982.
McCreight, Tim. Metalworking for Jewelry. Van Nostrand Reinhold, New York, 1979.
McRae, Bobbi A. Fiberworks Sourcebook. Betterway Publications, Inc., White Hall, Virginia, 1985.
Meilach, Dona. Decorative and Sculptural Ironwork. Crown Publishers, New York, 1977.
Metal Finishing Guidebook and Directory. 56th Guidebook-Directory Issue 1988. Volume 86, No. 1A. Metals and Plastics Publications, Inc. New Jersey.
Morton, Philip. Contemporary Jewelry. Revised edition. Holt, Reinhart, Winston, New York, 1976.
Neumann, Robert von. The Design and Creation of Jewelry. Chilton Book Company, Radnor, Pennsylvania, 1978.

4. VISUAL/AESTHETIC

The need for new visual stimulation and expansion of our ways of seeing, perceiving and developing ideas requires constant nurturing. The following titles may fulfill some of this requirement. Besides these titles, take a walk through your local library stacks or peruse the bookstore shelves to increase your visual knowledge. You never know what you might discover!

Albers, Josef. The Interaction of Color. Yale University Press, New Haven, Connecticut, 1963.

Beckwith, Carol. Maasai. Harry N. Abrams, Inc., New York, 1980.

Birren, Faber. Color Psychology and Color Therapy. Citadel Press, Secaucus, New Jersey, 1978.

Buzan, Tony. Use Both Sides of Your Brain. E.P. Dutton, New York, 1974.

Cartlidge, Barbara. Twentieth Century Jewelry. Henry N. Abrams, Inc., New York, 1985.

Chijiiwa, Hideaki. Color Harmony-A Guide to Creative Color Combinations. Rockport Publishers, Rockport, Massachusetts, 1987.

Color. Life Library of Photography. Time-Life Books, New York, 1970.

Contemporary Jewelry: The Americas, Australia, Europe and Japan. Exhibition Catalog published by the National Museum of Modern Art, Kyoto and Tokyo, Japan, 1984.

Courtney-Clark, Margaret. Ndebele-The Art of an African Tribe. Rizzoli International Publications, Inc., New York, 1986.

D'Allemagne, Henry Rene. Decorative Antique Ironwork-A Pictorial Treasury. Dover Publications, Inc., New York, 1968.

Dormer, Peter and Turner, Ralph. The New Jewelry, Trends + Traditions. Thames and Hudson, London, 1985.

Ebin, Victoria. The Body Decorated. Thames and Hudson, London, 1979.

Fisher, Angela. Africa Adorned. Harry N. Abrams, Inc., New York, 1984. Garner, Philippe, ed. Phaidon Encyclopedia of Decorative Arts, 1890-1940. Phaidon Press Limited, Oxford, England, 1978.

Gerlach, Martin, ed. Primitive and Folk Jewelry. Dover Publications, Inc., New York, 1971.

Gerritsen, Frans. Theory and Practice of Color. Van Nostrand Reinhold, New York, 1975.

Goethe, Johann Wolfgang von. Theory of Colors. 1810. Translated by Charles Lock Eastlake. M.I.T. Press, Cambridge, Massachusetts, 1973.

Graves, Maitland. The Art of Color and Design. McGraw-Hill, New York, 1951.

Gregory, R.L. Eye and Brain: The Psychology of Seeing. 3d ed. McGraw-Hill,New York, 1978.

Itten, Johannes. The Art of Color. Reinhold, New York, 1961.

Kuppers, Harold. Colour: Origins, Systems, Uses. Van Nostrand Reinhold, New York, 1974.

Lambert, Patricia and Staepelaere, Barbara and Fry, Mary G. Color and Fiber. Schiffer Publishing Ltd., West Chester, Pennsylvania, 1986.

Marx, Ellen. Optical Color and Simultaneity. Van Nostand Reinhold, New York, 1983.

Munroe, Alexandra, ed. Spectacular Helmets of Japan, 16th-19th Century. Exhibition Catalog, published by Japan House Gallery and the Association For the Research and Preservation of Japanese Helmets and Armor, Tokyo, Japan, 1985.

Musashi, Miyamoto. Harris, Victor-translator. A Book of Five Rings. The Overlook Press, Woodstock, New York, 1974.

Pirsig, Robert M. Zen and the Art of Motorcycle Maintenance. Morrow Publishing Company, New York, New York, 1974.

Raulet, Sylvia. Art Deco Jewelry. Rizzoli International Publications, Inc., New York, 1984.

Sharpe, Deborah T. The Psychology of Color and Design. Littlefield, Adams & Co., Totowa, New Jersey, 1975.

Stine, Richard. Non-Pointless Pencil Book. A Harvest/Harcourt Brace Jovanovich Book, 1978.

Verity, Enid. Color Observed. Van Nostrand Reinhold, New York, 1980.

Virel, Andre. Decorated Man-The Human Body as Art. Harry N. Abrams, Inc., Publishers, New York, 1979.

Willcox, Donald. New Design in Jewelry. Von Nostrand Reinhold, New York, 1970.

Willcox, Donald. Body Jewelry: International Perspectives. Henry Regnery Company, Chicago, 1973.

Supply Sources

THE FOLLOWING EQUIPMENT AND MATERIALS SUPPLIERS LISTING IS DIVIDED INTO THREE CATAGORIES:

CATEGORY A: is for most of us as we start out.

CATEGORY B: is the next big jump once we get more excited and need to expand the scope and scale of our work. Make sure you've got plenty of money, category B is not cheap! A suggestion, if you want to make larger anodized work, consider jobbing out the work to an anodizer who is set up to do bigger pieces. A large anodizing studio is prohibitive in cost for most of us! If you've got the bucks, do it.

CATEGORY C: includes other sources, some familiar to the jeweler/metalsmith, others not so familiar. Look closely.

This is not meant to be an exhaustive listing. For more complete local sources consult your telephone Yellow Pages or the Thomas Register. For a more complete national overview consult the back of the Metal Finishing Guidebook Directory.

A. SMALLER, LESS EXPENSIVE, MORE MODEST ANODIZING STUDIO SET-UP:

ALUMINUM
Hardware Stores
Lumber Yards
Home Improvement Stores
Hobby Stores
Scrap Metal Yards
Large aluminum retailers or other metals suppliers-check out their left-overs, end cuts, etc.

CLAMPS AND RACKING DEVICES
Discount and Home Improvement Stores
Hardware Stores-spring clamps
Stationary Stores-paper clamps in metal or plastic
Knitting/Weaving Stores-knitting needles for home-made racks
(Check the commercial racking companies near you-see Category B for a more expanded list.)

DYES
Weaving Stores-Fabric Dyes
Food Stores-Rit Dyes
EAST/WEST DyeCom--Dyes developed specifically for aluminum
P.O. Box 12294
Roanoke, VA 24024
703-345-4241

MET-L-ART, INC.--Dyes developed specifically for aluminum
P.O. Box 4611
Greenville, DE 19807
800-638-5278

Cerulean Blue, Ltd.--Fiber Reactive Dyes
P.O. Box 21168
Seattle, WA 98111
206-443-7744

RECTIFIERS FOR ANODIZING
Jewelry Supply Houses:

Allcraft Tool and Supply Company, Inc.
100 Frank Road
Hicksville, NY 11801
516-433-1660
212-895-0686
-or-
22 West 48th Street
New York, NY 10036
212-246-4740

Anchor Tool and Supply Company, Inc.
P.O. Box 265
Chatham, NJ 07928
201-887-8888

Borel & Frei
315 West Fifth Street
Los Angeles, CA 90013
213-689-4630

Jules Borel & Company
1110 Grand Avenue
Kansas City, MO 64106
816-421-6110
800-821-5686

William Dixon Company
Carlstadt, NJ 07072
201-935-0100

E. B. Fitler and Company
P.O. Box 242
Mifflinburg, PA 17844
717-966-2029

Otto Frei-Jules Borel
119 Third Street
P.O. Box 796
Oakland, CA 94604
415-832-0355
-or-
760 Market Street
San Francisco, CA 94102
415-421-8133

Gesswein and Company, Inc.
255 Hancock Avenue
P.O. Box 3998
Bridgeport, CT 06605
203-366-5400
800-243-4466
-or-
676 W. Wilson Avenue
Glendale, CA 91203
213-240-7113
800-232-2311
800-243-1204

Stanley O. Jacobs Company
315 West 5th Street
Los Angeles, CA 90013
213-624-6027
800-992-2029

Met-L-Art, Inc.
P.O. Box 4611
Greenville, DE 19807
800-638-5278

Marshall-Swartchild Company
2040 Milwaukee Avenue
Chicago, IL 60647
800-621-4767
-or-
109 N. Akard
Dallas, TX 75201
800-527-6390
-or-
1425 Fourth Avenue
Seattle, WA 98101
206-682-6158

Reactive Metals Studio
Box 425
Jerome, AZ 86331
602-634-3434

Rio Grande
6901 Washington, N.E.
Albuquerque, NM 87109
800-545-6566

I. Shor Company, Inc.
50 West 23 Street
New York, NY 10010
212-924-2200
800-221-8614

Swest, Inc.
10803 Composite Drive
Dallas, TX 75220
214-350-4011
-or-
5805-A Peachtree Corners East
Norcross, GA 30092
404-441-0404
-or-
1725 Victory Blvd.
Glendale, CA 91201
213-246-8385

BATTERY CHARGERS

Automotive Parts Houses
Discount and Home Improvement Stores
Hardware Stores

SEALANTS (SEALING SOLUTIONS)

East/West DyeCom
P.O. Box 12294
Roanoke, VA 24024
703-345-4241

MET-L-ART, INC.
P.O. Box 4611
Greenville, DE 19807
800-638-5278

Distilled or deionized water from food stores

FILTRATION SYSTEMS/PUMPS

Aquarium Stores
Local Water Pump Stores
Hardware Stores
Home Improvement Stores

PLASTIC TANKS, STAINLESS CONTAINERS-ANODIZING/DYEING

Restaurant Supply Houses
Hardware Stores
Discount and Home Improvement Stores

CHEMICALS (SULFURIC ACID, ETC.)

Check your yellow pages for chemical houses

LEAD FOR CATHODES

Check your yellow pages under sheet metal work
Scrap metal yards

HEATING DEVICES (SEALERS, DYES, ETC.)

Hardware stores
Flea markets, used stuff stores
Restaurant supply stores
Local natural gas/propane dealers

B. LARGER, MORE EXPENSIVE, MORE EXTENSIVE ANODIZING STUDIO SET-UP:

ALUMINUM

Check the Yellow Pages for local dealers
Try buying scrap from large companies in the form of end-cuts, over-runs, etc.
Scrap metal companies often purchase interesting surplus shapes, sizes, alloys, etc.
Be persistent, let your fingers do some walking, and get out to the scrap/junk yards and look around. Take a bone for the junk yard dog.

CLAMPS AND RACKING DEVICES

Apex Rack and Coating Company
739 Cottage Grove, S.E.
Grand Rapids, MI 49507
616-452-3139

Cosmos Minerals Corporation
906 Via Alondra
Camarillo, CA 93010
805-482-9825
800-235-5925

DBJ Industries
P.O. Box 4275
Hamden, CT 06410
203-281-4101

Duraclamp Company
1835 Whittier Avenue
Costa Mesa. CA 92627
714-645-1529

Moldsaver, Inc.
800 Kemp Street
Merritt Island, FL 32952
303-459-2766
800-327-2742

Rack Processing Company
2350 Arbor Blvd.
Dayton, OH 45439
513-294-1911

Sequel Corporation
119 Jonathan Blvd. N.
Chaska, MN 55318
612-448-3870

Servi-Sure Corporation
3553 N. Sheffield Avenue
Chicago, IL 60657
312-528-7575

Tilton Rack & Basket Corporation
66 Passaic Avenue
Fairfield, NJ 07006
201-226-6010

Vulcanium Corporation
3045 Commercial Avenue
Northbrook, IL 60062
312-498-3111
800-323-5348

DYES

Al-Tex Dyes Company
P.O. Box 2456
Charlotte, NC 28211
704-846-4117

Colonial Chemical & Dye Company
4041 Ridge Avenue
Philadelphia, PA 19129
215-438-2272

Duraclamp Company
1835 Whittier Avenue
Costa Mesa, CA 92627
714-645-1529

Lea Manufacturing Company
237 E. Aurora Street
Waterbury, CT 06720
203-753-5116

Okuno Chemical Industries Inc.
10-25 Hanatenhigashi 1 chome
Tsurumiku
Higashi-ku, Osaka 538, Japan
06-961-7781
(Sells products developed specifically for anodizing and dyeing die cast products)

C. H. Patrick & Company, Inc.
P.O. Box 2526
Greenville, SC 29602
803-244-4831

Pylam Products Company, Inc.
1001 Stewart Avenue
Garden City, NY 11530
516-222-1750

Sandoz Colors and Chemicals
4000 Monroe Road
Charlotte, NC 28205
704-372-0210

Specialty Chemicals and Services, Inc.
1615 Johnson Road, N.W.
P.O. Box 250089
Atlanta, GA 30325
404-799-1292

U.S. Specialty Color Corporation
2411 High Ridge Road
Matthews, NC 28105
704-846-4518

RECTIFIERS FOR ANODIZING

Aldonex, Inc.
2917 St. Charles Road
P.O. Box 148
Bellwood, Il 60104
312-547-5663

Brucar Process Systems Inc.
127M Brook Avenue
Deer Park, NY 11729
516-586-5800

Controlled Power Company
1955 Stephenson Hwy.
Troy, MI 48084
313-528-3700
800-521-4792

Kraft Dynatronix
105 Power Street
Amery, WI 54001
715-268-8118
800-826-7172

HBS Equipment Corporation
3000 Supply Avenue
Los Angeles, CA 90040
213-726-3033

LWD Scientific
132 Clement Street
San Francisco, CA 94118
415-387-4430

Rapid Power Technologies
Graysbridge Road
P.O. Box 291
Brookfield, CT 06804
203-775-0411
800-332-1111

Technic Inc., Equipment Division
485 Prospect Street
Pawtucket, RI 02860
401-728-8511

SEALANTS (SEALING SOLUTIONS)

Alcor Chemical Company, Inc.
9328 W. Bernice
Schiller Park, IL 60176
312-678-2880

Diversey Wyandotte Corporation
1506 John A. Papalas Drive
Lincoln Park, MI 48146
313-281-0930

Haviland Products Company
421 Ann Street, N.W.
Grand Rapids. MI 49504
616-361-6691
800-421-1134

Lea Manufacturing Company
237 E. Aurora Street
Waterbury, CT 06720
203-753-5116

Sandoz Colors and Chemicals
4000 Monroe Road
Charlotte, NC 28205
704-372-0210

Specialty Chemicals and Services, Inc.
1615 Johnson Road, N.W.
Atlanta, Georgia 30318
404-799-1292

U.S. Specialty Color Corp.
2411 High Ridge Road
Matthews, NC 28105
704-846-4518

FILTRATION SYSTEMS/PUMPS

Beaver Industries Inc.
Trimex Building, Route 11
Mooers, NY 12958
518-236-5489

Serfilco, Ltd.
1234 Depot Street
Glenview, IL 60025
312-998-9300
800-323-5431

Sethco Division,Met-Pro Corporation
70 Arkay Drive
Hauppauge, NY 11788
516-435-0530

PLASTIC TANKS

Belke Manufacturing Company
947 N. Cicero Avenue
Chicago, IL 60651
312-626-4606

Brucar Process Systems, Inc.
127M Brook Avenue
Deer Park, NY 11729
516-586-5800

Chem-Tainer Industries, Inc.
Division, County Plastics Corporation
361 Neptune Avenue
North Babylon, NY 11704
516-661-8300
800-645-5607

Grewe Plastics
123 S. 15th Street
Newark, NJ 07107
201-485-7602
800-227-0076

New England Plastic Coated Products, Inc.
350 Old Colony Road
Norton, MA 02766
617-222-3019

Udylite
2100 SE Main
P.O. Box 19601
Irvine, CA 92713
714-957-7800
-or-
21441 Hoover Road
Warren, MI 48089
313-497-9100

DROP-IN TANK LINERS

American Lining Corporation
5940 Reeds Road
Mission, KS 66202
913-362-9400
800-255-6245

DAG Plastics, Inc.
P.O. Box 147
Highway 91 N.
Belding, MI 48809
616-794-3960

Flexi-Liner Corporation
P.O. Box 1070
155 N, Aspan
Azusa, CA 91702
818-969-7702
800-423-4909

Specialty Plastic Fabricators
9624 W. Willow Lane
Mokena, IL 60448
312-479-5501

TESTING EQUIPMENT/OXIDE LAYER THICKNESS

CMI International Corporation
2301 Arthur Avenue
Elk Grove Village, IL 60007
312-439-4404

DeFelsko Corporation
410 Cedar Street
Ogdensburg, NY 13669
315-393-4450
800-448-3835

Electromatic Equipment Company, Inc.
606 Oakland Avenue
Cedarhurst, NY 11516
516-295-4300
800-645-4330

Kocour Company
4800 South St. Louis Avenue
Chicago, IL 60632
312-847-1111

Zorelco. Ltd.
8520 Garfield Blvd.
P.O. Box 25500
Cleveland, OH 44125
216-441-6100
800-321-4243

pH PAPER

Cerulean Blue, Ltd.
P.O. Box 21168
Seattle, WA 98111
206-443-7744

Fil-Chem Inc.
60 Franklin Avenue
Brooklyn, NY 11205
718-625-3400

VWR Scientific
P.O. Box 7900
San Francisco, CA 94120
415-467-4100

pH METERS

Analytical Measurements, Inc.
31 Willow Street
Chatham, NJ 07928
201-273-7500

Central Scientific Company
11222 Melrose Avenue
Franklin Park, IL 60131
312-451-0150

Hach Company
P.O. Box 389
Loveland, CO 85039
800-227-HACH

Lamotte Chemical Products Company
P.O. Box 329
Chestertown, MD 21620
301-778-3100

Nester Instruments
P.O. Box 1187
Millville, NJ 08332
609-825-0254

Presto-Tek Corp.
4101 N. Figueroa Street
Los Angeles, CA 90065
213-221-1178
800-421-8660

VRW Scientific
P.O. Box 7900
San Francisco, CA 94120
415-467-4100

SEE ALSO-GENERAL EQUIPMENT AND SUPPLIES LIST FOR OTHER LOCATIONS THROUGHOUT U.S.A.

GENERAL EQUIPMENT AND SUPPLIES

Central Scientific Company
111222 Melrose Avenue
Franklin Park, IL 60131
312-451-0150

VWR Scientific
P.O. Box 7900
San Francisco, CA 94120
415-467-4100
-or-
P.O. Box 66929
O'Hare AMF
Chicago, IL 60666
800-932-5000
-or-
P.O. Box 527
South Plainfield, NJ 07080
201-756-8030

and 24 other cities nationwide.

CHEMICALS

Bryant Laboratory, Inc.
1101 Fifth Street
Berkeley, CA 94710
415-526-3141

EM Science
A Division of EM Industries, Inc.
14714 Carmenita Rd, Suite 340
Norwalk, CA 90650
213-802-2908
800-821-9180
-or-
2909 Highland Avenue
Cincinnati, OH 45212
800-543-7152
-or-
3601 Kennedy Road
South Plainfield, NJ 07080
201-561-4000
800-526-7611

North Eastern Chemicals, Inc.
(Breatheasy, PMS solution for anodizing tanks)
240 Mary Street
Hackensack, NJ 07601
201-342-8774

Oakite Products, Inc.
50 Valley Road
Berkeley Heights, NJ 07922
201-464-6900
800-526-4473

Sargent-Welch Chemical Reagents
35 Stern Avenue
Springfield, New Jersey 07081
201-376-7050
-or-
9520 Midwest Avenue
Garfield Heights
Cleveland, OH 44125
216-587-3300
-or-
3125 7th Avenue, North
P.O. Box 10404
Birmingham, AL 35202
205-251-5125
-or-
5915 Peeler Street
Dallas, TX 75235
214-357-9381
-or-
4040 Dahlia Street
P.O. Box 7196
Denver, CO 80207
303-399-8220
-or-
1617 East Ball Road
Anaheim, CA 92803
714-772-3550

Check your local telephone Yellow Pages. Watch out, some chemical companies have high minimums. Shop around.

POLLUTION/WASTEWATER ASSISTANCE

Zerpol Corporation
1300 Schwab Road
Hatfield, PA 19440
215-368-0501

Haviland Products Company
421 Ann N.W.
Grand Rapids, MI 49504
616-361-6691
800-421-1134

RESISTS

Acu-Tape Division,
Foamtape Industries Inc.
658 Ridgefield Road
Wilton, CT 06897
203-762-0091

American Biltrite Inc.,
Tape Products Division
106 Gaither Drive
Mt. Laurel, NJ 18054
609-778-0700

Echo Supply
P.O. Box 32417
San Jose, CA 95152
800-621-0854

Flormel Adhesive Products, Inc.
P.O. Box 110
510 1st Avenue
West Haven, CT 16516
203-932-1345
800-356-7635

3M Company
Bldg. 223-6SE
St. Paul, MN 55144
612-733-7858

Sequel Corporation
119 Jonathan Blvd. N.
Chaska, MN 55318
612-448-3870

Tolber Division,
Pyramid Plastics, Inc.
220 W. 5th Street
Hope, AR 71801
501-777-3251

C. OTHER SOURCES:

SAFETY EQUIPMENT AND SUPPLIES

Bullard Safety
5586 E. Imperial
South Gate, CA 90280
213-771-8744

Conney Safety Products
3202 Latham Drive
P.O. Box 4190
Madison, WI 53711
800-356-9100

Lab Safety Supply
P.O. Box 1368
Janesville, WI 53547
608-754-2345
800-356-0783

PLASTIC CONTAINERS/PLASTIC PRODUCTS

Castino Restaurant Equipment and Supply
50 Utility Court
P.O. Box 2207
Rohnert Park, CA 94928
800-238-0404

Cambro Manufacturing Company
7601 Clay Avenue
P.O. Box 2000
Huntington Beach, CA 92647
714-848-1555
800-854-7631

Consolidated Plastics Company
1864 Enterprise Parkway
Twinsburg, OH 44087
216-425-3900
800-321-1980

Finegold's Restaurant-Hotel Equipment
9th and X Streets
Box 1113
Sacramento, CA 95805
916-441-6931

BASIC JEWELRY SUPPLIES

In addition to the listing above, consider the following:

Alpha Supply, Inc,
1225 Hollis Street
Box 2133
Bremerton, WA 98310
206-373-3302
206377-5629

ARE, Inc.
Box 8W
Greensboro Bend, VT 05842
800-451-5086

Bourget Bros.
1636 11th Street
Santa Monica, CA 90404
213-450-6556

D.J.S. & Company
1114 Greentree Road
Newark, DE 19713

T. B. Hagstoz & Son, Inc.
709 Sansom Street
Philadelphia, PA 19106
215-922-1627

C. R. Hill
2734 W. 11 Mile Road
Berkley, MI 48072
313-543-1555
800-521-1221

Komokin
(Traditional Japanese tools)
18-17, Minamiaoyama 3-Chome
Minato-Ku
Tokyo-107, Japan

Texas Jewelers Supply
7459 Airport Freeway
Fort Worth, TX 76118
817-595-0083

FINDINGS

B. A. Ballou and Company, Inc.
800 Waterman Avenue
East Providence, RI 02914
401-438-7000

Boston Findings Co.
387 Washington Street
Boston, MA 02108
617-357-9599
800-225-2436

T. B. Hagstoz & Son, Inc.
709 Sansom Street
Philadelphia, PA 19106
215-922-1627

Metal City Findings Company
456 Nordhoff Place
Englewood, NJ 07631
201-894-1155

Rio Grande
6901 Washington N.E.
Albuquerque, NM 87109
800-545-6566

Swest, Inc.
10803 Composite Drive
Dallas, TX 75220
214-350-4011

United Findings Company
2741 N. 29th Avenue, Suite 306
Hollywood, FL 33020
305-925-2233
800-327-0259

GENERAL

Arrow Star, Inc.
(Studio organizers, plastic containers)
661 William Street
Lynbrook, NY 11563
516-481-1414
800-645-2833
-or-
122 N. York Avenue
Dept. 61
Elmhurst, IL 60126
-or-
20425 So. Fordyce Avenue
Dept. 61
Carson, CA 90810

Eric Brand
(Musical instrument repair tools, supplies)
1117 West Beardsley Avenue
Elkhart, IN 46514

Brookstone Company
(Hard-to-find tools)
127 Vose Farm Road
Peterborough, NH 03458
603-924-9541

Brodhead-Garrett
4560 East 71st Street
Cleveland, OH 44105
216-341-0248
800-321-6730
-or-
P.O. Box 15528
Sacramento, CA 95852
800-824-7900
-or-
P.O. Box 6438
Macon, GA 31213
800-841-4967

Brownels, Inc.
(Gunsmithing supplies)
Route 2, Box 1
Montezuma, IA 50171
515-623-5401

C & H Distributors, Inc.
400 South 5th Street
P.O. Box 04499
Milwaukee, WI 53204
414-271-2250
800-558-9966

DRI Industries-Workshops Center
Box 28114
Warrensville Heights, OH 44128
216-831-0283
800-372-5282

The Eastwood Company
147 Pennsylvania Avenue
P.O. Box 296
Malvern, PA 19355
215-644-4412
800-345-1178

Jerry Company, Inc.
601 Linden Place
Evanston, IL 60202
312-475-8440

Leichtung, Inc.
4944 Commerce Parkway
Cleveland, OH 44128
216-831-7645
800-321-6840

The Graver Company
(Chisels, gravers,etc.)
879 Raymond Hill Road
Oakdale, CT 06370
203-848-8031

Post Tool
400 Oak Street
Oakland, CA 94607
800-367-8665
Other California locations:
Santa Rosa
Stockton
Fresno
Sacramento
Modesto

Sax Arts and Crafts
P.O. Box 2511
Allentown, PA 18001
215-395-8531
-or-
P.O. Box 2002
Milwaukee, WI 53201
414-272-4900
-or-
P.O. Box 95039
Arlington, TX 76005
817-640-0009

Small Parts, Inc.
6901 N.E. 3rd Avenue
P.O. Box 381736
Miami, FL 33138
305-751-0856
305-759-8167

Sporty's Tool Shop
Clermont County Airport
Batavia, OH 45103
800-543-7638

Tap Plastics, Inc.
12404 N.E. Halsey
Portland, OR 97230
503-256-3080
-or-
4538 Auburn Blvd.
Sacramento, CA 95841
916-481-7584
-or-
2945 S.W. Temple
Salt Lake City, UT 84115
801-484-7607

U.S. General Supply Corporation
100 Commercial Street
Plainview, NY 11803
516-349-7275
800-645-7077

Roper Whitney, Inc.
(Punches:hand/hydraulic)
2833 Huffman Blvd
Rockford, IL 61101
815-962-3011

PACKAGING CONTAINERS

The Mason Box Company, Inc.
Attleboro Falls, MA 02763
617-695-9381
800-225-2708

C. J. Fox Company
236 Aborn Street
P. O. Box 6186
Providence, RI 02940
401-421-2000
800-556-6868

Bradley's Plastic Bag Company
9130 Firestone Blvd.
Downey, CA 90241
213-289-0811
213-773-1478

Freund Can Company
198 West 84th Street
Chicago, IL 60620
312-224-4230

Conversion Chart

SOME HELPFUL CONVERSION FACTORS:

1 pint = 0.4732 liter.
2 pints = 1 quart = 0.9463 liter.
4 quarts = 1 gallon = 3.7853 liters.
128 fluid ounces = 1 gallon.
28.349 grams = 1 ounce.

Pints (pt.) multiply by 0.47 = liters (L.).
Liters multiplied by 2.11 = pints (pt.).
Quarts (qt.) multiply by 0.95 = liters (L.).
Liters multiplied by 1.06 = quarts (qt.).
Gallons (gal.) multiply by 3.78 = liters (L.).
Liters multiplied by 0.26 = gallons (gal.).
Grams per liter (g./L.) multiplied by 0.134 = avoirdupois ounces per gallon (oz./gal.).
Grams (g.) multiplied by 0.035 = ounces (oz.).
Avoirdupois ounces per gallon (oz./gal.) multiplied by 7.5 = grams per liter (g./L.).
Ounces (oz.) multiplied by 28.34 = grams (g.).
Fluid ounces per gallon (fl. oz./gal.) multiplied by 7.81 = cubic centimeters per liter (cc./L.).
Amperes per square decimeter (amp./dm.2) multiplied by 9.29 = amperes per sq. ft. (amp./ft.2).
Amperes per square foot (amp./ft.2) multiplied by 0.108 = amperes per square decimeter (amp./dm.2).

Ounces to Grams

Ounces	Grams
1	28
2	56
3	84
4	112
5	140

Gallons to Liters

Gallons	Liters
1	3.8
2	7.6
3	11.4
4	15.2
5	19.0

TEMPERATURE:

Fahrenheit (Degrees F) multiply fahrenheit temperature times 5/9 after subtracting 32 = degrees Celsius.
Celsius (Degrees C) multiply celsius temperature times 9/5 and then add 32 = degrees Fahrenheit.

Dye Color Chip Chart Information

SANDOZ ALUMINUM DYES

Sandoz Colors and Chemicals Corporation
4000 Monroe Road
Charlotte, North Carolina 28205
704-372-0210

NAME	CONCENTRATION GRAMS/LITER	OPTIMUM pH RANGE	LIGHTFASTNESS
Orminal 4N*	15.0	4.8-5.0	8
Gold S	2.0	6.5-7.0	5
Fast Gold L	2.0	5.5-6.0	7-8
Copper BF	2.0	6.0-7.0	4
Bronze G	2.0	4.5-5.0	6-7
Bronze 2LW	2.0	5.3-5.8	7
Fast Bronze L	2.0	5.5-6.0	7-8
Bordeaux R	2.0	5.8-6.3	3
Bordeaux RL	5.0	5.0-5.5	6-7
Red GLW	4.0	5.8-6.3	7-8
Fast Red B3LW*	5.0	5.2-5.7	8
Fiery Red ML	5.0	5.0-5.5	6-7
Orange 3A	2.0	5.8-6.3	6
Orange 2B	2.0	5.8-6.3	6
Orange RL	2.0	6.5-7.0	6
Golden Orange RLW	2.0	6.8-7.2	7-8
Yellow 4A	2.0	5.5-6.0	6
Yellow 3GL	3.0	5.7-6.2	8
Yellow D	2.0	5.5-6.0	5
Yellow G3LW	2.0	5.7-6.2	8
Green AEN	2.0	5.5-6.0	7
Green AX	2.0	5.5-6.0	6
Green LWN	2.0	6.0-6.3	7-8
Olive Drab No. 9	4.0	6.5-7.0	3
Blue A	2.0	5.5-6.0	8
Blue 4A	2.0	5.8-6.3	4
Blue 2LW	2.0	4.5-6.0	8
Fast Blue G*	4.0	5.5-6.0	7-8
Turquoise PLW*	5.0	6.5-7.0	8
Violet 3D	2.0	5.0-6.5	3
Violet CLW*	2.0	5.0-5.5	6-7
Brown BL	2.0	6.3-6.8	5
Brown GL	2.0	6.3-6.8	5
Brown GSL	5.0	4.5-5.5	6-7
Brown R	4.0	6.5-7.0	5-6
Olive Brown 2R	4.0	5.3-5.8	6
Yellow Brown 2G	2.0	4.5-5.0	5-6
Grey BL	1.0	5.5-6.0	6
Grey NL Liquid	1.0	5.5-6.0	6-7
Black BK	10.0	6.5-7.2	6-7
Black 2LW	10.0	4.2-4.7	8
Deep Black MLW*	10.0	4.2-4.5	8
Deep Black H3LW*	10.0	5.5-6.0	8
Fast Black GL Paste*	30.0	5.2-5.6	8
Fast Black OA	10.0	5.8-6.2	7-8

***SANODAL brand**

U.S. SPECIALTY COLOR CORP., INC.

542 N. Perry Street
Johnstown, NY 12095
or
P.O. Box 1417
Matthews, NC 28106
518-762-8031
518-762-8192

NAME	CONCENTRATION GRAMS/LITER	OPTIMUM pH RANGE	LIGHTFASTNESS
F.A.O.	15.0	4.5	8
SPECIALTY GOLD S	2.0	7.0	5
SPECIALTY COPPER BF	2.0	7.0	1
SPECIALTY BORDEAUX R	2.0	6.0	3
SPECIALTY BORDEAUX 2R	3.0	6.0	6-7
SPECIALTY ORANGE 3A	2.0	6.0	6
SPECIALTY GOLDEN ORANGE RLW	2.0	7.0	7-8
SPECIALTY YELLOW 4A	2.0	6.0	6
SPECIALTY GREEN AEN	2.0	6.0	7
SPECIALTY GREEN B CONC.	2.0	6.0	6
SPECIALTY OLIVE DRAB	2.0	7.5	3
SPECIALTY BLUE A	2.0	6.0	8
SPECIALTY BLUE 4A	2.0	6.0	4
SPECIALTY BROWN BL	2.0	6.5	5
SPECIALTY BROWN GL	2.0	6.5	5
SPECIALTY GREY BL	2.0	5.5	6
SPECIALTY BLACK BK/BKA	10.0	7.0	5-7
SPECIALTY BLACK 2LW	10.0	4.0	8
SPECIALTY DEEP BLACK MLW	10.0	4.5	8
SPECIALTY DEEP BLACK MLWA	10.0	4.6	8
SPECIALTY BLACK FLW	10.0	4.7	8

Brooch by David LaPlantz, fabricated, painted aluminum, 2 1/2" X 2", 1984.

PATCOALUM ALUMINUM DYES

C.H. Patrick and Company., Inc.
P.O. Box 2526
Greenville, NC 29602
803-244-4831

NAME	CONCENTRATION GRAMS/LITER	OPTIMUN pH RANGE	LIGHTFASTNESS
YELLOW X-3G	0.5-2.0	5.0-6.0	6-6/7
YELLOW 4A	0.5-2.0	5.0-6.0	6
GOLD FAO	7.5-15.0	4.5-5.5	8
GOLD S	1.0	5.0-6.0	5/6
GOLD EAN 200%	0.5	6.0-7.0	5
BRASS O	0.1-2.0	4.5-5.5	6
ORANGE 3A	0.5-2.0	5.5-6.0	6
COPPER BF	0.5-2.0	5.0-5.5	4
BORDEAUX R	0.5-2.0	5.0-6.0	4
MAGENTA X-F	0.5-2.0	5.0-6.0	4
NAVY X-R	1.0-5.0	5.0-6.0	6/7
BLUE 4A	0.5-2.0	5.0-6.0	4
BLUE A	1.0-2.0	5.0-6.0	8
TURQUOISE X-L	1.0-5.0	5.0-6.0	7/8
BRIGHT GREEN X-K	1.0-6.0	4.0-4.5	7
FOREST GREEN X-B	0.5-4.0	5.0-6.0	6
OLIVE GREEN OG-107	0.5-4.0	5.0-6.0	5/6
BROWN GL	1.0-2.0	5.0-6.0	5
DEEP BROWN X-RL	0.5-4.0	5.0-6.0	6
DEEP BROWN X-YLN	0.5-4.0	5.0-6.0	6
GREY BL	0.5-2.0	5.0-6.0	6
BLACK BK-N	10.0	5.0-5.5	6/7-7
BLACK CHP	10.0	5.0-5.5	6/7-7
DEEP BLACK MLW-N	10.0	4.5-5.0	8

Brooch by David LaPlantz, fabricated, painted aluminum, plexiglas, copper, waxed linen thread, bamboo colored with permanent marking felt tip pens, actual size, 1986.

Appendix

In the human body, the appendix is an unnecessary part. It is often forgotten or taken for granted until it ruptures. At that time, it becomes quite noticeable, and so is the scar, after the operation to remove the ruptured organ.

This appendix, however, serves as a vital place to house those items I forgot to place within the book or that have recently come to light.

TEMPERATURE INDICATOR

This temperature indicator is an alternative to a thermometer in the sealing tank. It is an inexpensive item found in the variety section of your local food or discount store. It is called a "Wet Willie" on the West Coast and I've seen them in Ohio under the name of "Pot Watcher." Whatever name it is called by, this glass disc is placed in the bottom of the sealing tank. As the solution comes close to the boiling point the "Wet Willie" will rattle, signaling that the boiling point is close at hand. If you like noises to warn you about temperature, instead of looking at a thermometer, this item is for you!

THERMALBAR

The Thermalbar Corporation sells an inexpensive thermometer that measures from 90-200 degrees F. A bright, liquid crystal display is surrounded by cast polyester resin. The price is under $15.00. Write to:

THERMALBAR CORPORATION
Box 14291
N. Palm Beach
Florida 33408
800-443-9333

TEMPILSTIKS

The following may be useful in torch annealing, when using a kiln without a pyrometer, or to predict certain results, in heat textures as Chuck Evans and Mike Edwards have discussed and demonstrated.

To measure the annealing temperature of aluminum, special chemical pencils may be purchased at welding supply houses. One such measuring device is called a Tempilstik. Each is rated for a specific temperature. The mark from the pencil will melt or show as a liquid smear at the specified temperature rating. Remember, aluminum is annealed between 650 to 800 degrees F. and air cooled.

PEELABLE LACQUERS

Peelable lacquers are available from:

BATAVIA COATINGS DIVISION/WHITTAKER
P.O. Box 428
Batavia, Illinois 60510
312-879-6800
(32R+ Red Lacquer Stop-off)

TOLBER DIVISION/PYRAMID PLASTICS, INC.
Millhorn Chemical and Supply
P.O. Box 460 Maywood, California 90270
213-771-8301

AN ALTERNATIVE TO PEELABLE LACQUERS

The peelable lacquers work well. The minimum purchase price or minimum size container, possibly a gallon size, however, may be more than you want to "have" around. Rubber cement works in a similar way, is cheaper, and can be purchased in small quantities.

FREE BOOK LIST

The Metal Finishing folks have an extensive book list on anodizing, plating, etc. Write to:

Metal Finishing
1 University Plaza
Hackensack, New Jersey
07601

VISION OR BLINDNESS

Don't forget! If you can see this page, watch out for those dangerous chemical splashes. **WEAR THE APPROPRIATE SAFETY GEAR!**

CLEAR VISION ALWAYS

Some of you may have noticed the goggles I was wearing in the three photos in the Safety chapter were **not** splash goggles, but impact goggles. There is a major difference. Wearing the wrong goggles might change your life. Splash goggles **do not** have holes, while impact goggles do have holes. I did this, on purpose, to see how many of you would notice. Did you notice? Good for you!

LEFT-HANDED PEOPLE COUNT ALSO

When I started to write this book, it occurred to me that I rarely see left-handed people demonstrating in books. So, let us all raise our left hands and salute the left-handed people of the world.

TATTOO

Yes, the left-handed demonstrator had a tattoo. The tattoo is the symbol for Artists Anodizing Aluminum. If you figured it out, congratulations.

EXPERIMENTS/RECORD KEEPING

When experimenting with different strengths of electrolytes, different electrolytes, aluminum alloys, dye concentrations, etc. I recommend that small portions, like quart size or less, be mixed. That way you won't waste valuable time and chemicals if the experiment fails. Keep exact records of the proportions and results, even if the results prove to be a failure. Don't throw away your samples. Tape them into your notebook. You never know when some new breakthrough will develop. Weeks or months later you mind may replay the scenes of that experiment in your subconscious, and see some new, more exciting relationships.

THE LAST DIGRESSION

Fear of failure, like our ever present shadows, is our constant companion. Fear may be used to help us to do better. It can also become a killer of spirit, a mangler of our faith, the dog that chews us up and spits out the remains. Not a pleasant picture, is it?

Of course not, but why let fear stop you from achieving your destiny? An old Chinese proverb says,"Fear knocked at the door, faith answered, but no one was there." This proverb helps to put life and its fears in some reasonable perspective.

In the face of fear, even a small bit of faith or courage will see you through to some degree of success.

A recent TV spot asked, "Can you recall what you were worrying about this time last year?" I couldn't remember. How about you?

After all, what is success? What is winning? Is winning something we try to do to impress others? If that is the only reason to win, I'd rather play another game. I'm not out to impress others, or puff up my own ego. I can do that by winning inside -- by honestly feeling good about myself, my skills, and my art. This inside type of knowledge we develop will help to make up the outward confidence, joy and spirited glow that most humans respect and emulate. Work for yourself, not to impress others. Keep focused on that which is special, unique and vital to your spiritual, intellectual and technical growth. These are the truths to attain and all the rest of life's riches, including respect from your peers will be an automatic given. Your spirit will thank you for discovering just who you are working for and who you wish to impress. Is that you? If not, why not?

DREAMS VERSUS REALITY REALITY VERSUS DREAMS

During the final "push" to finish this book, I had a most wonderful and "revealing" dream. Prior to this dream, I had not been dreaming at all, or at least nothing I could recall in the morning.

In the dream, I was watching a man working with his hands. He was making jewelry. I was fascinated by the way his hands looked, moved and worked. I kept looking at his hands and he kept looking at mine. We never spoke. I never actually saw his face.

I noticed how his hands, in comparison to mine, looked like hands that worked hard. He had callouses, cuts, and rough skin in all of those special places that grip tools and metal, with firm conviction and obvious direction of purpose. As I looked at my hands, I saw soft skin, no cuts, a lack of obvious callouses, in short, non-working hands. It's been nearly two months since I've last been able to make jewelry.

When I awoke, I was able to replay the dream in my mind and the message seemed obvious.

In the dream, I was watching myself working. The hands looked like my hands do when I am working. I believe my mind was telling me, "It is time to finish this book. Stop working with words and drawings! Get off that computer keyboard and get back to the bench where you belong!"

I feel my subconscious knew the end of the book was close and was signalling to me that soon, real "joy" would return to my life.

A new wave of jewelry is soon to be realized. The hands I was watching in my dream will be mine again!

Happy anodizing aluminum.

A long and healthy life to you! Until we meet again.....

HUMAN

David LaPlantz, a product of the World War II romance between Bernice and Milton LaPlantz, was born in 1944 in Toledo, Ohio. His formative years were spent growing in the joy and wonder of a nearby, farming community called Genoa.

His fondest memories were of riding his Monarch Silver King bicycle through the small town streets and alleys "checking" out other people's garbage barrels for "finds." He always seemed to be making objects from wood, plastic, cardboard, and any other material available. Going to the dump with his Father and Grandfather J.J. every Saturday was the highlight of each week. Their old red truck always seemed to bring more home than they took to the dump.

An athletic injury early in his high school years set the stage for a very "determined" attitude that became the foundation of his life. He learned from extreme frustration how to "win" on the inside, rather than the exterior winning that others see.

He likes chow dogs, old motorcycles, especially Triumphs, BSA's, and Nortons, rock'n roll music, jazz groups, working out to stay fit, going to the movies and making jewelry; not necessarily in any order.

His life-long goals are to find the good and the beautiful in all he sees, hears, touches and creates with his hands. He is a committed teacher and studio artist.

ARTIST

David LaPlantz also works under the pseudonym of David D. Deco or D. David Deco--the Three-D Man, or is it the 3-D Man? Who cares anyway?

His attitudes, like the mood swings of the true Gemini spirit, go from left to right and back with no rhyme or reason. He is "hot" and "cold," up and down, and never too predictable in any way. He hates things that remain the same and is always looking for a change, a restless spirit, but firmly grounded in his beliefs and faith.

He is prolific, making on the average of one piece per day each year. The energy output is fantastic and very tiring. He often finds it hard to relax.

Since the first champleve enamel/fine silver jewelry made in 1964, through the raised, fabricated and forged toys and other strange shaped vessels/containers of the late 1960's and 1970's, and continuing through the painted and anodized works of the 1980's, he defies being pinned down as to style or influence.

LaPlantz is represented in the following permanent collections:
Schmuckmuseum, Phorzheim, Germany.
Ralph Turner Collection, London, England.
The National Museum of Modern Art, Kyoto, Japan.
Crafts Council of New Zealand, Wellington, New Zealand.
American Craft Museum, New York, New York.

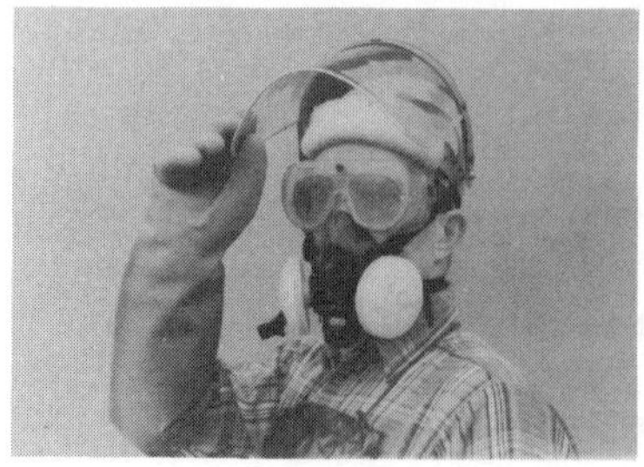

TEACHER

David LaPlantz never really "set out" to be a teacher. Like so many other events in his life, as he tells it, "I simply turned around one day and noticed I was a teacher of jewelry and metalsmithing!"

He received a BA Degree in Art from Bowling Green State University in 1966 and earned his M.F.A. Degree in metalsmithing in 1969 from Cranbrook Academy of Art.

He has taught at the following schools:
Institute of American Indian Arts, Santa Fe, New Mexico.
Flint Community Junior College, Flint, Michigan.
Colorado State University, Fort Collins, Colorado.
San Diego State University, San Diego, California.
Kent State University, Kent, Ohio.

Summer teaching:
Arrowmont School of Arts and Crafts, Gatlinburg, Tennessee.
California College of Arts and Crafts, Oakland, California.
Haystack-Hinckley School, Hinckley, Maine.
Oregon School of Arts and Crafts, Portland, Oregon.

In 1985, he was honored with a Fulbright Scholar Grant and traveled throughout New Zealand for six weeks teaching jewelry workshops.

LaPlantz currently teaches Jewelry Design in the Art Department of Humboldt State University, Arcata, California.